W9-BWG-081

This book discusses the political pressures that promote or inhibit particular policies for the management of hazardous waste. It integrates recent, original research on why and how hazardous waste policies are made and implemented, and it stresses the linkages among governmental agencies, public and private interests, political parties, citizen groups, and administrative-organizational factors in dealing with the hazardous waste problem.

To St. Francis of Assisi, patron of ecology

The Politics of Hazardous Waste Management

Edited by
James P. Lester
and Ann O'M. Bowman

Duke Press Policy Studies
Duke University Press   Durham, N.C.   1983

Printed in the United States of America on acid-free paper

Library of Congress Cataloging in Publication Data

Lester, James P., 1944-
  The Politics of hazardous waste management.

  (Duke Press policy studies)
  Bibliography: p.
  Includes index.
  1. Hazardous wastes—Government policy—United
States—Addresses, essays, lectures. I. Bowman,
Ann O'M., 1948-   . II. Title. III. Series.
HD4483.L48   1983   363.7'28   83-16595
ISBN 0-8223-0507-0
ISBN 0-8223-0523-2 (pbk.)

Contents

List of Tables and Figures

Tables

Figures

Preface and Acknowledgments

This book is concerned with the process and politics of hazardous waste regulation in the United States of America. The impetus for our collection is the fact that little work has been done, and very little written about, the political and institutional forces promoting (or inhibiting) policy formulation and implementation in this area. Although much has been written about the scope and nature of the hazardous waste problem, this particular issue has been the subject of very little systematic research that extends beyond physical or economic assessments of the problem. Our primary rationale for the book stems from the realization that governmental policy, more often than not, results from political considerations rather than prolonged assessments of the physical, technical, scientific, or social parameters of an environmental problem. Thus, by focusing upon the political process of hazardous waste regulation, we expect to contribute toward a systematic understanding of this complex, yet terribly important, public policy issue.

Our audience for the book is composed of three specific groups. First of all, we are directing this research collection to political scientists, sociologists, and other academic social scientists who are interested in the relationships between technology-intensive issues on the one hand, and the political processes of policy making and implementation on the other. Second, these readings are addressed to upper-division or graduate-level students in political science or public administration. For both professors and students alike, these readings are provided as a textbook for a course in environmental politics and policy. They may also accompany other courses in public policy more generally, and/or science and technology policy. Finally, we expect that this compilation of original research will provide additional insight for those currently involved in the implementation of a very controversial and complex public policy issue, including federal, state, and local administrative personnel, as well as other interested policy professionals.

The analytical approach used in this text stresses the linkages among intergovernmental relations, public-private relations, political parties, citizen participation, and administrative-organizational factors on the one hand, and public policy in this area on the other hand. It is not intended merely to provide a collection of interesting readings (although the readings are interesting in themselves). Rather, our purpose in this text is to collect and integrate recent and original research on this topic in order to aid the scholar, practitioner, and student in comprehending why and how hazardous waste policies are made and implemented. The quality of this research and its usefulness, therefore, rests largely with the intellectual efforts of the authors of the individual chapters. Each contributor demonstrated a strong commitment toward developing a systematic understanding of the forces influencing public policy in this complex area of decision making. They were also willing to pursue and sustain that commitment under the heavy editorial hand of the co-editors. To all of them, we extend our sincere gratitude for their efforts in the research which follows in this book.

We are also indebted to a number of individuals who (directly or indirectly) encouraged our work in this area of public policy. First, Lauriston R. King initially stimulated our interest in this project through our brief participation in Sea Grant activities at Texas A&M University. Kenneth Mladenka provided much encouragement in the early stages of the project in his role as Program Chairman of the 1982 Southwestern Political Science Association Meetings. In addition, Samuel A. Kirkpatrick, Charles W. Kegley, Norman Luttbeg, and Walter A. Rosenbaum offered both their advice and encouragement throughout various stages of the project. Texas A&M's Office of University Research, the Department of Political Science at Texas A&M, and the University of South Carolina's Department of Government and International Studies provided funds which enabled us to conduct some of this research and to prepare the manuscript. In addition, Lisa Grubbs, Edna Lumpkin, Robin Scott, and Theresa Abel provided cheerful and extremely competent clerical assistance for the preparation of the manuscript.

We also wish to thank Richard C. Rowson, Director of Duke University Press for his early (and sustained) interest in the project, Reynolds Smith, our editor, and Ed Haynes, for their help in the production process, and Mary Lou Back for her assistance in the marketing phase of the book.

Finally, to each other, we express our sincere thanks since this book is better as a result of our collaborative efforts.

James P. Lester
Ann O'M. Bowman

April 25, 1983

I. Introduction and Approach

1. The Process of Hazardous Waste Regulation: Severity, Complexity, and Uncertainty

James P. Lester

> I believe that it is probably the first or second most serious environmental problem in the country. One of the difficulties is that we really do not know what the dimensions of the problem are. Essentially, there is very little downside risk to anybody who illegally disposes of chemicals in such a way as to be harmful to the public health.
>
> We do not know where the millions of tons of stuff is going. We feel that the things that have turned up like the Love Canal and Kin-Buc situation are simply the tip of the iceberg. We do not have the capacity at this time really to find out what is happening. In my view, it is simply a wide open situation, like the Wild West was in the 1870s, for toxic disposal
>
> The public is basically unprotected. There just are not any lawmen out there, State or Federal, policing this subject.
>
> James Moorman, Assistant Attorney General for Land and Natural Resources, U.S. Department of Justice, May 16, 1979

Dramatic incidents such as kepone dumping in the James River, the "Valley of the Drums" in Kentucky, and the damage caused by the Love Canal chemical disposal sites have served as catalysts for mobilizing public interest in the responsible management of hazardous wastes. Consequently, the hazardous waste issue has recently become the focus of much congressional interest[1] and scholarly research.[2] Indeed, the solid and hazardous waste problem is often described as "the single most threatening environmental issue facing the country," or the "environmental problem of the century."[3]

The recognition of this environmental problem has provided a basis for the passage of such federal legislation as the Solid Waste Disposal Act of 1965, the Clean Air Act of 1970 (and subsequent amendments), the Federal Water Pollution Control Act of 1972, the Federal Environmental Pesticide Control Act of 1972, the Safe Drinking Water Act of 1974, the Toxic Substance Control Act of 1976, the Resource Conservation and Recovery Act of 1976, and, most recently, the Comprehensive Environmental Response, Compensation, and Liability Act of 1980 (or "Superfund," as it is more commonly called). Table 1.1 provides a comprehensive list of existing legislation in this area of public policy.

Table 1.1. Federal laws and agencies affecting toxic substances and hazardous waste control

Statute	Year enacted	Responsible agency	Toxic substances sources covered
Comprehensive Environmental Response, Compensation, and Liability Act[a]	1980	EPA	Hazardous waste spills and inactive or abandoned hazardous waste sites
Toxic Substances Control Act[a]	1976	EPA	Chemical substances, except tobacco, nuclear materials, alcohol, pesticides, and foods, food additives, drugs, cosmetics, and devices covered by FDA authority
Resource Conservation and Recovery Act[a]	1976	EPA	Solid wastes, including hazardous wastes
Clean Air Act	1970; amended 1977	EPA	Hazardous air pollutants
Clean Water Act (formerly Federal Water Pollution Control Act)	1972; amended 1977	EPA	Toxic water pollutants
Safe Drinking Water Act	1974; amended 1977	EPA	Drinking water contaminants
Federal Insecticide, Fungicide, and Rodenticide Act	1947; amended 1972, 1975, 1978	EPA	Pesticides
Act of July 22, 1954 (codified as Section 345 (a) of the Food, Drug, and Cosmetic Act)	1954; amended 1972	EPA	Pesticide residues in food
Marine Protection, Research and Sanctuaries Act	1972	EPA	Ocean dumping

Act	Year	Agency	Coverage
Federal Food, Drug, and Cosmetic Act	1938	FDA	Basic coverage of food, drugs, and cosmetics
Food additives amendment	1958	FDA	Food additives
Color additives amendments	1960	FDA	Color additives
New drug amendments	1962	FDA	Drugs
New animal drug amendments	1968	FDA	Animal drugs and feed additives
Medical device amendments	1976	FDA	Medical devices
Federal Meat Inspection Act	1967	USDA	Food, feed, and color additives and pesticide residues in meat and poultry products
Poultry Products Inspection Act	1957	USDA	
Egg Products Inspection Act	1970	USDA	
Fair Packaging and Labeling Act	1976	FDA	Packaging and labeling of food and drugs for humans or animals, cosmetics, and medical devices
Public Health Service Act	1944	FDA	Sections relating to biological products
Occupational Safety and Health Act	1970	OSHA, NIOSH	Workplace toxic chemicals
Federal Hazardous Substances Act	1960	CPSC	Hazardous (including toxic) household products (often the same as consumer products)
Consumer Safety Product Act	1972	CPSC	Hazardous consumer products
Poison Prevention Packaging Act	1970	CPSC	Packaging of hazardous household products
Lead-Based Paint Poisoning Prevention Act	1973; amended 1976	CPSC, HEW, HUD	Use of lead paint; on toys or furniture, on cooking, drinking, and eating utensils, in federally assisted housing
Hazardous Materials Transportation Act	1975; amended 1976	DOT (Materials Transportation Bureau)	Transportation of toxic substances generally

Table 1.1. (continued)

Statute	Year enacted	Responsible agency	Toxic substances sources covered
Federal Railroad Safety Act	1970	DOT (Federal Railroad Administration)	Railroad safety
Ports and Waterways Safety Act Dangerous Cargo Act	1972 1952	DOT (Coast Guard)	Shipment of toxic materials by water
Federal Mine Safety and Health Act	1977	Labor (Mine Safety and Health Administration), NIOSH	Toxic substances and other harmful physical agents in coal or other mines

a. Primary legislation in this area of public policy.
Source: Environmental Law Institute, An Analysis of Past Federal Efforts to Control Toxic Substances (Springfield, Va.: National Technical Information Service, 1978), and information provided by agencies.

In spite of this legislative activity by the federal government, it is clear that an effective solution to the problem will require substantial cooperation that extends well beyond the federal government to include state and local governments, the private sector, and individual citizens. In fact, the effective control of hazardous waste is an issue that transcends all levels of government, the private sector, and all lines of socioeconomic status, age, race, and lifestyles.

Even with the extreme vigilance of all affected parties, however, it is not certain that the critical problem of hazardous waste can be (or will be) managed effectively. Recent studies suggest that virtually all clay containment mechanisms for the "safe" disposal of chemical wastes are eventually vulnerable to chemical penetration; hence, groundwater supplies are in greater jeopardy than once believed.[4] Moreover, governmental resources (both fiscal and technical-administrative) for dealing with this problem are declining as a new administration challenges the necessity of environmental regulation and shifts its priorities in directions other than the environment. These new governmental priorities and diminished resources now threaten to affect adversely the progress that has been achieved thus far. Citizen groups, on the other hand, continue to express strong support for the maintenance of existing regulations for environmental pollution. Indeed, a recent survey of public opinion by Louis Harris found that "not a single major segment of the [American] public wants the environmental laws made less strict."[5]

The present study cannot resolve the debates and conflicts that characterize this issue since they are inherent to a pluralistic political system in which no individual or interest group has a monopoly on the definition of the problem and the proper policy response to it. This study will, however, provide an analytical framework for identifying and discussing some of the primary determinants of hazardous waste policy making and implementation at each level of government. In the process, we lend some insight and perspective to the uncertainty and complexity in this area of environmental policy.

In this book, we contend that the events surrounding the hazardous waste issue may be viewed in terms of three broad dimensions or themes—*severity* of the public policy problem itself, *complexity* in the policy process for dealing with this issue, and *uncertainty* in terms of an effective and prompt policy response to it—which are further explored in chapters to follow.

Technology: Severity of the Problem

It is presently agreed among most scientists and policy makers alike that the problem of hazardous waste is a very serious one. Indeed, in its report on hazardous waste disposal, the Subcommittee on Oversight and Investigations of the House Interstate and Foreign Commerce Committee summarized this particular problem:

> The hazardous waste disposal problem cannot be overstated. The Environmental Protection Agency (EPA) has estimated that over 77,140,000,000

pounds of hazardous waste are generated each year, but only 10 percent of that amount is disposed of in an environmentally sound manner. Today, there are some 30,000 hazardous waste disposal sites in the United States. Because of years of inadequate disposal practices and the absence of regulation, hundreds and perhaps thousands of these sites now pose an imminent hazard to man and the environment. Our country presently lacks an adequate program to determine where these sites are; to clean up unsafe active and inactive sites; and to provide sufficient facilities for the safe disposal of hazardous wastes in the future.[6]

Despite the seriousness of the situation, EPA is experiencing great difficulty in obtaining accurate measurements of the magnitude of the problem. For example, there is substantial uncertainty about how many firms are actually involved in hazardous waste management, about the total volume of chemical wastes that are generated in the United States each year, and about the total number of sites used for hazardous waste disposal.[7]

Although estimates vary widely as more data are collected and made available to the public, the estimates (as of September 1981) suggest that over 400,000 firms are involved in handling hazardous wastes.[8] However, the data base for this figure is incomplete and the EPA therefore has no reliable estimate of the total number of firms involved.[9] Precise figures on waste generation or disposal sites are also uncertain. The EPA estimates that anywhere from 41 million to 57 million metric tons of industrial hazardous wastes are generated each year. Similarly, the estimated number of hazardous waste sites in the United States ranges from 32,000 to 50,000.[10]

In October 1981, the EPA identified 115 sites with the potential to pose imminent threats to health and the environment;[11] however, the methodology used to identify these priority sites for remedial action under the Superfund program was soundly criticized by the Congressional Office of Technology Assessment (OTA).[12] Other estimates suggest that the number of waste sites posing a significant threat to public health and/or the environment is anywhere from 1,200 to 34,000, depending on the estimation technique used.[13]

In any case, EPA Administrator Anne Gorsuch, in October 1981, identified 24 of these 115 sites as posing a "greater potential danger" to public health than New York's Love Canal because they threaten to pollute drinking water supplies in densely populated areas.[14] On July 23, 1982, the EPA added 45 more hazardous waste dumps to its list of the 115 worst sites in the country, making them eligible for cleanup under the $1.6 billion Superfund program.[15] Finally, on December 20, 1982, the EPA released a list of the 418 worst toxic sites that are eligible for cleanup under the Superfund program (see appendix A for a list of these sites). Most of these sites, as well as EPA's earlier list of 2,000 "potential hazardous waste disposal sites" (released on January 1, 1981), are found in the states of New Jersey, Michigan, Pennsylvania, New York, and Florida, all of which are classified as high waste-generating states.[16]

Quite aside from various estimates of the severity of the problem, the economic costs of cleaning up the hazardous waste problem are also staggering. For example, it is reported that nearly $100 million has been spent on the Love Canal area alone.[17] Cleaning up all the abandoned sites nationwide is expected to cost anywhere from $4 to $50 billion.[18] These figures, however, say nothing about the costs of yet undiscovered sites which may be hazardous, nor does it allow for additional costs brought about by delays in cleanup action. In this sense then, the $50 billion figure is a *conservative estimate* of the actual costs involved.

Furthermore, it is often argued that the hazardous waste problem is largely the result of the failure of the free market to impose the cost of proper waste disposal on the original chemical manufacturers. Thus, the costs of proper waste disposal options pose complications for the policy process since waste disposal industries have few economic incentives for adhering to safe (and legal) waste disposal practices.[19] Table 1.2 provides estimates of the costs of alternative waste disposal technologies. At best, the profit motive and other incentives to lower the cost of disposal dictate that industries rely on landfill disposal (where it is allowed), which is the least desirable method from an ecological perspective.[20]

Taken as a whole, the situation discussed above illustrates both the severity of the hazardous waste problem and the substantial uncertainty associated with estimating the physical and economic parameters of this public policy issue.

Politics: Complexity in the Policy Process

A second aspect of this issue is that the policy process for decision making and implementation is characterized by extreme complexity. As Getz and Walter note, the Resource Conservation and Recovery Act (RCRA) mandate draws at least four different groups of actors into its comprehensive regulatory system.[21] The complexity of this situation stems from the observation that "each set of actors responds to different incentives. It is still an open question whether their divergent and incompatible interests can be meshed through a series of continuous mutual adjustments into an effective regulatory scheme."[22]

For example, at the federal level, the EPA is obliged to: (1) identify "hazardous" wastes; (2) establish a manifest system for tracing hazardous wastes from generation, to transporter, to disposal facility; and (3) set federal minimum standards for hazardous waste disposal, enforced through permits for disposal facilities. However, EPA rulemaking under RCRA is complicated by the fact that the "hazardous waste regulations are the most complex and voluminous ever promulgated by the federal government."[23] Specifically, EPA's objective is to develop a consensus among the various affected interests as a necessary condition in its rulemaking. Consequently, this strategy has led to extensive intervention by both environmental groups and affected industries as each attempts to gain concessions vis-à-vis the RCRA regulations. While it is not certain that this extensive intervention has produced a more "rational" (i.e., a more comprehensive) and legitimate (in terms of majority support) hazardous

Table 1.2. Costs of environmentally sound methods for disposal of hazardous wastes

Method	Cost (dollars per metric ton)
Land spreading	2–25
Chemical fixation	5–500
Surface impoundment	14–180
Secure chemical landfill	50–400
Incineration (land based)	75–2,000
Physical, chemical, or biological treatment	Varies

Source: U.S. Environmental Protection Agency. Office of Water and Waste Management, *Everybody's Problem: Hazardous Waste* (Washington, D.C.: U.S. Government Printing Office, 1980), 15.

waste policy, it is clear that the implementation of RCRA has been delayed by this activity.[24]

The next group of actors in the policy process are the states themselves. To qualify for full EPA authorization, the states are first given the opportunity of developing detailed legislative programs for licensing waste facilities that are "equivalent to" EPA's minimum guidelines, for developing their own version of the manifest forms, and for enforcing compliance and prosecuting violators. Residual authority, it should be noted, rests with the EPA; if any portion of a state plan fails to measure up to minimum agency criteria, the EPA is authorized to intervene and operate the state program on its own.

As Carnes notes, there is substantial variation among the states in terms of their response to RCRA. While some states have proceeded to enforce their existing hazardous waste regulations or have adopted new ones, other states have delayed formulating new legislation which would bring them into compliance with RCRA. For the latter group of states, it is argued that "slow and erratic federal funding, delays in EPA's adoption of RCRA rules, and the inflexibility of those rules have constrained the formulation of state programs."[25] This partial explanation, of course, assumes that intergovernmental (especially federal-state) relations are dominant factors in shaping the various states' policy response. However, a counterargument suggests that *internal* characteristics of the states themselves are dominant influences upon state policy action. Within this perspective, the states are constrained by the availability of fiscal and staff resources to monitor and enforce compliance, by administrative fragmentation and confused lines of authority, by excessive interest group influence upon policy-making institutions (principally upon the state legislature and the state bureaucracy), and by a lack of political leadership at the executive level in this area of public policy.[26]

Third, local communities are also drawn into the issue since their views must be considered before any waste disposal facility can be sited in areas under their jurisdiction. The EPA requires some mode of "public participation" before granting a permit for a new installation or when recertification for an existing facility is requested. In addition, Section 7004 of RCRA requires the EPA and the states to encourage, provide, and assist public participation in the development, revision, implementation, and enforcement of any regulation, guideline, information, or program under RCRA. The mechanisms available to local citizens include public hearings and meetings, citizen suits, public notification requirements, and public education programs. Thus, local government officials and private citizens are actively involved in attempting to influence siting decisions for hazardous wastes as well as specific regulations for the safe transportation and disposal of hazardous wastes. However, local citizen groups express strong dissatisfaction with siting waste disposal facilities in their jurisdictions.[27] Indeed, it is suggested that public participation vis-à-vis the hazardous waste siting issue leads to extensive polarization among citizens and heightened political conflict.[28]

Finally, the fourth group of actors includes the private firms generating the hazardous wastes, as well as those in the waste disposal industry. The chemical industry has lobbied extensively at both the federal level (i.e., through the Congress and EPA), as well as at the state level (i.e., through the state legislatures and appropriate state agencies), to influence the development of regulations that would be more acceptable to their laissez-faire position. Specifically, they have argued against the enactment of new (and more stringent) regulations or for regulations that reflect actual management conditions (i.e., the degree of hazard) as opposed to uniform national standards that reflect a worst-case scenario.[29] However, one of the lessons of the Love Canal experience is the need for public-private cooperation. Without strong cooperation and financial commitment by the private sector, protection of the environment from chemical pollution is an elusive, if not altogether impossible, goal.[30] Ultimately, both public and private actors will have to achieve a workable consensus.

This need for public-private cooperation is made difficult, however, by the growing involvement of organized crime in the hazardous waste disposal industry. Evidence continues to suggest that organized crime is involved, although the extent is presently unknown. A superintendent of the New Jersey State Police testified to that effect in late 1980:

> There is organized crime activity in toxic waste disposal, toxic waste hauling. We do not deny that. It has to be made very clear; it is not something we just learned today. [However] . . . we have no reliable intelligence base to tell us that organized crime toxic waste in our State. I think we have credible intelligence data to tell us that organized crime controls solid waste. We find ourselves hung up on this, and I think we are in some way being portrayed outside our agencies as not really believing that there is this involvement. That could not be further from the truth. It is accurate, and what has been said here today is accurate.[31]

Beyond the severity of the problem and the complexity stemming from multiple actors' involvement (and their divergent and often incompatible interests), a number of additional considerations further complicate the process.

Public Policy: Uncertainty of Outcomes

A third characteristic of this issue is uncertainty in outcomes across several areas. For example, there is some uncertainty associated with society's overall commitment to environmental protection. As one analyst notes:

> Groups whose immediate agendas are at odds with governmental activity concerning the environment and wish a change in the order of priorities perceive environmentalists as either too radical (for example, in advocating utopian solutions) or too conservative (by restricting commercial development of wilderness areas, for example). The postponement of deadlines and relaxation of environmental rules reflects pressure from energy and industrial interests, which contend that the stringent application of environmental regulations poses obstacles to energy development and threatens industrial growth and employment. They argue that in order to meet current energy demands, readjustments must be made to previous commitments to protect and clean up the environment. They propose environmental trade-offs without seriously evaluating existing patterns of energy supply and usage or considering the consequences of continued high-level energy use on the stability of delicate ecological systems.[32]

The loss of confidence in the ability of government to respond effectively to environmental problems has led some to advocate deregulation and a return to self-regulating market mechanisms. These "neoconservative regulatory reformers," as they are often called, advocate reduced environmental regulations because, they argue, such regulations involve increased costs to business and so contribute to high inflation rates so characteristic of the 1970s and 1980s.[33]

This current mood of "environmental retrenchment" may ultimately affect society's commitment to the responsible management of hazardous wastes. For example, with the arrival of the Reagan administration in 1981, and increased support for deregulation, it is argued that recent policy changes are adversely affecting the laws for controlling hazardous wastes. Specifically, the EPA has cut its staff and budget approximately in half and its research efforts even more. Of the $452 million that has accumulated in the Superfund, the EPA has spent only $88 million (as of December 1982) to finance cleanup of abandoned chemical waste dumps. In addition, the administration's 1982 budget cut the preliminary investigation and cleanup capability from 1,300 to 900 waste sites.[34] Furthermore, after announcing that it intended to make major concessions to industry, the Reagan administration, in February 1982, proposed a reversal of rules that banned (under RCRA) the burying of hazardous waste at landfills.[35] Then, in March 1982, EPA also suspended its previous requirement that manufacturers of hazardous wastes report each year on what happened to those wastes. Instead, it was proposed that EPA take an annual survey of 10 percent of the companies involved.[36] Shortly thereafter, EPA announced its "national contingency plan" for dealing with hazardous waste sites. Essentially, this plan proposed to replace the Carter administration's stringent blanket licensing procedures for all disposal sites to controls based on the "degree of hazard" posed by chemicals stored or disposed of at the sites. In effect, these actions represented important concessions to the chemical industry. Moreover, a refusal by EPA Administrator Anne Gorsuch to turn over documents in compliance with a subpoena led to an unprecedented contempt of Congress charge against her in December 1982. Apparently, her refusal fueled suspicions in Congress that the EPA was allowing hazardous waste polluters to "buy out" of potential damage suits too cheaply.[37]

Earlier, in late March 1982, it appeared that the EPA was taking a more conciliatory approach to hazardous waste management by making several major concessions to the environmentalists. For example, EPA reversed its three-week-old decision to permit burial of drums of toxic wastes in landfills and issued interim rules to prohibit burial of toxics.[38] Second, EPA reversed a preliminary decision made in October 1981 and required an estimated 10,000 hazardous waste facilities to obtain liability insurance to protect citizens against possible contamination.[39] Finally, EPA retained the cleanup standards originally proposed by the Carter administration for all incinerators that burn hazardous wastes and it imposed more stringent standards for new hazardous waste landfills which become effective on January 26, 1983.[40] Thus, these recent policy shifts by EPA, as well as delays in finalizing RCRA regulations and delays in undertaking cleanup action under the Superfund program, have added to the extreme uncertainty characteristic of this public policy issue. Specifically, it is uncertain whether we are now entering an extended period of "environmental retrenchment" or whether government and private actors will reach a mutual understanding and work toward effective solutions to this issue.

A second area of uncertainty concerns the availability (and accuracy) of scientific and health information on which to base public policy. For example, "little is known about health hazards associated with exposure over a long period of time to small amounts and combinations of chemicals in ground water or elsewhere."[41] Moreover, as noted by Professor Beverly Paigen in her testimony before the House Subcommittee on Oversight and Investigation, "epidemiological studies can never prove cause and effect; these studies only show an association of disease with geographical location."[42] Thus, a dilemma exists in the sense that we simply do not have enough solid scientific information to identify with certainty the "right" level of regulation.[43]

This scientific uncertainty in turn presents new challenges to our political institutions, including the legislative, executive, and judicial branches in terms of formulating, implementing, and interpreting governmental policy in this area. For example, the fact that this issue-area is itself a highly technical and complex one underscores the need for technical information and expertise as a basis for formulating rational policy choices. There is recent case study evidence suggesting that a state's legislative capacity to provide this technical information and expertise to state legislators is instrumental in facilitating state policies for solid and hazardous waste management.[44] Unfortunately, many states lack this technical advisory capability.

Second, when the available scientific evidence is conflicting (or nonexistent) disputes may arise over questions of scientific "fact" and thus delay responsiveness by appropriate bureaucratic authorities. This problem of confusion and delayed administrative responsiveness is illustrated by Congressman John J. LaFalce's testimony on the Love Canal situation:

> From day one, Mr. Chairman, I have been attempting to get the Federal Government and the State government to realize the significance of the problem there and to conduct themselves in an appropriate, responsible, and responsive manner, and both levels of government have failed. They have not adequately addressed themselves to the health concerns of the people. They have not conducted adequate testing. Right now there still is a question mark about what the true health conditions are at the Love Canal site.
>
> Countless times I have asked them to come and do comprehensive testing. We have witnessed inertia, ineptitude, indifference. We have not witnessed cooperation between the Federal Government and the State government. We have witnessed confrontation between the Environmental Protection Agency and the New York State Department of Health, confrontation between the White House and the Governor's mansion.
>
> Mr. Chairman, I have to point out to you and underscore that these studies were not undertaken for the right reasons, nor were they undertaken, we have subsequently found out, in the right way. They should have been undertaken years ago—3 years ago, 2 years ago, 1 year ago.
>
> They were undertaken in January of 1980 because the Department of Justice at the end of December 1979 brought a lawsuit against the Hooker Chemical Co. The Department of Justice said, "We'd better have further testing done." Therefore, in order to gain evidence for a lawsuit, rather than fulfill the high responsibility to protect the health and safety of the citizenry, the Federal Government at long last embarked on a test.[45]

Third, the scientific uncertainties discussed above present new challenges to the judicial system which currently places the burden of proof in trials involving exposure to toxic chemicals upon the plaintiff. In such instances, however, it is difficult to prove that a certain illness resulted from a particular chemical. Thus, it is uncertain how the courts will manage the highly complex issues associated with the responsible management of hazardous waste. Doubts have been expressed over the courts' ability to deal with this and other "transcientific" issues.[46]

In addition to these institutional problems, there are physical uncertainties associated with this issue. For example, there is a shortfall in landfill disposal capacity due to citizen opposition to new sites coupled with growing amounts of wastes being generated every year.[47] Given this situation, two possible outcomes are suggested. This shortfall in disposal capacity may encourage the use of alternative methods of disposal (e.g., resource recovery); alternatively, it may lead to more (not less) illegal disposal of hazardous wastes. Moreover, as noted above, recent research suggests that clay liners in existing landfills are inadequate in containing hazardous wastes.[48] This research was a factor in EPA's decision to require synthetic (plastic) liners in its new landfill regulations issued on July 26, 1982. However, synthetic liners used in disposal operations are not always efficient in containing wastes.[49] Even EPA in its new landfill regulations recognized that there is no such thing as a "secure landfill"—that liners will eventually leak—and that landfills may only be viewed as temporary collection facilities or "controlled releases" into the environment.[50] Thus, protection of the public's health and safety is uncertain whenever and wherever chemical leachates penetrate this "safety barrier," which some research suggests is highly probable, if not inevitable.

Finally, there is much uncertainty associated with the effectiveness of past governmental responses to the problem at Love Canal. The EPA released its assessment of the current situation at Love Canal on July 17, 1982.[51] The report concluded that the "Ring Three" area was "habitable" although it drew substantial criticism from both of New York's U.S. senators, Daniel P. Moynihan and Alfonse D'Amato, from several environmental groups, and from some of the scientific consultants who were involved in the study. On the other hand, the report was praised by New York's Governor Carey and many local officials in Niagara Falls who favor redevelopment of the Love Canal neighborhood. Essentially, the EPA study claims that the clay cap on the Love Canal and the drainage system around it have halted the flow of chemicals from the canal. Critics, however, charge that EPA did not expend enough effort in attempting to determine the accuracy and reliability of certain tests.[52]

Thus, substantial uncertainty continues to pervade this public policy problem. That uncertainty is associated not only with present and future governmental and private sector actions, but also with previous actions, such as the Love Canal cleanup operation.

Purpose and Method

In view of these three aspects of this public policy problem, our purpose in this book is to provide insight into the problem in terms of untangling some of the complexity in the political process of decision making for toxic substances and hazardous wastes. The empirical evidence around which the chapters revolve is an examination of the economic, technical, political, and perceptual factors that shape governmental responses to this public policy problem at three levels of government. For example, some of the following questions continue to provoke interest among scholars of environmental politics and provide the foci for further exploration within the context of hazardous waste policy:

1. What are the determinants of hazardous waste policy at the federal, state,

and local levels? Are technological factors (i.e., the technical need for policy) dominant or conditioning factors in their relationship to hazardous waste policy formulation and implementation? To what extent do political (especially administrative-organizational) factors exert a strong influence upon policy formulation or implementation?

2. To what extent does "intergovernmental tension" constrain policy implementation at the subnational levels in this area?

3. How do federal (or state) decision makers process technical information on the hazardous waste issue? Are risk assessments used and, if so, what factors are associated with their use and impact?

4. What is the nature of public-private relations in this area (i.e., consensual or nonconsensual), and how have these relations promoted (or inhibited) policy making and implementation?

5. What kinds of institutional capabilities in the fifty American states are associated with greater responsiveness to the problem of hazardous waste, e.g., legislative reform, consolidation (or fragmentation) of the state environmental bureaucracy, etc.?

6. What role (if any) does citizen participation play in the formulation and implementation of hazardous waste policy?

Despite the compelling nature of these questions, very little exists by way of an answer. Recent research on this issue has attempted to document the scope and nature of the problem, but little has been done to describe and analyze the political process for dealing with it. Although we are indeed concerned with acquiring an understanding of the severity of the hazardous waste issue and the nature of regulatory outcomes, our primary focus in this book is on describing and analyzing the *political processes* associated with governmental responses to it.

This study is not, however, concerned with purely descriptive accounts of this important topic. As students of the public policy process, we are committed to the systematic accumulation of knowledge. Thus, in the following chapters, our evidence is oriented around a conceptual framework which guides our understanding of the forces promoting or inhibiting hazardous waste policy making or implementation. In the chapters which follow, we draw upon *both* case study and aggregate approaches. We have adopted both approaches here because we believe that the ultimate goal of those conducting case studies is the development of propositions that are useful across a number of cases. Consequently, we have used the case studies for providing insight and rich detail as well as for generating hypotheses suitable for testing by ourselves or by others who follow in this research area.

In general, our model for analyzing the determinants of hazardous waste policy is derived from previous conceptual frameworks advanced by Donald Van Meter and Carl Van Horn, by Paul Sabatier and Daniel Mazmanian, and more recently by George Edwards.[53] Figure 1.1 represents an adaptation of their earlier work and it details the factors believed to promote (or inhibit)

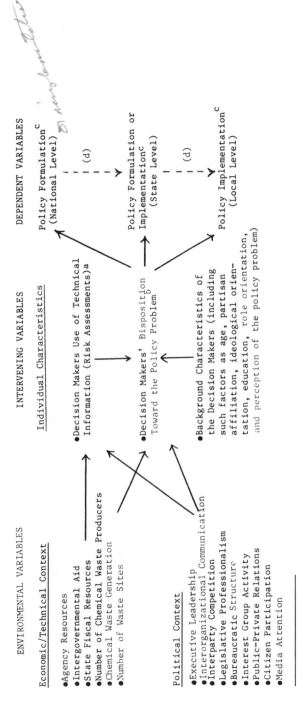

ENVIRONMENTAL VARIABLES

Economic/Technical Context

•Agency Resources
•Intergovernmental Aid
•State Fiscal Resources
•Number of Chemical Waste Producers
•Chemical Waste Generation
•Number of Waste Sites

Political Context

•Executive Leadership
•Interorganizational Communication
•Interparty Competition
•Legislative Professionalism
•Bureaucratic Structure
•Interest Group Activity
•Public–Private Relations
•Citizen Participation
•Media Attention

INTERVENING VARIABLES

Individual Characteristics

•Decision Makers Use of Technical
 Information (Risk Assessments)[a]

•Decision Makers' Disposition[b]
 Toward the Policy Problem

•Background Characteristics of
 the Decision Makers (including
 such factors as age, partisan
 affiliation, ideological orientation, education, role orientation,
 and perception of the policy problem)

DEPENDENT VARIABLES

Policy Formulation[c]
(National Level)

(d)

Policy Formulation or
Implementation[c]
(State Level)

(d)

Policy Implementation[c]
(Local Level)

[a]Includes the use of cost/benefit analyses, risk/benefit analyses, and/or risk assessments (both
qualitative and quantitative).

[b]Refers to decision makers' policy support, neutrality, or opposition.

[c]Includes hazardous waste policy objectives, standards, and enforcement activities by various authorities.

[d]Refers to "intergovernmental conflict," defined as attitudinal distance in support for hazardous waste
policies between (or across levels of) decision makers.

SOURCE: Compiled by the author.

Figure 1.1. A model of the determinants of hazardous waste policy at three levels of government

hazardous waste policy making and implementation. While these authors are primarily concerned with policy implementation, many of their variables influence policy formulation as well. In addition, we include other factors believed to affect policy formulation. As the following chapters show, we assume that three major groups of factors affect public policies for the regulation of hazardous waste: the external economic/technical context, the external political context, and the internal characteristics and perceptions of the policy actors themselves.

The Economic and Technical Context

The economic and technical context is described as those factors at the national, state, or local levels which provide pressure for, or constraints on, the development or implementation of public policies for dealing with this issue. At state and national levels, for example, the availability of financial resources to the implementing agency is necessary to hire the staff and to conduct the technical analyses involved in the development of regulations, the administration of permit programs, and the monitoring and enforcement of compliance.[54] Similarly, the availability of economic resources within the implementing jurisdiction (states or cities) obviously sets limits on, or provides opportunities for, the provision of public goods and services by a government on behalf of its constituents.[55] Finally, in the event of resource shortages, the availability of intergovernmental aid provides additional resources to both state and local governments for policy implementation purposes.[56]

Further, it is suggested that actual environmental conditions (e.g., the severity of the pollution problem) will directly influence support for regulatory (e.g., hazardous waste) policy formulation and/or implementation. Essentially, it is argued that the greater the severity of the problem (or the greater the scope of those affected), the more likely the demands for regulatory controls.[57] However, Mazmanian and Sabatier suggest that policy implementation is constrained by difficulties in measuring the seriousness of the problem, the diversity of behavior being regulated, the percentage of population within a political jurisdiction whose behavior needs to be changed, and the extent of behavioral change required among the target groups.[58]

As the following chapters indicate, the availability of fiscal resources and measurement of problem severity are significant constraints at both federal and state levels.

The Political Context

The political context is similarly described as those factors at the federal, state, or local levels which exert strong influences on the type of policies adopted or implemented at each level of government. For example, changes in executive leadership obviously influence the nature of intergovernmental (or intragovernmental) relations which, in turn, affect support for (or opposition toward) the formulation or implementation of environmental policies.[59]

Second, a high degree of interorganizational communication is believed to be essential for effective policy implementation. Specifically, "if different sources of communication provide inconsistent interpretation of standards and objectives or if the same source provides conflicting interpretation over time, implementors will find it even more difficult to carry out the intentions of policy."[60]

Third, the characteristics of the implementing agencies, or bureaucratic structure, is another crucial factor influencing policy implementation. By bureaucratic structure, we mean the degree of hierarchical integration within and among implementing institutions.[61] A lack of hierarchical integration or organizational fragmentation may inhibit the coordination necessary to implement successfully a complex policy. It may also waste scarce resources, delay policy changes, create confusion over program responsibilities, and lead to increased jurisdictional jealousy and conflict.[62]

In addition, the institutional capabilities of the state legislatures (or the degree of legislative professionalism) are also believed to exert a strong influence on policy making.[63] Many state legislatures provide only a minimum of services to legislators, meet relatively infrequently, and in general are considered organizationally ineffective in facilitating public policies. Thus, it is logical to assume that even the most salient environmental issues will not have a firm and consistent impact on state policies within the context of weak (or nonprofessional) legislative institutions.[64]

Other political conditions believed to influence successful policy formulation or implementation include partisan support (or opposition) for the policy within the implementing jurisdiction; issue saliency and support among the public at large; the level of interest group mobilization; the amount of media attention focused on the issue of hazardous waste; and the nature of public-private relations.[65]

Individual Level Factors

Although these contextual factors are important, they are not the sole (or necessarily the most important) forces acting upon decision makers as they go about formulating or implementing policies in this area. Several authors suggest that the disposition (or commitment) of the implementors themselves is a critical factor in understanding public policy and implementation.[66] Indeed, it is suggested that each of the components of the model discussed above is filtered through the perception of the implementor within the jurisdiction where the policy is delivered. According to Van Meter and Van Horn, three elements of the implementors' attitudinal response may affect their ability and willingness to carry out the policy: their understanding (or comprehension) of the policy problem; the direction of their response toward it (i.e., acceptance, neutrality, or rejection); and the intensity of that response.[67] In addition, some recent research indicates that perceptual forces, such as decision makers' perception of the policy problem, influence not only their level of support for a particular policy,[68] but even their receptivity to risk assessments of the severity of the problem.[69]

Furthermore, the background characteristics that decision makers bring to bear on a public policy problem further influence their policy preferences.[70]

In summary, the linkages in figure 1.1 suggest hypotheses which may be examined empirically in the following chapters, assuming that operational measures of each phenomenon can be developed and the necessary empirical data collected. While this discussion suggests that static relationships exist between these contextual/individual factors and hazardous waste policy, it should be emphasized that the policy process in this area is a dynamic one. Thus, in the chapters which follow, we will endeavor (at least conceptually) to untangle the complex interrelationships that exist between contextual level factors and individual level factors on the one hand, and public policies for the regulation of hazardous waste on the other.

Organization of the Study: Alternative Perspectives

The data and insights that are essential for an application of this conceptual framework will be provided in the following chapters, which are organized around two major perspectives. These two alternative (but nevertheless complementary) viewpoints include the federal perspective and the subnational (i.e., state and local) perspective.

Part II of the book is concerned with the federal perspective. It assumes that the major constraints and/or factors promoting hazardous waste policy are conditions at the national level. For example, Richard Riley, in chapter 2, suggests that the Environmental Protection Agency (EPA) itself is a major constraining influence upon implementation of the Toxic Substances Control Act (TSCA) and the Resource Conservation and Recovery Act (RCRA). Specifically, he describes and analyzes relations between the EPA on the one hand and Congress, the private sector, and outside agencies on the other and how these relationships inhibit policy implementation. He concludes that such variables as communication between agencies, the availability of fiscal resources, statutory factors, and several other economic and political factors at the national level are delaying implementation of TSCA and RCRA.

However, an argument is also made that the EPA has moved slowly, albeit carefully, in terms of its responsibilities for hazardous waste management. Ultimately the EPA's careful planning may, in fact, promote policy implementation. This is precisely the argument that Steven Cohen and Marc Tipermas make in chapter 3. They argue that while Superfund implementation has been delayed by the Reagan administration, the preimplementation project's outputs will ultimately be responsible for minimizing the delay.

Harvey Lieber, in chapter 4, on the other hand, returns to the thesis that the EPA's delays have frustrated implementation of RCRA. Specifically, he argues that the EPA-directed implementation process has served to confuse the states seeking to follow federal guidelines and assume control over hazardous wastes. Scientific and technological uncertainties, as well as a minimal commitment by

EPA, have resulted in a program which will take at least a decade from its passage to be implemented as intended by its legislative sponsors.

In part III, we begin with an additional view of the forces promoting or inhibiting the regulation of hazardous waste. The subnational perspective assumes that factors internal to the states themselves (as opposed to forces external to the states, e.g., the EPA) are major influences upon policy implementation. For example, chapter 5, by Bruce Williams and Albert Matheny, discusses Florida's attempt to implement RCRA. The authors suggest that responsible hazardous waste management in Florida is constrained by private sector market incentives against proper disposal, by the inability (or unwillingness) to adequately assess the risks associated with hazardous waste by government officials charged with implementation, and by symbolic legislation that fails to provide for adequate enforcement.

John Worthley and Richard Torkelson, in chapter 6, examine intergovernmental conflicts (i.e., conflicts between state agencies and conflicts between state and local governments) and public-private relations in the state of New York as problems for the successful implementation of policies for the management of hazardous waste. They suggest that intergovernmental tensions and public-private sector conflicts are significant barriers to policy development and implementation in New York State.

Focusing on the state of Texas, Kenneth Kramer, in chapter 7, argues that internal political conditions such as strong petrochemical interest groups, weak political institutions (especially a lack of formal gubernatorial strength and a highly fragmented bureaucracy), and diffused media attention inhibit a more aggressive state policy in Texas for the responsible management of hazardous waste.

At the same time, however, there are forces at the state or local levels which are encouraging stringent policies for the regulation of hazardous waste. For example, David Morell, in chapter 8, suggests that the response taken by California to the hazardous waste problem emerged as a direct result of gubernatorial leadership in creating an unusual government agency to formulate innovative policy alternatives. In addition, chapter 9, by Walter Rosenbaum, argues that while steadily decreasing federal and state fiscal resources and personnel have adversely affected publicly funded citizen participation programs for hazardous waste management, the role and impact of citizen participation on government policy may actually increase as a result of privately funded activities by citizen action organizations.

However, chapter 10, by Susan Hadden, Joan Veillette, and Thomas Brandt, explores the complexities involved in the role of public participation in the siting process for hazardous waste. They conclude that the actual impact of public participation upon hazardous waste policy implementation is, at best, uncertain. The complexity of the siting process, as well as opportunities for public participation (and the potential for local opposition to new hazardous waste sites) may encourage waste generators to use alternative forms of waste disposal. These complex procedures may also encourage more (not less) illegal waste disposal by firms who seek either to avoid these procedural complexities or the strong possibility of citizen opposition to new sites.

Finally, chapter 11, by James Lester, James Franke, Ann Bowman, and Ken Kramer, draws from the four case studies of the states and tests several relationships that are identified (or suggested) in these prior analyses. It pays particular attention to the role and influence of fiscal resources, technological factors, and politics (e.g., the role of partisanship and administrative/organizational factors) as forces shaping the fifty American states' policies for the regulation of hazardous waste during the period 1976–79. Essentially, it concludes that technological factors (especially technological factors relating to the severity of the problem) and political factors (especially legislative professionalism and consolidation of the state environmental bureaucracy) encourage state policies for the regulation of hazardous waste.

In part IV, the concluding chapter by Ann Bowman draws together broad and specific determinants of hazardous waste policy discussed in previous chapters. In this final chapter, Bowman synthesizes, interprets, and supplements the findings of our contributors.

II. Federal and Intergovernmental Perspectives

2. Toxic Substances, Hazardous Wastes, and Public Policy: Problems in Implementation

Richard Riley

During the twentieth century, the federal government has often intervened to meet public demands for goods and services which the private sector has been either unable or unwilling to fulfill. Washington's various public policy responses to technological and demographic changes, as well as socioeconomic deprivation, have resulted in an enormous increase in the regulatory, distributive, and redistributive policy initiatives and the accompanying bureaucracies to implement these policies.

The 1970s witnessed increasing criticism of this growing "administrative state." Political candidates as well as incumbents reinforced public disenchantment with rising costs and declining effectiveness of modern government to achieve social goals. Fortunately, scholarly critiques of the policy process accompanied this public outcry against big government. The professional literature dealing with policy implementation has generally been of two types: macro-oriented model building, and micro-oriented case studies. The first, more theoretical body of implementation literature emphasizes certain variables deemed essential to program success: (a) policy standards and objectives; (b) interorganizational communications and enforcement activities; (c) program resources; (d) attitudes and resources of constituency groups; (e) commitment, skills, and disposition of implementing officials; and (f) the prevailing economic, social, and political conditions.[1] This more macro-oriented literature has in most instances followed earlier attempts by social scientists conducting case study analyses of specific program implementation which have attempted to explain why some policies fall short of mandated objectives.[2] Both forms of implementation literature point to the same basic reasons for program failure— defects in the original legislation, an inadequate bureaucratic response to a perceived problem, complicated intergovernmental and interagency communications, and a multitude of exogenous variables which detract from program performance.

This chapter focuses upon some important environmental legislation passed in 1976, the Toxic Substances Control Act (TSCA) and the Resource Conservation and Recovery Act (RCRA). Both statutes resulted from major efforts in the early 1970s to deal with what has become a severe problem—the production and improper disposal of toxic and hazardous materials. The legislation gave the Environmental Protection Agency (EPA) extensive authority to protect the public from various hazardous substances in the environment, and to effect comprehensive regulation of toxic substances posing a public health hazard. As will become apparent, many of the same variables discussed above by policy analysts—e.g., vague and unrealistic program goals, complicated interagency and constituency communications, a changing economic and political environment, and uncertainty over whether the EPA will move ahead with full implementation of TSCA and RCRA—have afflicted the EPA in its program mission.

This chapter is organized in the following manner. Because they have had a similar fate before the EPA, the chemical industry, and the public, each statute will be examined individually by looking at some of the more controversial provisions of each act and early EPA experience with implementation. The primary source materials used for this analysis are annual reports of the EPA, transcripts of two House and Senate subcommittees that conducted hearings on the legislation in 1979 and 1980, several oversight reports by the General Accounting Office, and selected comments by the mass media on the fate of TSCA and RCRA. After examining the implementation history of the legislation, a short discussion will follow of the "Superfund" bill passed in December 1980 and more recent implementation activities of TSCA and RCRA under the Reagan administration. The chapter then concludes with an assessment of how this survey contributes to our knowledge of policy implementation generally, TSCA and RCRA implementation specifically, and the outlook for government, industry, and the public reaching some consensus on a severe pollution problem of the 1980s. By focusing upon TSCA and RCRA, we shed much light on the role and influence of interagency relations, public-private relations, agency-industry relations, and agency-congressional relations upon policy implementation for the regulation of toxic substances and hazardous wastes.

Early Activity Under TSCA

In 1976, the Congress passed the Toxic Substances Control Act (TSCA) after a long debate over the growing problem of industrial wastes. By intention, the act was supposed to protect public health and the environment from excessive risk resulting from the production, use, and improper disposal of chemical substances. In signing TSCA into law, President Ford referred to the legislation as "one of the most important pieces of environmental legislation that has been enacted by the Congress."[3] The basic objectives of TSCA are fairly broad, some might say too broad:

1) *Adequate data* should be developed with respect to the effect of chemical substances and mixtures on health and the environment and . . . the development of such data should be the responsibility of those who manufacture and those who process such chemical substances and mixtures;

2) *adequate authority* should exist to regulate chemical substances and mix-

tures which present *an unreasonable risk of injury to health or the environment*, and to take action with respect to chemical substances and mixtures which are *imminent hazards*; and

3) authority over chemical substances and mixtures should be exercised *in such a manner as not to impede unduly or create unnecessary economic barriers to technological innovation while fulfilling the primary purpose of this Act* to assure that such innovation and commerce in such chemical substances and mixtures do not present an *unreasonable risk of injury* to health or the environment.[4]

To fulfill these objectives, the EPA was to require testing of chemicals identified as possible risks,[5] scrutinize new chemicals prior to manufacture,[6] regulate chemicals known to present a health risk,[7] and report and maintain data on all chemical substances.[8]

Major problems with TSCA have dealt with delay in the rulemaking process, several substances exempted from regulation, the hesitancy of industry to comply with premanufacture notification (PMN), and overall delay incurred by the agency in exercising the authority intended by Congress in passing the act. Under Section 4 of TSCA, the agency sets rules only when it can substantiate statutory findings about chemical substances involved and when the industry has failed to develop necessary data to prove the safety of the substance to the public. The agency must find that a chemical presents an unreasonable risk, that there are inadequate data already available to perform a "reasonable risk assessment," and that testing is necessary to provide such data. A testing rule may also be based on an agency finding of "substantial production and exposure to humans or the environment."

Under Section 4(a), the EPA tried to inventory all chemicals produced, imported, or processed for commercial purposes since January 1, 1975. The intention was to begin building a data base on chemical substances and to distinguish between existing and new chemicals. The agency published an Initial Chemical Substance Inventory of approximately 44,000 chemical substances on June 1, 1979, two and one-half years after TSCA went into effect. A Revised Chemical Substance Inventory added another approximately 11,000 chemicals to the original list in July 1980.

The EPA has often been somewhat defensive about its role and responsibilities under TSCA. In its second annual report on the administration of the act, the agency noted:

> With TSCA, as with any major new legislation, it has been necessary to cope with many administrative and managerial difficulties, including those associated with building and housing a new and expanding organization, identifying and hiring qualified professionals, and developing efficient operating procedures and effective working relationships.[9]

Aware of the needs and interests of industry, EPA notes that another problem is "the tension between EPA's needs for scientific and other data for use in identifying and assessing human health and environmental risks and the chemical industry's concerns about the costs of providing such data."[10] Finally, the agency emphasizes the conflict between industry's interest in protecting trade secrets and the public's wanting access to data submitted under TSCA.

In the 1980 report on TSCA administration, the agency devoted one page in a twenty-nine-page report to "problems," alluding only to difficulties encountered under Section 4(a)(1)(A)(ii). This section requires manufacturers and processors to test chemical substances for health and environmental effects.[11] A complicated testing procedure was being used to determine the sufficiency of existing data on chemicals: "This process of collecting and evaluating relevant information . . . has been extremely resource-intensive and time-consuming. Therefore, the Agency is reevaluating its approach based on the length of time it is taking to develop the first testing rule."[12] The agency had just published its first testing rule, three and one-half years after TSCA had taken effect. The proposed rule would require testing of chloromethane and selected chlorinated benzenes, but the agency declined to require additional health effects testing of acrylamide, as recommended by the Inter-agency Testing Committee.[13]

In 1979, the EPA was also involved in several lawsuits brought under TSCA, three of which involved polychlorinated biphenyls (PCBs).[14] Section 6(e) of TSCA requires the agency to establish rules governing the marking and disposal of PCBs, and to prohibit (with certain important exceptions to be discussed below) their manufacture, processing, distribution, and nontotally enclosed use. The marking and disposal rules were issued by the agency in February 1978, and in May 1979 a final rule banning the manufacture, processing, distribution, and nontotally enclosed use of PCBs was issued. This rule lowered, from 500 parts per million (ppm) to 50ppm, the minimum PCB concentrations in materials covered by the regulation.

Of the three PCB-related cases, the most important was *Environmental Defense Fund* v. *EPA*, wherein the EDF had petitioned the Federal Court of Appeals for the District of Columbia Circuit for review of the agency's May 31, 1979, rule defining minimum concentration levels. On October 30, 1980, two major portions of the agency's PCB Ban Regulations were invalidated by the Court.[15] Both the definition of PCBs as those in concentrations of 50ppm or greater and the designation of intact, nonleaking PCB-containing transformers, capacitors, and electromagnets as totally enclosed uses of PCBs (and therefore exempt from regulation) were found to be unsubstantiated by proper evidence. Since the October 1980 decision, EPA has had to consider how to proceed with new rulemaking efforts regarding these two issues. The agency's 1980 report notes:

> The most serious problem presented by the Court's decision is that the Agency has little factual information about the manufacture, processing, and distribution and use of PCBs in concentrations below 50ppm, but has reason to believe that a ban on these activities would disrupt major industries in the United States.[16]

Congress Looks at TSCA

Many of the difficulties experienced by EPA in implementing TSCA became apparent in April 1980 during four days of hearings before the House Subcommittee on Consumer Protection and Finance, of the Committee on Interstate and Foreign Commerce. The main purpose of the hearings was to reauthorize the act for fiscal year 1982 and to modify the agency's ability to deal more responsibly with PCBs. During three days of testimony before the subcommittee, witnesses from the EPA, the Food and Drug Administration, the U.S. Department of Agriculture, a chemical manufacturing trade organization, an environmentalist group, and the hazardous waste disposal industry all cited several reasons which they felt had delayed implementation of TSCA.

Section 5 requirements of the act deal with premanufacture notification (PMN). Beginning thirty days after publication of the chemical inventory, anyone intending to manufacture or import a new chemical substance must notify EPA at least ninety days prior to manufacture or processing concerning the substance name and any test data on effects of the substance. The agency then can take action, if necessary, to limit exposure or delay production until more information is provided. These requirements trouble industry personnel, as the testimony of one witness on April 15, 1980, indicates. Robert Polack, Chairman of the Toxic Substances Task Force of the Synthetic Organic Chemical Manufacturers Association (SOCMA), complained about the appropriateness of the tests then being required by the agency. Polack testified that "one cannot and should not require a chemical whose [sic] volume, use and exposure patterns are limited, to spend lots of time and . . . money on testing prior to commercialization."[17] Polack also complained about the 15–45 page PMN form then being proposed by the agency, which he said, if adopted, would cost a company anywhere from $1,100 to $15,000 to complete properly. Because of the high cost of toxicological and environmental testing, Polack argued that EPA should allow most new low volume chemicals to be produced without extensive testing.[18] "EPA has spent almost 2 years . . . regarding the development of a premanufacture notification program. So for 2 years they have been struggling with how to implement a program that does not need a single rule to be implemented."[19] Polack also noted that "we don't like uncertainty and we have been immersed in uncertainty with respect to the PMN program." One might safely assume that his sentiment was probably representative of much of the chemical industry at the time regarding PMN requirements under TSCA.

Ms. Janie A. Kinney, Chairperson of the Toxic Substances Advisory Committee of EPA, then testified and expressed concern about the agency's delay in issuing testing rules and regulations. When asked by Rep. Richardson Preyer (D-NC) to what she attributed the delay in acting quickly, Kinney responded in a revealing manner:

Well, I think there are not any simple answers. . . . [T]he statute is very complicated. EPA was faced with having to undertake a massive effort simply to get procedural guidelines and regulations in place because they had no program at all and all of a sudden they had a major new act to implement. . . . I think part of it has been the fact that EPA tries to develop absolutely perfect rules before it even issues a proposed rule. I think the Agency would be better served, and the public would be better served, if EPA used the rulemaking process that is in the statute to develop some of the information it may need in issuing its final regulations.[20]

Subcommittee Chairman James Scheuer (D-NY) then attributed agency delay to ignorance of the industry: "When EPA officials . . . admittedly don't know a great deal about the chemical industry, they tend not to act."[21]

Another subcommittee witness was Dr. Louis Slesin of the Natural Resources Defense Council (NRDC), an organization that had filed suit in 1979 against the EPA challenging that agency's delay in initiating testing rules as required by TSCA Section 4(e).[22] When asked by Scheuer to what he attributed the delay in inventorying chemicals and reviewing PMNs, Slesin mentioned three reasons: agency timidity in exercising its discretion and promulgating necessary rules; "scientific perfectionism," or the testing, retesting, and further delay in promulgating test rules on chemicals known to be hazardous; and agency lack of knowledge of the industry. According to Slesin, "EPA does not . . . have at the moment people on the staff that have the intimate knowledge of the way the industry works. . . . If you want to go and regulate, you have to know what is going on." Chairman Scheuer then asked how government could attract such expertise, and Slesin noted that money and a personal belief that the agency can in fact do something are essential. "At the moment everything is so wishy-washy that you are not sure if you go into EPA whether you will ever get anything done. We see 3 years later very little has been done."[23]

On April 16, the subcommittee examined H.R. 7003 and its proposal to eliminate PCB-containing transformers, capacitors, and electromagnets in food-related areas, a statutory change that had been motivated in part by PCB contamination of foodstuffs in several states in August 1979.[24] Chairman Scheuer again was critical of EPA, since it had in effect banned only about 1 percent of PCBs in commercial production. "Given the EPA's failure to date to carry out the congressional mandate, we think that the Congress is not only justified but really under a moral obligation to take some specific actions in regard to the enormous dangers imposed by PCBs."[25] Additional testimony on April 16 involved officials from the U.S. Department of Agriculture and the Food and Drug Administration who recommended necessary changes in H.R. 7003 to regulate better the production and distribution of PCBs.[26]

On April 17, Scheuer charged that there was "a very deep feeling of dissatisfaction in Congress on the way EPA has implemented the congressional ban on PCBs." Under Section 6, "totally enclosed" PCB facilities are exempted from regulation. The EPA has determined that PCB transformers, capacitors, and electromagnets, which constitute over 99 percent of the PCBs now in use, should be classified as "totally enclosed" uses, and thereby exempt from regulation. Scheuer considered that agency ruling to be a "clear and flagrant violation and a contradiction of the congressional mandate" contained in Section 6(e) of the act.[27]

The subcommittee also heard from a representative from the hazardous waste disposal industry, Melvyn Bell of Energy Systems Company (ENSCO) of El Dorado, Arkansas. Bell spoke about some recurring problems in the PCB-incineration process and the delicate issues of facility siting, community relations, and financial responsibility. He insisted that neither TSCA nor the Resource Conservation and Recovery Act (RCRA) adequately addressed the dual problems of public acceptance of disposal facility siting and of financial responsibility in case of catastrophic damage to humans and/or to the environment.[28] Hoping that the then-pending Superfund legislation would allay some of these fears, Bell also insisted that the public should be made aware that "because of the things they want and need in some cases, they are generators of those [chemical waste] materials." A December 1981 report by the General Accounting Office criticized the EPA for its failure to regulate PCBs under TSCA, noting that the agency "has made slow and limited progress in implementing the mandate and has little assurance that industry is complying with its regulations." It attributed this lackluster record to EPA delay in issuing rules on the marketing and disposal of PCBs, ineffective enforcement actions that compel compliance, and the lack of acceptable disposal facilities for handling PCBs.[29]

In 1980, the General Accounting Office published a report focusing on TSCA implementation, the major findings of which tend to confirm what several witnesses before Congress have stated regarding poor organization, inadequate staffing, and overall delay in implementing TSCA.[30] According to the GAO, the Office of Toxic Substances—later reorganized as the Office of Pesticides and Toxic Substances (OPTS)—as the primary office of responsibility for TSCA, was from the beginning poorly organized and understaffed. EPA assistant administrators could not reach an overall consensus on an appropriate structure for implementing new TSCA authority. Although the recruitment of senior management staff was initially delayed pending the approval of an agreed-upon organization structure, key management and staff positions were either vacant or lagging behind authorized levels after three years in OPTS, as table 2.1 indicates.

The comptroller general's report also faults the agency for its failure to anticipate passage of TSCA in 1976. There was no strategic planning reflecting a comprehensive approach to improve long-term regulatory effectiveness or to anticipate major problems and generate appropriate responses. On October 22, 1976, one day after passage of the act, the deputy administrator finally established a TSCA Strategy Work Group to develop an implementation strategy. After public meetings in December of 1976, the agency published a draft plan outlining possible implementation approaches in February 1977, more than

Table 2.1. OPTS staffing of permanent positions (1977–79)

Staffing level	FY77	FY78	FY79	FY80[a]
Authorized	88	196	382	510
Actual	88	171	313	423
(Shortfall)	0	(25)	(69)	(87)

a. As of May 2, 1980.
Note: The position of assistant administrator was filled in October 1977; three positions for deputy assistant administrators were filled from January to June 1978; eight division directors were appointed June–October 1978; and sixteen (of an authorized twenty-five) branch chiefs were appointed from January to October 1979.
Source: Comptroller General of the United States, Report to the Congress, *EPA Is Slow to Carry Out Its Responsibility to Control Harmful Chemicals* (October 28, 1980), 10–11.

one month after TSCA was to have taken effect. The agency then published a revised strategy in the *Federal Register* on October 28, 1978, entitled "Proposed Implementation Approach: Request for Public Comment," wherein it discussed such objectives as developing necessary organization and staff; establishing procedures for testing and evaluating chemical hazards; initiating information gathering; establishing mechanisms for premanufacture notification of new chemicals; and developing a coherent agencywide approach to toxic substances. By September of 1979, the direction of the program was still in doubt, as evidenced by comments from a deputy assistant administrator:

> There is as yet no coherent OTS-[Office of Toxic Substances]wide view of TSCA's essential purpose, and hence no agreement except by coincidence as to how we should implement it. Rather, there is a Section 5 program, a Section 4 program, a Section 6 program, a Section 8 program, etc., each apparently driven to a greater or lessor [sic] degree, by the inevitable bureaucratic imperatives of 1) self-preservation, and 2) exercise of an authority because it is there.[31]

On March 7, 1980, the Assistant Administrator of OPTS proposed a reorganization of the Toxics program, the major purpose of which was to eliminate major organizational and administrative problems derived from having TSCA responsibilities divided among three separate deputy assistant administrators. Functional changes occasioned by this reorganization were to consolidate the entire chemical exposure assessment program, all regulatory, and all information-rulemaking authority into one division. The comptroller general's report welcomed this change, viewing it as having a "positive effect on the future course of the Toxic Program."[32]

The report also criticized the agency's failure to fulfill its responsibilities under Section 8 of the act, dealing with the collection of basic information on existing chemicals. When the act was passed, little or no data existed on the number of chemicals in commerce, how they were being used, who was being exposed, and which chemical substances posed health hazards. Despite four years of activity, in 1980 the agency had not collected the necessary information requested by Congress in the act. Although the agency had published its Revised Inventory by July 28, 1980, these data consisted only of production-related information as to *name* and *quantity* of substances produced. There were no data concerning *use*, *exposure*, and possible *health effects* of identified substances.[33]

The final area of agency responsibility under TSCA commented upon by the GAO concerned premanufacture notification (PMN). It criticized the agency's program to control new chemicals, for as of July 1980, the EPA still had not issued final implementing regulations, strategies to implement all TSCA authorities affecting new chemicals, or operating procedures to determine what agency action to take on new substances. In a speech before the American Association for the Advancement of Science on January 5, 1980, the assistant administrator of OPTS noted:

> From EPA's perspective, . . . by far the most disturbing result observed to date concerns the general lack of toxicity testing data submitted with notices on new chemicals. . . . [V]ery little health and environmental effects testing has been performed on the new chemicals brought to EPA's attention thus far, . . . virtually all of which involve some degree of human or environmental exposure.[34]

He stated that unless this condition were corrected, "our review of notices will be based upon a fundamental lack of information and data." Through July 16, 1980, of the 199 PMNs received by OPTS, 60 percent of them (120) contained absolutely no toxicity data.

Two other issues that had troubled agency personnel attempting to comply with TSCA concerned how much information was needed to review adequately new chemicals, and how to preserve confidentiality of the reported information so as to prevent industrial sabotage. Because industry had complained about costs incurred in providing new chemical substance data, the EPA revised its form for PMN on October 16, 1979, and in so doing, reduced industry costs by approximately 149 percent (to approximately $8,900 per chemical). But still, public interest organizations and industry spokespersons objected to the proposed forms, though for different reasons. Whereas environmentalists believed that *more* information should be submitted with the PMN, industry personnel raised questions concerning the agency's authority to require any information via letterwriting rather than case-by-case rulemaking, a noticeably longer process.[35] Concerning the confidentiality issue, the agency had experienced much difficulty in obtaining the actual chemical identity of new substances. In the 183 PMNs submitted to the agency through July 16, 1980, wherein the company claimed confidentiality, two-thirds of them (123) did not contain the chemical identity of the substance.[36] The agency's difficulty in obtaining this information is in part the result of Section 5(d)(2) of TSCA, which allows the company to use the generic name, indicating Congress's recognition of the importance of confidentiality prior to commercial manufacture of a chemical substance.

In summary, recent experience with TSCA has revealed some fundamental problems that must be resolved if the program is to be a success. From a *statutory* standpoint, the law calls for "adequate data" on the effects of chemical substances and mixtures on health, and increased regulation of those substances which present "an unreasonable risk of injury" or an "imminent hazard" to health or the environment. These directives require a complete chemical inventory and, from the outset, EPA was slow to compile this list of toxic chemicals. The agency also experienced several problems relating to *communications*: (a) a lack of basic knowledge of existing chemicals; (b) difficulty in getting the chemical industry to supply data on new substances entering the marketplace (PMN requirements); and (c) an aroused public that wanted government to act but not if it disturbed the local social and economic environment. From congressional testimony and investigations by the General Accounting Office, the *disposition* of EPA toward increased regulation was marked by caution and a quest for "the perfect" rule acceptable to all parties concerned. Finally, from the perspective of *resources*, both finances and staffing were inadequate as late as three years after TSCA took effect in January 1977. And as if the problems surrounding full implementation of TSCA were not enough, the EPA was simultaneously having to administer new legislation governing the disposal of hazardous wastes under the Resource Conservation and Recovery Act.

Congressional Reaction to Hazardous Waste Disposal

Like its handling of TSCA, the EPA's implementation of the Resource Conservation and Recovery Act (RCRA) of 1976 has confronted similar problems concerning the actors involved, the scope of the act, agency discretion, public opposition to certain provisions, and inadequate agency resources. Recently, Getz and Walter observed that effective implementation of RCRA would involve successful interaction of four different actors—the agency, firms generating hazardous wastes, firms charged with disposal of such wastes, and local communities having to condone the siting of any disposal facilities within separate jurisdictions. They argue that EPA presumes the existence of hazardous substances in the environment, which compels generating firms to prove their innocence, rather than the agency to establish industry guilt. This approach minimizes agency costs and maximizes social costs to industry and the public. Getz and Walter also note that RCRA actually increases disposal and administrative costs among generators and disposers alike, while at the same time affecting a suspicious local public having to consent to disposal facility location. No one wants such facilities in his backyard! Getz and Walter are skeptical of EPA's ability to accommodate diverse interests in implementing these "cradle-to-grave" regulations on hazardous wastes.[37]

Not until 1976 did Congress recognize the need to protect land, drinking water supplies, and the food chain from hazardous wastes. Whereas TSCA deals with the identification of existing and new chemical substances in commerce that pose health hazards, RCRA deals with the disposal of such wastes. Passed on October 21, 1976, RCRA's purpose was to increase environmental protection. Subtitle C[38] is the heart of the legislation and has caused the most difficulty over the past few years, in that it requires the establishment of federal standards for defining and dealing with "hazardous wastes."[39] Subtitle C creates a comprehensive control system for managing these wastes, including appropriate monitoring, recordkeeping, and reporting. EPA was to have developed the regulatory framework of RCRA by April 21, 1978 (eighteen months after enactment, in accordance with 42 U.S.C. 6922), and to operate a hazardous waste regulatory program in each state,[40] although each state could assume responsibility for its own state program if it developed and implemented a program equivalent to the federal plan.

By mid-1978, several months after the regulatory program was to have been in place, the EPA was experiencing several difficulties with RCRA implementation. Subtitle C of the act requires EPA to develop national standards for the protection of health and the environment from solid waste disposal facilities. Under this section, the agency was to inventory all disposal sites and publish a list by October 1977 of all sites (municipal and industrial) not conforming to the minimum standards. By October of 1983, all sites not in full compliance ("open dumps") were to be either closed or upgraded to established standards. In its report of June 1978, the GAO emphasized the serious threat that improper waste disposal activities posed to the nation's groundwater supply.[41] It also noted that the EPA had already postponed its site inventory publication until October 1978, and that the October 1983 target for upgrading or closing all open dumps was in jeopardy. The GAO observed that the relationship between waste disposal practices and groundwater had generally been ignored by responsible officials. And even worse was the lack of basic data about the scope of the problem: "The extent of the damage done to this important water resource has not been determined and the total number of sites which may be contaminating groundwater has not been established by the responsible Federal or State agencies."[42] By late 1978, eight months after the regulatory program was put in place, the agency was experiencing much trouble. According to the GAO, the development of environmentally sound disposal facilities was essential to the successful implementation, and such capability did not then exist:

There is currently a shortage of suitable disposal sites and the problem will become even more acute as additional wastes are determined to be hazardous, existing sites are closed because they do not meet environmental requirements, and wastes . . . are taken to offsite facilities. *Effective implementation of the program cannot be accomplished unless additional treatment, storage and disposal capacity can be developed.*[43]

As of December 1978, four of the ten agency regions[44] reported that their offices could not monitor the disposal of wastes already being generated, and officials in four of the ten states visited by GAO (New Jersey, North Carolina, Ohio, and Pennsylvania) reported a capacity shortage in their states.

The GAO report also discussed public opposition to expanding disposal capacity. A 1976 EPA-funded study which surveyed about 80 percent of the hazardous waste disposal industry found that 42 percent of the firms interviewed saw public opposition as a constraint in obtaining new sites or expanding old ones. Officials in five of the seven EPA regions visited indicated that proposed disposal sites were stopped primarily because of public opposition. Selected states wherein such conflict emerged included Minnesota, New York, Washington, and Pennsylvania.[45] GAO recommended that the EPA monitor more closely state waste management plans in order to identify siting problems early in the process. It also recommended legislation to create a national trust fund, supported by fees assessed on the disposal of hazardous wastes, to cover all postclosure financial liability and to prevent further contamination. Passage of the Superfund act of December 1980 implemented this recommendation.

In early 1979, the GAO again commented upon the effectiveness of state hazardous waste programs. Reporting that American industry would generate an estimated 56 million metric tons of hazardous waste in 1980 alone, the GAO stated that current state programs were not adequate to fulfill RCRA requirements: "At present, operational hazardous waste regulatory programs or programs to be implemented in the near future . . . cannot adequately safeguard public health and the environment."[46] It said EPA was unable to provide necessary technical assistance to monitor state programs, nor could the agency assume responsibility for operating those programs if the states were unable to do so. Because of staffing and funding limitations, state agencies and the EPA were delinquent in completing the initial stages of implementation. The agency had not only failed to meet the April 21, 1978, deadline for defining "hazardous waste" and implementing a viable program, it also had not developed the necessary environmental standards for use by disposers, processors, and storers of such wastes. Of the twenty-six states visited or contacted by GAO, all but two of them had either no existing hazardous waste program or the project was in the very early stages of development:

Most states have not carried out even some of the basic first requirements of a hazardous waste program. In addition, they had only general estimates on the volumes of hazardous wastes being produced within their states and were not certain where they were being disposed of. None of the 26 states had fully identified waste generators within their state jurisdictions, and none had adequate enforcement programs.[47]

Though most states recognize the necessity to control the handling and proper disposal of such wastes, GAO maintained that ". . . few if any states have hazardous waste programs which provide the control intended by RCRA."[48] By October 1978, only seventeen states had enacted specific legislation for the management of hazardous wastes.

To make matters worse, GAO concluded that by EPA's own admission, the ten EPA regional offices would not have adequate staff to execute certain essential provisions under RCRA: six regional offices said they could not review state disposal sites to verify if they are environmentally sound; eight could not provide assistance to industry and the public on RCRA requirements; eight could provide no help to states in developing state regulations; and none of the ten regional offices could provide technical assistance to states in initiating their programs.[49]

By 1979, RCRA implementation had attracted much criticism on Capitol Hill. During July and August 1979, the Senate Subcommittee on Oversight and Management held hearings on the implementation of the act. In March 1980, the subcommittee released its report wherein it was very critical of agency regulatory revisions identifying the characteristics of and listing selected hazardous wastes. According to the subcommittee, the agency's proposed regulations to implement Section 3001 of RCRA were "not sufficient to insure protection of the public and the environment from problems associated with the disposal of hazardous waste."[50] The agency's tendency to ignore toxicity characteristics of various products left a major gap in the regulatory scheme.

The subcommittee also criticized EPA's delay in promulgating regulations under Section 3001. EPA had allotted a disproportionately small share of funds to ensure land protection, with a mere three percent of the total agency budget for FY 1980 devoted to hazardous waste programs, as opposed to air and water programs.[51]

Other problems alluded to by agency personnel, and mentioned by the subcommittee in its report included the need to encourage public participation without hopelessly delaying implementation, the desire to devise regulations that will survive judicial scrutiny by federal courts, public uncertainty about hazardous waste regulation and disposal facility siting, and reduced morale and confidence among agency employees at the program level.[52] Overall, the report maintained that

> [T]he current organizational structure of the EPA Office of Solid Waste has resulted in mismanagement in the implementation of RCRA in several ways. These include a) inadequate budget allocation to solid waste programs; b) inadequate application of human resources to solid waste programs; c) continued delay in the promulgation of RCRA regulations; and d) internal conflicts of interest and infighting throughout the implementation of RCRA.[53]

Remedial Action Under RCRA: Fighting to Catch Up

The investigation and evaluation of hazardous waste disposal sites only recently became a major issue for the EPA. Not until mid-1979 did the agency begin to develop a strategy for addressing problems posed by the uncontrolled sites. Until that time, there was no coordinated, agencywide approach to the problem.

Through 1981, the GAO continued to monitor hazardous waste management programs mandated by the RCRA. In a study, requested by the House Subcommittee on Oversight and Investigations, on alternative waste disposal and cleanup methods, which was released in November 1980, the GAO concluded that: (1) solid waste disposal on land is the cheapest and therefore predominant alternative, but it is also the most dangerous to the environment; (2) deep well injection and high temperature incineration both offer effective alternative means of disposal; (3) basic research on these three methods has already been completed and as of mid-1979 EPA had revised its research strategy to emphasize hazardous waste identification, uncontrolled waste site problems, hazardous waste technology and risk assessment; and (4) regional or areawide disposal facilities are being planned for the future. As a forecast of future trends, the GAO noted EPA's belief that "the States, either separately or in regional groups, must assume the prime responsibility for . . . hazardous waste disposal."[54] But the GAO also criticized EPA on two particular points: the failure of the agency to undertake a comprehensive, national inventory of closed and abandoned waste disposal sites, the extent of danger they pose, and the estimated total cleanup cost; and the yet-to-be-devised standards for the design and operation of hazardous waste disposal facilities.

The task of locating hazardous waste disposal sites has consumed much of EPA's time and the results have been frustrating. To indicate how imprecise the data are, the agency estimated in November 1978 that there were 838 disposal sites containing significant amounts of hazardous wastes, 103 of which were judged to be potentially dangerous. Three months later, in February 1979, an EPA-contracted study estimated that between 32,000 to 51,000 sites nationwide may contain hazardous waste. But since early 1979, the agency and the various states have begun to identify and document potential hazards. Whereas 4,098 sites had been identified by December 31, 1979, that number had increased by 112 percent to 8,677 potentially hazardous sites by December 31, 1980, with more than 400 new sites being discovered each month.[55]

In July 1981, the GAO cited further evidence that EPA implementation of RCRA was not proceeding smoothly. As of June 1981, EPA had not approved a single state solid waste management program. Subtitle D of RCRA had originally intended to see the several states adopt their own management programs with federal assistance and advice, but such plans had not materialized.

The GAO attributed this delay in part to the agency not publishing necessary guidelines for developing and implementing state solid waste management programs until July 31, 1979, fifteen months after the date required by RCRA.[56] The report also noted that the "open-dump" inventory was still incomplete and did not fulfill the original purposes for which it was intended. And last, future loss of federal funding after FY 1982 to the states for solid waste management would have an adverse effect upon the later planning and operation of such programs. In summary, the GAO was not optimistic about the states' capacity to continue such waste management programs without federal aid.

Given the severity of the hazardous waste problem, recent passage of the Comprehensive Environmental Response, Compensation, and Liability Act (CERCLA) of 1980 is a significant development in the continuing battle to preserve the environment. Signed by President Carter on December 11, 1980, Superfund, as it is known, is a revolving trust fund that will finance the cleanup of hazardous waste sites first, and then try to recover the incurred costs from responsible parties. It is intended to complement existing laws governing hazardous wastes and to encourage responsible parties to reduce voluntarily the damage incurred by this waste. The fund provides for a budget of $1.6 billion over fiscal years 1981–85 for the purpose of cleaning up dumpsites where the perpetrator is unknown, cannot be located, cannot afford cleanup costs, or has abandoned the site. It also provides for another new agency within the Public Health Service, the Agency for Toxic Substance and Disease Registry, which reports directly to the Surgeon General and maintains a literature and data inventory on the health effects of toxic substances.

It is still too early to assess the full impact of Superfund, and the Reagan administration has sent mixed signals concerning its commitment to protecting the environment. On July 29, 1981, EPA administrator Anne Gorsuch testified before the House Subcommittee on Commerce, Transportation and Tourism. She insisted that, in spite of several delays in implementing the act, "I have made the full implementation of Superfund the *highest priority* of the Environmental Protection Agency."[57] However, the proposed FY 1983 budget for the EPA called for a 20 percent cut in funds, to $975 million, and for FY 1984, the agency may request only $700 million, one-half the FY 1982 budget.[58] Morale within the agency supposedly is declining, and its task of inventorying thousands of chemicals now in the market, as required by TSCA, is being threatened, as well as its having to promulgate safety standards for chemical dumpsites.

Apart from supporting CERCLA, the administration's statements and certain EPA actions in 1981–82 indicated some confusion about its commitment to upholding environmental policy. In May 1981, there had been reports that the administration intended to make some major concessions to the chemical industry by relaxing certain regulations and thereby reducing the cost of disposing of hazardous wastes.[59] Apparently, standards devised by the Carter administration requiring blanket licensing procedures for all hazardous waste disposal sites would be replaced by a more lenient licensing standard which depended upon the "degree of hazard" posed by chemicals stored or disposed of at the sites. The modification appealed to the administration on both philosophical and budgetary grounds, and it was attractive to a chemical industry which had advocated the strategy change for years.

Several events in 1982 indicated the erratic pattern of hazardous waste policy within the Reagan administration. On February 25, the EPA proposed to reverse previous rules (to have become effective on November 19, 1981) that prohibited the burying of hazardous liquids in drums at waste disposal landfills,[60] thus permitting disposal sites to accept up to 25 percent of their capacity as barrels of toxic liquids. The agency said that it was suspending the ban for a ninety-day period, in part because it was unworkable and because industry had complained about the cost of complying with the previous regulations.

This apparent policy change prompted several protests from environmental groups and members of Congress supporting RCRA and other hazardous waste policy mandates. Commenting upon the agency actions during the preceding year, Rep. James Florio (D-NJ), Chairman of a House subcommittee with jurisdiction over hazardous waste policy, characterized EPA actions as "a wholesale retreat" from the efforts to clean up hazardous waste disposal practices.[61] And Khristine Hall, an attorney for the Environmental Defense Fund, noted that during the Carter years there was a commitment to cleaning up the environment, but "any commitment now is with regard to loosening regulations."[62]

On March 19, 1982, the EPA reversed its three-week-old decision and established an interim rule prohibiting the burial of any container in which toxic liquids are standing in observable quantities.[63] Gary N. Dietrich, director of EPA's Office of Solid Wastes, stated at the time that "it was a mistake to suspend the rule."[64] Rep. Toby Moffett (D-CN), chairman of the House Government Operations Committee's Environmental subcommittee, praised the EPA reversal: "I am glad EPA has finally come to its senses. It was clear from the beginning that the 90-day suspension and proposed regulations were ill-conceived and a danger to public health. This is a clear victory and the first of its kind during this administration for those of us who truly care about environmental protection."[65]

On March 12, in response to the CERCLA mandate of 1980, the EPA announced a "national contingency plan" for dealing with abandoned hazardous waste disposal sites.[66] But, because the plan dealt with merely administrative procedures, it was criticized by environmentalists and some members of Congress. On March 14, 1982, the chemical industry received another concession from the administration when the EPA suspended its requirement that manufacturers of hazardous wastes report annually on the disposal of such wastes.[67] The agency indicated that henceforth it would survey only 10 percent of the companies annually. Agency officials stated that the change would allow for a more cost-effective accumulation of data required by RCRA and thus impose a lesser burden on chemical companies and other waste generators.[68] Whereas

Dietrich, EPA director of solid wastes, supported the decision to reduce annual reporting requirements, Hugh B. Kaufman, assistant to the director of the hazardous site control division in the EPA and a frequent critic of the agency's enforcement work, declared that without the annual report the EPA would not have sufficient information on which to monitor compliance. Noted Kaufman, "What this means is that the only mechanism giving the agency information to prevent midnight dumping of toxic wastes is now being abandoned."[69]

The barrage of criticism from environmental groups, industry, state officials, and members of Congress, stemming from these policy proposals in March, led to a further reassessment of ongoing EPA policy in late spring and early summer of 1982. On April 12, the EPA reversed a much-criticized decision made in October 1981, and said that an estimated 10,000 hazardous waste disposal facilities would now have to obtain liability insurance protecting people from chemical contamination.[70] The agency said that disposal facilities would have ninety days to obtain minimum coverage of $1 million per accident and $2 million per year. Large companies could satisfy the insurance coverage requirement by showing that they have sufficient financial assets to meet potential claims, including a net worth of at least $10 million. EPA Assistant Administrator Rita Lavelle said that earlier the agency had felt that insurance liability was a state, rather than a federal, issue, "but, frankly, the testimony came in otherwise."[71] Public comment on the insurance subject was nearly unanimous in favor of requiring insurance for hazardous waste sites. EPA Administrator Gorsuch said in a written statement accompanying the agency release that the insurance requirements "have been asked for by state agencies, the regulated industry, and the general public."[72]

Actions by the EPA in the summer of 1982 were rather surprising, given the agency's past series of reversals and ambivalent policy announcements. On July 13, after six years of public and private debate and numerous court battles, the EPA released approximately 600 pages of final operating requirements for 2,000 existing landfills and storage ponds handling toxic wastes, thereby completing the body of rules needed to supplement RCRA.[73] The regulations set design and construction standards for containing contaminants at landfills, impoundments, and other land disposal facilities, as well as requirements for monitoring the operation of the sites and for undertaking corrective action if they leak. The rules would take effect in January 1983 and could cost industry as much as $500 million a year to comply, although that amount could increase if many of the 2,000 existing waste sites are required to undertake extensive cleanup programs.

Conclusions and Policy Implications

This survey of two major pieces of environmental legislation in the 1970s focuses upon many of the variables identified by previous policy analysts who have attempted to explain why there is often a large gap between policy and performance. Like earlier studies of implementation, it has discussed such variables as communications, resources, statutory factors, and several economic and political factors that often delay implementation.

First, both TSCA and RCRA contain very broad, encompassing goals which demand extensive cooperation among the EPA, industry, state government, and the public. For this to occur, some consensus must be reached on the severity of the problem, how best to resolve that problem, and what constitutes successful performance. Under both statutes, these matters have frequently been left to the EPA and target groups, and the result has been delay.

Second, problems of communication among the many actors involved in the implementation of TSCA and RCRA have beset government's attempt to deal with toxic substances and hazardous wastes. EPA has begun a massive survey to catalog chemical substances and hazardous waste disposal sites. And identification is only the first of its many tasks, since extensive rulemaking is then necessary to monitor, contain, and reduce public exposure to harmful substances. Unfortunately, responsible offices within the EPA were delinquent, until recently, in publishing rules required for effective implementation of TSCA and RCRA. This delay has resulted from several factors—alleged perfectionism and timidity of EPA in response to hostile publics (both industry and consumers), inadequate staff and budgetary resources, and declining morale within the agency. In time, EPA personnel could lose their sense of commitment amid criticism from industry, from the public that is affected and alarmed by the presence of hazardous substances in the environment, from congressional staff and committee members, and from presidential politics. All these detractors have their own unique motivations to explain their behavior, but to the EPA they all seem accusatory in nature.

Third, developments in implementation of RCRA guidelines in 1982 indicate ambivalence within the Reagan administration on which goals are more important—reducing the burden and cost of enforcing environmental regulations or maintaining faith with past legislative mandates. Some of this hesitancy may be attributed to the administration's having to contend with multiple voices from Congress, state government, industry, and environmental groups. Not only has the administration had to be sensitive to these different demands, but some reports have noted that fissures are developing in the administration's facade of unity, the most notable contrast being the stance of high ranking officials within the EPA and the Office of Management and Budget.[74] To date, it has been difficult to determine which will prevail. These internal struggles between policy makers within the administration will likely continue until one or the other camp emerges with a clear advantage to dominate future decisions. In the meantime, it will continue to send mixed signals to industry and the states as to what should be done.

Finally, financial resources required for protecting the environment from toxic substances and hazardous wastes are limited. Given the severity of the problem and the recurring cries to reduce government spending, waste, and inefficiency, Washington and the several state governments face major tasks ahead. And yet there is reason to believe that the goals of TSCA and RCRA may eventually be realized in spite of the several problems the legislation faces. States have begun to deal with hazardous waste management at the urging of EPA, and as some recent research indicates, the states' capacity to respond may well be determined by political, organizational, and technological rather than mere economic variables.[75] There is hope that as both the public and private sectors become more aware of the grave ecological problems accompanying the production and disposal of chemical substances and by-products, their resolve to do something about these problems will stiffen. Though the cost of taking remedial action may be high, the ultimate costs of inaction could be fatal.

3. Superfund: Preimplementation Planning and Bureaucratic Politics

Steven Cohen and Marc Tipermas

The federal government usually does not plan how to implement a new program until after implementation has already begun. It is even rarer for an agency to plan for program implementation prior to enactment of legislation authorizing a new program. This chapter examines one of the rare exceptions—EPA's Superfund program for response to hazardous waste incidents.

The "preimplementation planning project," as it came to be known at EPA, was originally premised on the notion that a relatively rational look at program implementation issues prior to enactment of legislation would help EPA meet statutory deadlines called for in the Superfund legislation. This was something EPA had infrequently been able to do in other environmental program areas. While the project had some value as a planning tool, this chapter suggests its true value may have been due to more to its use as a bureaucratic tactic for establishing an area of distinctive competence.

What Is Superfund?

In 1976, the U.S. Environmental Protection Agency (EPA) and the New York State Department of Environmental Conservation began to investigate chemical contamination at Love Canal in Niagara Falls, New York. In August 1978 New York's Commissioner of Health declared a health emergency at Love Canal. Before the year was over, 237 families had been evacuated from the area. By July 1980 a significant number of Americans had heard the story of Love Canal, and an ABC News-Harris Poll found that 86 percent of the American public favored "giving the problem of toxic chemical dumps and spills a very high priority for federal action."[1] On December 11, 1980, President Carter signed into law the Comprehensive Environmental Response, Compensation, and Liability Act of 1980 (CERCLA)—the $1.6 billion Superfund—which had been passed by an overwhelming congressional majority the week before.

In many respects Superfund was a path-breaking piece of legislation. Under Superfund "those that pollute, pay"—upfront—into a trust fund to be used for cleaning up hazardous waste sites and responding to emergency incidents. The trust fund is financed by $1.38 billion in taxes on the production of certain organic and inorganic chemical feedstocks and crude oil, and $220 million in general appropriations. The legislation is not a traditional EPA regulatory program in which EPA sets standards, issues permits based on those standards, and then takes enforcement actions for failure to comply with permits. To be certain, CERCLA does require private parties to report: (1) to EPA the existence of hazardous waste sites they own or operate and (2) to the Coast Guard's National Response Center certain releases of designated hazardous substances. It also establishes the liability of private parties for the costs of response to a release or substantial threat of release of hazardous substances from waste sites, facilities, or vessels they own or operate. On balance, however, Superfund should not be seen as a regulatory program. According to former EPA Assistant Administrator Eckhart C. Beck, the philosophy behind the program was one of "shovels first, lawyers later." Of course, where financially viable responsible parties can be found the government must first request that such parties take response actions. But when hazardous waste sites are truly abandoned, or corporate recalcitrance results in delayed response, Superfund is a direct-action program enabling the government to move swiftly and in advance of protracted litigation to protect public health and the environment.

Models of Decision Making and Superfund

Rational and Incremental Decision Making

Table 3.1 compares the rational and incremental modes of policy making. In 1955 the use of the rational model of decision making was tested empirically by Martin Myerson and Edward Banfield.[2] After conceiving of planning as the rational selection of a course of action Myerson and Banfield evaluated the decision-making activities of the Chicago Housing Authority regarding the placement of housing projects—in terms of this rational model. Although accepting the caveat that pure rationality was impossible, they maintained that a decision could be considered either more or less rational and stated that "decisions may be made with more or less knowledge of alternatives, consequences, and relevant ends and so we may describe some decision making processes as more nearly rational than others."[3]

In *A Strategy of Decision* David Braybrooke and Charles Lindblom note that "Martin Myerson and Edward Banfield make a persuasive case against [the existence of] a thorough going examination of the problem situation, alternative courses of action, and of consequences in their case study of policy making with respect to the location of public housing in the city of Chicago."[4]

The main reason we do not make decisions as the rational model prescribes is that we simply cannot; or as Lindblom and Braybrooke observe, "The synoptic [rational] ideal is not adapted to man's limited problem solving capabilities."[5] They go on to point out that the synoptic ideal is not suited to: (1) the inadequacy of information; (2) the costliness of information; (3) the closeness

Table 3.1. Rational and incremental models: A comparison

Rational process	Incremental process
1. The analysis of the situation	1. The policy maker works directly on agreement on specific projects, policies, or programs, and *not* toward agreement on abstract goals.
2. End reduction and elaboration	2. The policy maker is concerned with the comparison and evaluation of increments only.
3. The design of courses of action	3. The policy maker considers only a restricted number of policy alternatives.
4. The comparative evaluation of consequences in light of ends	4. Ends are adjusted to means, as well as the other way around. The problem is constantly redefined. Policy objectives are derived largely from an inspection of our means.
5. The selection of the preferable alternative	5. Many alternatives are attempted in a series of "attacks" on the problem of concern.
6. The assessment of the action taken in light of both ends and means	6. Assessment relies on experience and feedback because policy making is remedial. In Lindblom's view: "public problem solving proceeds less by aspiration toward a well defined future state than by identified social ills that seem to call for remedy."[a] In short, ultimate ends are not of great concern.

a. Charles Lindblom, quoted in Galloway (see below), 179.
Source: Thomas D. Galloway, *The Role of Urban Planning in Public Policy-making: A Synthesis and Critique of Contemporary Procedural Planning Thought*, Ph.D. Dissertation (1971), Ann Arbor, Mich.: University Microfilms, 1974. The figure is drawn from material on pp. 71 and 175–80.

of observed relationships between fact and value in policy making (it may be more difficult to separate ends and means than rational theory assumes and demands); and (4) the openness of the system of variables with which it contends (the difficulty of ever adequately understanding all the relevant variables). The root causes of incrementalism are found both in the inability of human decision makers to act according to the rational ideal, and in the fact that decision makers are satisfied with normal decision-making processes. Of course, in some cases decision makers have *no choice* and must accept nonrational decision processes because they lack the resources needed to establish rational processes. In order to divert resources to a "rational" planning exercise, the Superfund organization had to be successful enough at bureaucratic power politics to obtain extra resources. Nevertheless, it is possible to identify rational "islands" in the incremental "sea." If management can be convinced that extraordinary planning processes are required in a priority situation, resources will be diverted, and a measure of rational decision making can be achieved.

Incrementalism can be seen to have risen out of some sort of understanding of the limits to human capability. As John Steinbrunner indicates in *The Cybernetic Theory of Decision*, something akin to incremental decision making literally abounds in nature. Steinbrunner gives the example of the tennis player returning his opponent's serve. The player does not rationally collect all the data relevant to the decision-making process and rationally calculate a response to the situation. The player does not figure out the response in terms of air currents, the exact velocity of the ball, measurement of the angle of approach, and so on. Rather, the player scans the situation for the one or two relevant variables needed to return the ball. Summarizing, Steinbrunner notes that:

> We have already seen the secret of the ability to handle variety. The simplest cybernetic mechanisms do not confront the issue of variety at all, for they make no calculations of the environment. The mechanisms merely trace a few feedback variables and beyond that are perfectly blind to the environment. Hence degrees of complexity in the environment are of no concern within the decision making mechanism itself, and the burden of calculation which the analytic (rational) paradigm seems to impose is not a problem for cybernetic assumptions.[6]

Decision-making processes acting as cybernetic mechanisms deal only with what they must, reacting only when the environment changes the variables the mechanism deems critical (as a thermostat would react to the temperature and turn on the heat). In this sense, incremental decision making is simply the survival of those procedures and institutions best suited to deal with critical environmental variables. It is the normal way that humans deal with complexity, and the rational ideal is never even considered.

For the most part government makes decisions and policies in the manner outlined by Braybrooke and Lindblom. Normally, government establishes many decision-making processes (roughly analogous to cybernetic mechanisms) that deal with complexity by only addressing a few variables.

In noncrisis situations, the environment is stable and "decomposable" (i.e., able to be separated into segments in order to act on it), although we understand very little of it. Still, we have enough understanding to proceed. When the amount of change needed is small and our understanding of a situation is low we pursue a strategy of disjointed incrementalism. This strategy is "remedial," "serial," and "exploratory": moving *away* from the problems rather than *toward* solutions. Braybrooke and Lindblom note that "analysis and evaluation are socially fragmented, that is they take place at a very large number of points in a society. Analysis of any given problem area and of possible policies for solving the problem is often conducted in a large number of centers."[7] The concept of overlap, or "where one decision maker leaves off, another will pick up," is central to cybernetic theory, and is central to the incremental model's response to the need for comprehensiveness addressed by rational theory. In our view incrementalism is an accurate description of normal decision making. However it is less useful in explaining decision making during revolution, crises, and what Braybrooke and Lindblom termed "grand opportunities." In these situations Yehezkel Dror's "extra rational processes,"[8] Steinbrunner's "cognitive based inference machines,"[9] and Charles O. Jones's nonincremental (and we would argue "cognitively based") "speculative augmentation"[10] are core elements of the decision-making process. In these situations normal decision-making procedures cannot produce needed outputs. Focusing decisions on a few obvious information sources that can trigger appropriate cybernetic mechanisms does not permit solution of the problem at hand. In crisis situations humans infer from available theory and data and make a decision. Essentially we guess. We speculate in order to augment our normal ability to produce decisions. It is in these situations that incrementalism cannot describe decision processes.

Superfund and the Rational Ideal

The Superfund planning operation can be explained as being the result of a "somewhat grand opportunity," or crisis, when incremental processes were unable to produce acceptable results. The planning project was the result of a political-bureaucratic decision to set aside a small amount of resources to permit the bureaucracy to conduct an exercise in organized guessing. Although the Superfund preimplementation planning project proceeded in stages consistent with the rational process described earlier in table 3.1, the project was *not* an effort to rationally plan. In the programmatic area, it was an effort to "rationally" speculate. The Superfund planners guessed all the time. They never had all the information needed to make rational decisions. Moreover, the project staff often had an incomplete sense of their goals (as indicated by the lack of a single piece of legislation for which to plan). Nevertheless, the project cannot be seen as a normal piece of bureaucratic business: it was clearly nonincremental. An effort to plan the program in seven functional workgroups was in part an attempt to follow the stages of the rational method, but because of its limited information base, truncated time frame, and bureaucratic turf objectives, cannot be characterized as mainly rational.

In addition, the large number of non-workgroup preimplementation activities can be seen as *mainly* incremental. The non-workgroup activities were basically the "firefighting" efforts required to create a new organization. In sum, the preimplementation project was an eclectic collection of incremental, rational, and speculative processes and products. The project had a bit of everything: organized and systematic guessing, the codification and reiteration of previous operational experience at hazardous sites and spills (an incremental process), and the rational design of a workgroup process as well as the rational design of the financial management and management information systems. We turn now to a more detailed discussion of this process.

Agency Involvement in Superfund Activities: A Chronology

The enactment of Superfund was the culmination of a four-year process that began in earnest at Love Canal. As the magnitude of the environmental problem became more obvious, and the potential for political trouble became clearer, EPA's top management began to see the cleanup of existing hazardous waste sites as a critical environmental problem. Although the agency was slow to move at first, EPA staff conceptualized and wrote the initial hazardous waste site Superfund bill and the agency eventually became the prime mover behind the legislation.

Several EPA and congressional studies, and the public outcry being voiced at hazardous waste sites throughout the United States,[11] made hazardous waste management a top priority of EPA and convinced agency management to: (1) develop and work for the enactment of Superfund legislation; (2) establish a National Hazardous Waste Enforcement Task Force that would increase use of existing enforcement authority under the Resource Conservation and Recovery Act (RCRA) and the Clean Water Act (CWA); and (3) establish a new Hazardous Waste Site Control Branch to assist in the discovery and investigation of waste sites and to foster the use of existing cleanup authorities at such sites. On June 11, 1979, President Carter submitted his administration's Superfund proposal (developed at EPA by a task force under the direction of Assistant Administrator Thomas C. Jorling) to Congress and called for immediate action. Shortly thereafter, in a memorandum dated June 27, 1979, EPA Deputy Administrator Barbara Blum established an interim Hazardous Waste Site Enforcement and Response System within EPA and termed the system the "highest priority effort for the Agency." On the surface, then, by the end of June 1979 EPA had done all it could to establish a substantial hazardous waste response program pending passage of the legislation.

Origin of the Preimplementation Planning Project

In the opinion of EPA staff who played a key role in the development of the Superfund legislation and in the establishment of the interim response and enforcement system, however, an additional step was needed: a planning effort to ensure rapid and effective implementation of legislation once it was enacted. Even though the Superfund legislation called for a large federal trust fund, it was clear that the amount of money being discussed (between one and four billion dollars) would be inadequate for the total cleanup of the nation's hazardous waste sites. If Superfund dollars were to be carefully targeted to the worst waste sites, a well-planned program would be required. In addition, in the summer of 1979, EPA was being criticized for slow implementation of its toxic substances and hazardous waste programs. EPA staff felt that a similar situation could not be allowed to develop regarding Superfund. This was due both to the public pressure for federal action and the fact that EPA, itself, was highlighting the great magnitude of the problem as part of its legislative strategy. If EPA were unable to implement the legislation shortly after enactment, then the agency's overall credibility and effectiveness might well have been severely undermined.

In meetings during September 1979 with Eckhart C. Beck, Jorling's successor as assistant administrator, his staff advocated the establishment of a Superfund preimplementation planning project. Specifically they recommended the creation of (1) a new senior-level management position with responsibility for overseeing both the agency's existing emergency response program and a Superfund planning effort and (2) a new staff office to assist this official in focusing on

> the development of national policies and strategies for the Agency's hazardous waste site discovery, response, [and] containment [program] . . . whether or not the new Superfund is passed and signed into law. However, because passage of Superfund is expected by Spring or Summer, 1980, national policy and strategy development is specifically needed now to assure timely and cost-effective Superfund program development and implementation.[12]

Over the next several months, Beck acted on virtually the entire staff recommendation. In October he decided to create a new Associate Deputy Assistant Administrator (ADAA) for Environmental Emergency Response and Prevention. By February 1980 after the lengthy EPA personnel process worked itself out, Michael B. Cook was selected as ADAA. A reorganization proposal for a new Superfund planning office was developed in November 1979 and finally approved under EPA's complicated internal reorganization process five months later in March 1980. The new planning office was called the Office of Analysis and Program Development (OAPD). OAPD and EPA's emergency response organization, the Oil and Special Materials Control Division (OSMCD), both reported directly to Cook.

Early History: The Working Group Structure

The preimplementation planning project began in March 1980 when the short-handed OAPD engaged a consulting firm to provide analytic support for Superfund planning. The first activity was to prepare a draft workplan for the project. This draft was reviewed by the ADAA, Michael Cook, and OAPD, and then on April 2, 1980, it was circulated to relevant offices throughout EPA. Comments were requested from OSMCD, and the Offices of Water and Waste Management, Legislation, General Counsel, Research and Development, Planning and Management, and Enforcement.

As the draft workplan noted, planning and implementation of legislation that had not yet been enacted presented EPA with an uncertain set of planning parameters:

> Implementation planning for Superfund must allow for significant changes in program design based on major alterations introduced during the markup of the various bills, or during a conference committee deliberation if different bills are approved by the Senate and the House. . . .
> Since a Superfund program does not exist at this point, and since various options are still being debated on The Hill, the Superfund program is actually the "dependent variable" in the planning process, and there are many "degrees

of freedom" as to what might happen in the next two years. The purpose of this first planning task, therefore is to ensure that the process design takes all of the changing parameters into account, but yet recognizes that some things will remain constant and need to be done regardless of what happens to the Superfund proposal.[13]

In order to deal with legislative uncertainty, the approach taken was to reduce the various legislative proposals to components common to each. The goal was to plan an enhanced environmental emergency response program that could be implemented under any Superfund proposal. Perhaps the most significant element of the draft workplan, however, was the establishment of agencywide workgroups in the following seven areas: (1) Administration, organization, and management; (2) financial management and cost recovery; (3) data management and support; (4) program operations, oil and chemical spills management; (5) program operations, hazardous waste site management; (6) research and development; and (7) state programs, local government, and community relations. In addition, in order to ensure that the overall workgroup effort was consistent and timely, a coordinating "synthesis group" was established.

On May 12, after receiving comments on the draft workplan and completing revisions, Cook issued a "Final Work Plan to Prepare a Superfund Implementation Strategy." The memo outlined: (1) the major areas of responsibility for each workgroup, (2) the major policy issues each group needed to analyze, and (3) the major budget issues that each group needed to examine. Each workgroup was to develop a formal planning report for its area of responsibility.

By mid-May, then, all the pieces of an effort to plan rationally for Superfund were in place: issues were to be examined, options were to be developed and weighed, and agencywide workgroups were to carry out the analyses.

The focus of the first workgroup reports was "early decision issues." Early decision issues were defined as those issues that had to be identified, analyzed, and decided at the earliest possible date, either because of their influence on other issues, or because their resolution required a long lead time. The May 12 memorandum set forth a schedule for workgroup activities throughout the summer of 1980.

The tasks undertaken by the workgroups were taken quite seriously by the workgroup chairs and membership. Over $500,000 in contract funds and considerable EPA staff time was devoted to workgroup activities. A veritable blizzard of paper was produced, the value of which will be discussed later in this chapter.

In June, the synthesis group convened a two-day retreat to make preliminary decisions on early decision issues. Many of the policy and operational issues still faced by the Superfund program were first identified during this planning exercise. The workgroup reports, and particularly the integrating summaries included in the several synthesis group reports, began to familiarize EPA's management and Superfund staff with the complex set of problems the agency would face when attempting to implement the Superfund program.

Planning Activities Outside the Workgroup

In addition to the activities centered in the workgroups, a number of critical preimplementation tasks were undertaken by OAPD and OSMCD outside the workgroup structure. OAPD devoted substantial effort to: (1) staff support for the agency's legislative activities regarding passage of Superfund (OAPD staff helped analyze the various Superfund proposals); (2) beginning agency review of Superfund regulations (EPA has a lengthy process of internal review prior to the promulgation of regulations often taking more than a year to complete); (3) beginning the process of letting a multimillion dollar contract for Superfund policy and management analytic support (even combined and expanded, OAPD and OSMCD would never have enough government employees to manage a billion dollar federal program, and therefore extensive contractor assistance would be needed to augment scarce staff); (4) developing and pushing through the bureaucracy a formal proposal to reorganize OSMCD and OAPD into a new Superfund Deputy Assistant Administrator's Office (this process normally took from six months to a year, but only took three months to complete); and (5) developing an innovative and complex community and local government relations program for Superfund (it was obvious to EPA management that Superfund would have to deal with the most difficult public relations problem of any environmental program). OSMCD expended considerable time and funding for: (1) beginning the process of letting national contracts to perform cleanup work, undertake engineering studies, and conduct field investigations of hazardous waste incidents; (2) assisting OAPD in developing the Superfund reorganization proposal; and (3) keeping the existing emergency response program operating while awaiting passage of Superfund (when Superfund enactment was delayed, the maintenance and enhancement of the ongoing program enabled the Superfund organization to maintain a sense of mission and momentum).

By the time the Superfund legislation was enacted, the preimplementation planning project achieved a number of programmatic objectives. In the area of *program development and management support* EPA had completed (1) a master calendar of key program events and milestones; (2) a reorganization of the EPA Headquarters Superfund Program Office into the Office of Hazardous Emergency Response (OHER); (3) plans for the development of major Superfund regulations; (4) a feasibility study of near-term data processing and management information system requirements; and (5) an analysis of the staff and financial resources required to implement Superfund.

In the area of *fund management* the preimplementation planners had completed or nearly completed (1) an assessment of allowable uses of the trust fund; (2) an analysis of near-term financial management systems required for Superfund; (3) a plan for delegating the authority to obligate trust fund monies to on-scene response personnel; and (4) an interagency agreement with the Coast Guard on trust fund expenditures.

In the area of *on-scene response* the preimplementation planners had completed or nearly completed (1) a draft of the National Contingency Plan (NCP) —the major document guiding Superfund response actions; (2) a hazardous site response management plan—detailing the phases and decision points required for long-term cleanup actions; (3) a hazardous waste site investigation manual; (4) action plans for emergency response at twenty critical hazardous waste sites; (5) criteria for setting emergency and long-term response priorities; and (6) identification of priority waste sites for long-term remedial action.

In the area of *intergovernmental coordination* EPA had completed (1) a survey of state capabilities to participate in Superfund implementation; (2) an analysis of citizen-government interaction at hazardous waste incidents; (3) a draft community relations program for Superfund; and (4) a survey of Superfund-related federal agency authorities and capabilities.

Institutional Issues Influencing Preimplementation Planning

The preimplementation planning project was not simply an exercise in rational planning. It quickly developed two objectives that were distinctly non-rational. One was to stake out bureaucratic turf. Cook and his new planning office would attempt to preempt other EPA units from obtaining Superfund functions to the extent possible. It was not that other organizations wanted Superfund, rather it was that Cook needed speedy concurrence and noninterference from other EPA units if Superfund were to be rapidly implemented. By jumping the gun on other units Cook could create a sense of momentum that would give him greater discretion and help Superfund avoid some of EPA's exhausting internal review processes. A second nonrational objective was to meld the new planning office with EPA's existing emergency response organization. The preimplementation planning project was to foster the working relationships necessary to form a single Superfund organization.

Internal Issues

As noted earlier, the new Superfund planning unit was formally titled the Office of Analysis and Program Development (OAPD) and reported directly to Cook (see figure 3.1). OAPD started small. At its inception it had five full-time professional employees and one secretary. Nine months later, by the time the Superfund legislation was enacted and OAPD was reorganized into the permanent staff office for a new Superfund Deputy Assistant Administrator, it had grown in size to twenty-five employees. As figure 3.1 indicates, OAPD was not the only organization reporting to the ADAA for Environmental Emergency Response and Prevention. The other unit was the seventy-person Oil and Special Materials Control Division (OSMCD), a decade-old organization that had been headed by Kenneth Biglane since its establishment. According to its functional statement in EPA's organizational manual, OSMCD was "responsible for developing national policy . . . [for] programs of prevention and response in the areas of oil and hazardous substances." In concert with the Coast Guard, OSMCD was responsible for the nation's oil and chemical spill programs, and for the pre-Superfund hazardous waste site response program.

When OAPD was established, its primary interorganizational relationship was with OSMCD. The fact that most OAPD staffers were younger and less technically oriented did not exactly endear them to OSMCD's more seasoned bureaucrats, many of whom had been with Biglane since the division's birth. Although inadequate authority and insufficient funds had made it difficult to respond to the nation's burgeoning hazardous waste site problem, OSMCD was *the* organization responsible for implementing the EPA's more limited emergency response program upon which Superfund was designed to expand. If enacted, the proposed Superfund legislation would set in motion a program that was considerably more complex than that overseen by OSMCD. If a successful transition was to be made from the old program to the new, the "policy types" in OAPD and the "operations types" in OSMCD would have to be merged into a single cohesive unit. The experienced hands in OSMCD would have to educate the newcomers in OAPD about real-world constraints on program implementation. OAPD would have to devise methods for transforming old institutional routines into the new standard operating procedures required to implement Superfund.

External Issues

An additional critical task undertaken by OAPD was to educate the rest of the agency about the need for Superfund and obtain broad intra-agency consensus behind the program. Before Superfund, EPA's emergency response program was a small—at most $10 million a year—program, of little interest to the other, primarily regulatory units in EPA. With Superfund, this function would be expanded to a $400 million a year program that would be the agency's highest priority and would require agencywide support to succeed.

This education function was facilitated by the fact that each workgroup had a representative on it from OAPD and the synthesis group was made up exclusively of these OAPD staff members. This arrangement enabled OAPD and Cook to influence and interpret workgroup outputs. It insured that the workgroup structure would serve, rather than dominate, the new Superfund organization. By participating in this institutional arrangement, other units in EPA essentially conceded the primacy of Cook's organization with regard to the agency's Superfund activities.

A key role in the planning project was played by the "synthesis" group. Under Cook's persistent prodding, the synthesis group dominated the overall project and was able to meet deadlines and deliver on schedule an analysis of early decision issues and a rough implementation plan.

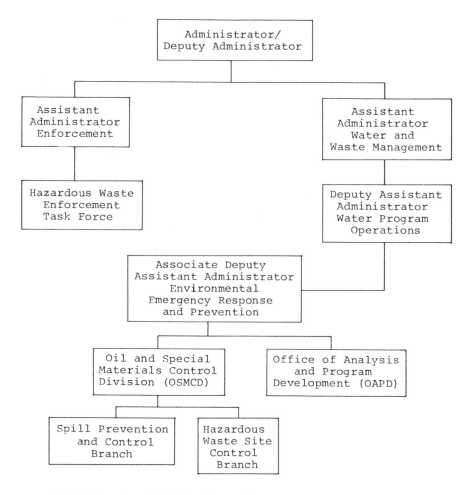

Source: Compiled by the authors.

Figure 3.1. Organizational chart for Superfund-related programs at U.S. EPA headquarters, 1980

In most cases, the workgroups took on the character of the Chairperson's formal organization: the spills operations workgroup (number four) was controlled by the spill response branch of OSMCD; the sites operations workgroup (number five) was an appendage of OSMCD's hazardous site control branch; and OAPD dominated the workgroups on administration and program management, financial management, data management, and state programs.

As the planning project progressed, it appeared that more and more of the critical tasks were performed outside of the workgroup structure. After an initial emphasis on workgroups, lead responsibility for tasks that should have fallen under the jurisdiction of the workgroups was assigned to individual staff or managers under Cook's direct control. This occurred in part because work-group members were drawn from Superfund-related offices throughout EPA. The amount of time that group members outside of Cook's organization could devote to workgroup activities was fairly minimal, leaving most staff work to OAPD, OSMCD, or contractor staff. As a result, once Cook's office had staked out the Superfund turf and taken full advantage of the turf-legitimizing function played by the agencywide workgroups, the workgroup structure collapsed. By the time the Superfund legislation was enacted in December 1980, the work-groups had disappeared. A second schedule for workgroup activities was never developed. Functions once performed by the workgroups were assigned to the newly reorganized Office of Hazardous Emergency Response (OHER). Never-theless, by December 1980 the preimplementation planning project had accom-plished an impressive array of political and bureaucratic (as well as program-matic) objectives, and had completed a number of critical activities. OAPD and OSMCD had been moved up a notch in the EPA hierarchy, reported directly to a deputy assistant administrator (DAA), and formed the core of the Office of Hazardous Emergency Response. OHER became the "Superfund Office." It possessed the Superfund jurisdiction, and was able to claim hazardous waste cleanup as its area of distinctive competence. With that claim to distinctive competence came the spoils of bureaucratic war: more staff, more office space, more funding, and more power.

The Value and Limitations of Preimplementation Planning

Value

Because of Congress's tendency to enact vast quantities of symbolic environ-mental legislation, and the related tendency by Congress and OMB to deny EPA the additional resources needed to implement new legislation, most environmental programs are very slow in getting off the ground. EPA usually misses the statutory deadlines for promulgating regulations. In the case of Superfund, at least some of EPA's top management shared the perception that hazardous waste was a highly salient political issue, and that EPA could not afford to allow Superfund program implementation to be delayed. The im-mediacy of the threat to environment and public health was so obvious that agency management decided that it was essential to create the image and the reality of an action-oriented, dynamic program ("shovels first and lawyers later").

The Superfund preimplementation project was established to give the agency a head start on Superfund implementation. Although the value of specific project outputs can be questioned, the project was responsible for establishing new organizational capacity in EPA, an event of unquestioned value. Particular outputs may have been unsuccessful, but the total effort constituted a valuable learning experience for EPA. In that sense the whole was greater than the sum of the parts. The preimplementation project allowed the Superfund organization to develop its area of distinctive competence before operational pressures began. It enabled EPA to minimize the effects of mistakes always made by new organizations. In addition, the Superfund organization was able to preempt turf battles over Superfund by working in the area at a time when risks were high (the absence of legislation) and rewards for involvement were low (due to the absence of significant new resources to participate in Superfund).

The strategy of planning through agencywide workgroups allowed OHER to claim Superfund turf, not because it was able to deal with every Superfund issue, but because it was a "managerial innovation" that created the image of an organization that was "out front" of Superfund. Its effect was roughly analogous to that found by Harvey Sapolsky in his study of the Polaris missile system. Regarding the much touted Program Evaluation and Review Technique (PERT) Sapolsky observed that "the program's innovativeness in management methods was . . . as effective technically as rain dancing. It was, nevertheless quite effective politically."[14] The Superfund workgroups were more effective than rain dancing, but were at least as important a tool for bureaucratic politics as they were for rational planning.

The absence of *major* turf battles and the partial freedom from operational pressures enabled the Superfund organization to become familiar with relevant policy issues. The project permitted greater development of issues, and more refined problem definition than would have otherwise been possible. This early issue development insured that operational decisions would be made with a greater than usual amount of forethought. Operational mistakes would still be made, of course, but these errors would tend to be errors of commission rather than errors of omission. Mistakes that result from deliberate choices will tend to be identified earlier due to a greater understanding of the trade-offs involved in making those choices.

Another value of the project was that the "head start" made it possible to meet statutory deadlines, such as the promulgation of key regulations and the publication of a hazardous waste site response priority list. Although the Reagan administration halted these efforts, they would have permitted the rapid implementation of Superfund.

Other benefits obtained by the project are more concrete and measurable. The Superfund organizational structure was firmly in place when legislation was enacted. Recruitment of staff was well underway. Year-long contract procurement processes had already begun. Internal regulation review processes had been underway for several months. The Superfund office was lodged in relative comfort in the basement (formerly the garage) of EPA's headquarters building in southwest Washington. These open-space quarters even acquired a special name among Superfund staff—"The Blue Lagoon"—after the football field–size turquoise carpeting covering the floor. Office machinery had been purchased, a softball team was established, and T-shirts with the Superfund logo were already in use. Although these mechanical processes and material possessions may seem rather pedestrian, they are the lifeblood of the organizational organism. In fact, an examination of OAPD's correspondence files shows that nearly twenty percent of the memoranda written during the preimplementation period related to such mundane matters as office space, staffing, furniture, contracting, and the like. It was unquestionably useful to have these "routine," but absolutely necessary, matters taken care of before January 19, 1981, when Jimmy Carter delegated Superfund implementation to EPA under the first Superfund Executive Order.

Limitations

While the preimplementation project had great value it also had significant limitations. A primary measurable limitation was the amount of effort spent planning for provisions of the legislation not included in the final bill. The legislation passed by Congress was a last-minute compromise. Although it retained many of the features in the original Carter administration proposal, its third party liability provisions and several other sections of the act were somewhat weakened in the final bill. Efforts aimed at planning the implementation of the more stringent administration proposal were a bit wasteful.

The change of administration from Carter to Reagan caused a second area of misdirected effort. The original Superfund proposal and the legislation finally enacted were strong, federally oriented programs. The states had a significant role in setting priorities and in sharing costs but overall Superfund was designed to be a federal program. Unlike other environmental programs, Superfund did not have a "state grant" provision; states would participate through contracts or cooperative agreements. These mechanisms permit a greater measure of federal control. When the Reagan administration came into power, priorities shifted, and EPA was instructed to quickly delegate as much of Superfund to the states as possible. As a congressman, OMB Director David Stockman had been a vociferous opponent of the Superfund proposal, instead favoring a smaller program of grants to the states. Stockman and other Reagan officials found that the legislation could be interpreted in such a way as to transform Superfund into more of a state-dominated program. This shift in emphasis had the effect of leaving unplanned much of the fund-financed cleanup program. EPA would find that, in late FY 1981 and through much of FY 1982, the only way to obligate Superfund money for long-term cleanup was to enter into cooperative agreements with the states. The plans for a strong federal program were set aside and the agency began a poorly thought out strategy of state delegation. A primary determinant of early funding decisions was the ease of entering into cooperative agreements with particular states. Although the agreements all pertained to important hazardous waste sites, EPA would probably acknowledge that the first sites acted on were not the nation's highest priority hazardous waste sites. The lack of a coordinated program was a result of the preimplementation project's justifiable failure to anticipate that Superfund would become a state-oriented program. The other side of that coin is that some of the planning for a federally centered program was wasted effort.

Another potential limitation of the project was what could be termed "vacuum planning"—planning with no basis in reality. The Superfund planners in OAPD were frequently criticized by the "operations-types" in OSMCD for unrealistic planning goals and processes. However, as the organizations established increased formal and (particularly) informal working relationships, the two units began to merge and the problem of "vacuum planning" became less pronounced. It is easy to see how planning could become divorced from reality if the planning unit did not have a close working relationship with the operations unit. On the other hand, too close a link with the operations unit could distort the planning unit's mission and force it to become involved in overly short-range planning for current operations.

As noted earlier, the project can be seen as a partially nonincremental exercise in organized guessing. Hence, as in any game of chance, the key is to increase the odds of guessing right. It would, of course, be optimal if one could amass so much information and analyze so many options that it would be possible to come up with a menu of implementation plans to fit any political eventuality. Certainly, as in gambling, the more times you play, the more times they pay. Unfortunately each play expends resources, and the option of comprehensive contingency planning is usually only available to military planners. For the rest of the federal bureaucracy it is usually necessary that the time, staff resources, and contract dollars available for planning be targeted carefully.

The Superfund organization decided to use its headstart to clarify and force management decisions on "technical" implementation issues. Typically, these routine but absolutely necessary decisions absorb enormous amounts of time and energy. Issues related to organization structure, management strategy, decision rules, and funding criteria take a good deal of time to decide. By having internal agreement on some of these issues *before* enactment, agency management thought, it might be possible to take on operational responsibilities at a more rapid than usual pace.

The reorganization that created the Office of Hazardous Emergency Response is an example of a time-consuming, routine process that was carried out during the planning project. Reorganization proposals in EPA are subject to a lengthy negotiation and review process. The units being reorganized must negotiate with the personnel department and agree to specific staffing patterns, grade levels, and position descriptions. Prior to the Reagan administration, the agency's Management Organization Division's approval was required on the proposed organization structure, and on specific, detailed statements of organization functions. Although the process was streamlined under Administrator Gorsuch, in the past, high level officials throughout the agency were asked to concur on any reorganization proposal after it was approved by the Management Organization Division, and before it was submitted to the administrator's approval. The OHER reorganization was approved by the administrator before the Superfund legislation was enacted. The dislocation that naturally follows any reorganization (new reporting relationships, new duties, etc.) did not affect

Superfund implementation. The benefit of having the new organization in place is beyond question. No staff time was diverted from operational responsibilities to negotiate the details of the reorganization, and staffers had the security of knowing who their new bosses were.

The preimplementation project's attempts at policy analysis are less obvious successes, but were still worth undertaking. The process of analyzing key policy issues always involves a great deal of individual and institutional learning. The "percolation" of ideas and concepts takes time. Even though many of the initial attempts at formulating policy were not ultimately utilized in decision making, asking these questions at an early stage in the implementation process enabled the staff to spend an extra measure of time sorting through the implications of various options.

In our view, advancing both the mechanics and analysis necessary to implement the program was valuable. However, where choices must be made between devoting resources to policy development or "mechanics" we would advocate devoting resources to bureaucratic housekeeping. The benefits of these activities are more certain and tangible. However, at a minimum, some amount of policy analysis is needed both to stake out a legitimate claim to turf and to keep talented staffers interested in the project.

In sum, the chief lessons of the Superfund project were: (1) under conditions of uncertainty the first task in the planning process is to identify whatever certainty exists; (2) once essential functions are identified, planning should be limited to those subjects; and (3) focus efforts on time-consuming bureaucratic necessities such as contracting, personnel actions, securing office space and equipment, and reorganization.

4. Federalism and Hazardous Waste Policy

Harvey Lieber

Intergovernmental problems have hindered the implementation of legislation designed to regulate hazardous wastes. Of particular significance are the United States Environmental Protection Agency's (EPA) delays in issuing regulations and inadequacies in state laws and programs. Using an intergovernmental focus, this chapter traces the evolution of federal legislation, especially Subtitle C of the Resource Conservation and Recovery Act (RCRA) of 1976, as it is currently being carried out by EPA in the state program authorization process.

Federal-state relations in this area are explored first with a view toward the impact of changes in presidential administrations on these relationships. The chapter then concentrates on an analysis of intergovernmental factors which have constrained RCRA's implementation at the state level.

Also briefly described are the aforementioned state-level legislative and programmatic inadequacies and EPA's regulatory delays. The agency's implementation process has served to confuse states seeking to follow federal guidelines and assume control over hazardous waste management. Scientific and technological uncertainties, as well as minimal commitment until recently by EPA, have resulted in a program which may take at least a decade and a half from its initial passage to achieve implementation as its legislative sponsors intended.

National-State Environmental Relationships

Hazardous waste regulation under the Resource Conservation and Recovery Act is one of a half-dozen major EPA programs. In terms of the degree of national control and direction vis-à-vis state and local governments, it ranks in the middle between such basically national, almost preemptive, laws as the Toxic Substances Control Act and the Federal Insecticide, Fungicide, and Rodenticide Act on the one hand and the EPA's noise, drinking water, and solid waste programs, which have generally consisted of technical assistance and funding support to local and state programs. Thus hazardous waste, like air and water pollution, is a federal program with considerable tension between federal regulators and state program administrators.

The Resource Conservation and Recovery Act of 1976 (RCRA), P.L. 94-580, attempts to provide for strict regulation of hazardous waste through the establishment of a permit and manifest program. However, it should be noted that the act's main focus is on nonhazardous or solid waste. It requires unacceptable municipal waste sites to be either closed within five years or improved according to strict standards, authorizes funds for state solid waste management planning, and gives EPA the authority to investigate various solid waste management alternatives and make recommendations for resource recovery and recycling.

Subtitle C of the act directs EPA to promulgate regulations to protect human health and the environment from improper management of hazardous wastes. It requires EPA to define criteria and methods for identifying hazardous waste and provides for the establishment of a manifest tracking system and record-keeping to assure that hazardous wastes are managed properly and monitored carefully from the point where they are generated to the place where they are sent for final disposal. Besides giving standards for generators and transporters, it requires facilities that store, treat, or dispose of hazardous wastes to obtain permits which must be periodically renewed upon inspection and which specify procedures to protect public health and safety.

Thus for both solid and hazardous waste programs, EPA, in conjunction with state and local governments, was to develop and implement programs through planning grants, research, and technical assistance. However, EPA can retain or delegate considerable enforcement responsibility for hazardous waste. Its solid waste authority is very limited; the federal agency cannot even ban open dumps except in the never yet invoked case of an imminent hazard. Almost from its inception in 1965, the national solid waste program has been vulnerable to reductions and, at various times, to the danger of abolition. It suffered from "malignant neglect" and was an "unwanted orphan," if not an "outcast."[1]

In contrast, hazardous waste, originally a minor part of RCRA, has evolved into the keystone federal effort, especially after Love Canal and similar incidents across the country, which culminated in the 1980 Superfund legislation, increasing even more the national government's role in cleaning up abandoned hazardous waste sites.

While relatively skimpy, the legislative history of Subtitle C indicates that Congress did not intend to preempt the role of the states;[2] rather it attempted to encourage institutionalization of RCRA's goals at the state level with the aid of implementation grants to the states. This was to be a federal effort, administered at least initially by EPA, which would authorize state programs to operate in lieu of federal programs. But EPA would still retain ultimate oversight responsibility.

Superfund legislation, P.L. 96-510, the Comprehensive Environmental Response, Compensation, and Liability Act of 1980 (CERCLA), also continued the mixed (or ambiguous) federal-state partnership in dealing with hazardous waste. State and local governments are responsible for remedial action at the abandoned site until a cooperative agreement or contract is reached with the federal government on cleanup terms which include federal funding. Even here, the states share 10 percent or more of the remedial cleanup costs. Eventually the state is to take on full responsibility and funding for long-term operation and maintenance of the site; but in the event that waste materials are to be removed, the recipient facility must have already been judged acceptable under Subtitle C federal regulations.

As in the case of water and air pollution, federal solid waste legislation appears to have encouraged the growth of state programs from what had been exclusively a local function. Thus in 1965, at the time of the enactment of the first Solid Waste Disposal Act which provided federal assistance to states, primarily to replace open dumps with sanitary landfills, there were fewer than ten state employees with solid waste functions. By 1978 more than 1,000 were employed in state solid waste programs.[3] Similarly, in 1965 there were only two state solid waste management programs; by 1975 all states had undertaken such activities.[4] And as of late 1982 all but four states have submitted solid waste management plans to EPA for review and approval.

In a similar manner, concern for hazardous wastes has surfaced only in the last decade. Even in 1979 a National Wildlife Federation survey asserted that only seven states had specific toxic substance legislation, only seven maintained a list of toxic substances or hazardous wastes, and only thirteen had a designated official or agency for hazardous wastes.[5]

Legislative History and Implementation Through Five Administrations

In reviewing the history of solid and hazardous waste legislative programs, three major patterns emerge. The first is that of a growing federal role, from a modest program of research, technical assistance, and limited funding to one of national government regulatory leadership often at variance with state perspectives. The second is that of congressional initiative and agenda setting as well as vigilant, often critical oversight of the implementation process. As in other ecological areas, Congress's role was that of "environmental protector" with the ideas and initiative emanating from Congress, especially its Democratic members.[6] Finally, the Executive Branch, whether Democratic or Republican, has consistently given the solid and hazardous waste programs a low priority, resulting in a confused and delayed regulatory issuance and implementation process.

The first law, the Solid Waste Disposal Act of 1965, was tacked on to air pollution legislation and was initiated to create a "national research and development program for new and improved methods of proper and economic solid waste disposal."[7] States, and especially local governments, retained responsibility for maintaining and operating disposal sites with federal technical and financial assistance. However, a key White House adviser candidly admitted that the succeeding "Nixon administration did its best to restrict the federal government's role in solid waste management."[8]

Broader legislation with a more active national government role in recovery of materials and energy and encouraging recycling, the Resource Recovery Act of 1970, was opposed and almost vetoed. The two main reasons were budgetary and ideological. There was a fourteenfold increase in funding over a two-year period to $239 million by FY 1973. Second, "the bill was seen as a first step toward an all-out program that would shift responsibility for solid waste disposal construction from the local to the federal government, just as sewage construction funding gradually had been shifted from a state and local responsibility to a federal one."[9]

However, attention was slowly focusing on toxic waste materials. The 1970 act required EPA to study hazardous waste storage and disposal. Its report to Congress, delivered almost a year late, declared that the management of hazardous residues was generally inadequate and that the threat to public health and welfare was increasing. Federal, state, and local laws were found to be spotty or nonexistent. EPA therefore proposed a regulatory program in cooperation with the state and private sector, including the creation of "regional processing facilities" for hazardous waste disposal.[10]

EPA had submitted a proposed Hazardous Waste Management Act of 1973 which would authorize a regulatory program for treatment and disposal of EPA-designated hazardous wastes: "States would implement the program subject to federal standards in most cases."[11]

By 1975, in a theme often repeated later, EPA's associate administrator told Congress that the agency would not repropose the bill "in light of President Ford's decision to hold the line against all new spending programs in order to fight inflation and keep the budget in control." Congressman Fred Rooney, Chairman of the House Subcommittee on Transportation and Commerce, immediately criticized his response as "chintzy in an area that means so much to the health and welfare of this nation. . . . Here we are trying to solve a problem and you say just because we are trying to battle inflation we can't spend another nickel."[12]

Such congressional attitudes finally led to the passage of the Resource Conservation and Recovery Act in the last days of the 94th Congress, during the presidential election campaign. The Senate committee report on the bill was therefore critical of administration attempts "to restrict federal activities to control hazardous wastes" through administrative reorganizations, reduced budget requests and delays in submitting reports to Congress."[13]

After passage of the act, the new Democratic administration, in the tradition of its predecessors, continued to give low priority to its implementation. Promulgation of regulations, mandated to be issued eighteen months after enactment, was considerably delayed by a decision to have the hazardous waste permit rules issued as a consolidated (and complicated) permit, stapled together with program requirements governing three other major permit programs under the Safe Drinking Water Act, the Clean Water Act, and the Clean Air Act.

The implementation process will be discussed in greater detail in the next section. However, the Carter administration initially deemphasized the program too—for example, planning to phase out solid waste planning grants by FY 1984. Several developments, however, forced it to concentrate more resources on hazardous waste problems. The unfolding of the Love Canal tragedy, and then other cases around the country such as Kentucky's Valley of the Drums, focused public attention on hazardous waste disposal amid allegations of EPA inaction. For example, such criticisms were highlighted in an ABC News television program, "The Killing Ground," aired on March 29, 1979, and in a *Time* cover story, "The Poisoning of America," which was replete with graphic photographs of abandoned dumps.[14]

This national and local media attention, as well as congressional oversight activities and criticisms forced EPA to upgrade the hazardous waste program and also led to the passage of the Superfund legislation. Hearings held in the late 1970s by House and Senate oversight committees gave agency critics, both outside environmental groups and internal dissidents, an effective forum to highlight implementation deficiencies.

Dissatisfied lower level EPA career officials claimed that the agency was under tight White House and Office of Management and Budget constraints and so was not aggressively searching out problem areas. These knowledgeable and outspoken civil servants charged that they had been instructed "to reduce the scope of the hazardous waste management regulations, which they were developing . . . in order to accommodate the President's cutting back on the federal budget to fight inflation."[15] Their charges were well publicized in congressional committee testimony and by the media and eventually resulted in the agency adopting a more militant public image.

After a series of such hearings, House and Senate committees issued critical reports. The House Interstate and Foreign Commerce Committee scored EPA's failure to formulate adequate regulations in time "or to conduct a comprehensive search for hazardous waste sites and to pursue enforcement actions vigorously."[16] The Senate Government Operations Committee similarly charged that EPA had "failed to make full use of its authority" to implement and enforce RCRA, and faulted EPA for not recognizing the seriousness of the problem and for not using all the money Congress had appropriated for the program.[17]

The General Accounting Office (GAO) has similarly issued a stream of investigative reports highly critical of EPA's program with titles such as these: "Hazardous Waste Management Programs Will Not Be Effective: Greater Efforts Are Needed," "Hazardous Waste Sites Pose Investigative, Evaluation, Scientific and Legal Problems," "Solid Waste Disposal Practices: Open Dumps Not Identified, States Face Funding Problems," and "Hazardous Waste Facilities With Interim Status May Be Endangering Public Health and the Environment."[18]

Finally, the Reagan administration began by accepting a Carter administration decision to phase out federal solid waste activities; however, it accelerated from FY 1984 to FY 1982 the denial of all funds for nonhazardous waste activities. It also proposed to cut by 10 percent Subtitle C grants to states. Initially it deferred the issuance of landfull regulations, postponed insurance requirements for operators of hazardous waste sites, and proposed partially lifting the ban on landfill of containerized liquid wastes.

However, an outcry by environmentalists, as well as by the trade association of companies engaged in destruction and detoxification of hazardous wastes, reinforced by a court suit, forced EPA and its new leadership to pull back from its proposed modifications and to issue landfill rules earlier than it had originally wished. Thus the Reagan administration too is under considerable pressure to aggressively pursue the control of hazardous wastes.

The Implementation Process: Delayed Federal Regulations and Uncertain State Program Assumption

RCRA contains numerous detailed provisions, subtitles, and sections, as is common in current environmental legislation. And also as is usual, Congress failed to provide explicit implementation instructions on how the act was to be carried out in terms of sequence or priorities. It was left to EPA to work out the details and attempt to ascertain congressional intent and conflicting public sentiments. And it is surely naive to believe that "issuance by the Environmental Protection Agency of final and interim final rules in November 1980 *has ensured strict regulation* of these substances."[19] Promulgation of rules is but one intermediate step in a torturous administrative process which often involves protracted court proceedings as to their general validity and specific application. Achieving voluntary compliance and/or agency enforcement of regulations thus takes many years and is far from being automatic and self-enforcing.

Subtitle C regulations were mandated to be issued eighteen months after passage of the law. However, they were finalized only by November 1980, four years after passage of RCRA and typically only under the gun of court-ordered deadlines, imposed after the agency had been sued for delay by environmentalists and others. The remaining rules were finally issued in July 1982, again under court order to prevent further delay.

The regulations deal with identification and listing of hazardous wastes (Section 3001); standards for generators and transporters of hazardous waste (Sections 3002, 3003); performance, design, operating requirements, and permits for facilities that treat, store, or dispose of hazardous wastes (Sections 3004, 3005); and notifications of hazardous waste activities (Section 3010). Most of these regulations are controversial and currently involved in litigation.

The focus of this chapter will be on Section 3006 regulations, which are guidelines describing conditions under which states can be authorized to carry out their own programs.

Section 3006 provided for two stages in establishing state programs that would meet minimum federal requirements. The first is interim authorization under which states can operate their own less exacting programs for two years while upgrading them; the second is full authorization. The act states only that to qualify for interim authorization the state program should be "substantially equivalent" to federal programs. Full or final authorization, a more stringent stage, has three main criteria: a state program must be "equivalent" to federal programs, consistent with federal or other state programs, and provide adequate enforcement. While state requirements in fully authorized programs may not be less strict than those of the federal government—i.e., they must meet federal minimum standards—a 1980 amendment to Section 3009 did allow states to impose requirements which are more stringent than those of EPA.

However, even after a state program has been fully authorized, it can be withdrawn by EPA upon a determination that the state is not administering and enforcing the program in accord with conditions of approval. Thus even when states are administering hazardous waste programs, EPA retains an oversight and inspection role including recall authority of state-issued permits.

The House report stated that the states were given "primary enforcement authority" and the "primary option of implementing federal minimum standards relating to hazardous waste." Federal minimum standards were justified as providing regulatory uniformity for states and industry as well as serving to prevent driving business out to states which decide to be dumping grounds.[20] Thus RCRA was not to preempt the states' role; only in the case of their reluctance to act would the federal government administer the program.

The interim authorization stage was intended, first of all, to avoid disrupting states with hazardous waste laws and programs already in operation. It would also give them time to bring their programs into conformity with federal minimum standards. Thus the interim two years was to be both a grace period and an orderly transition time frame for states to build up or modify their programs in accord with federal regulations.

In recognition of these understandings, EPA's draft implementation strategy paper encouraged state assumption of hazardous waste programs. Federal officials voiced their desires at several hearings: "Our desire is to turn this program over" to the states.[21] "We are very, very heavily dependent for the success of this whole act and its implementation on state and local agencies. . . . The actual implementation of the program must be done at the state level."[22]

Federal officials hoped that the desire of states for control of programs affecting their constituents and the business interest in avoiding the possibility of duplicate programs and requirements, further encouraged by federal funding, would induce the states to apply for interim and then full authorization of their programs in lieu of federal enforcement.

Regulations elaborating upon the law's provision providing for withdrawal of state approval would include the following cases: inadequate legal authority, unsatisfactory program operations, or enforcement programs failing to comply with requirements; and modification of a state program in such a way that it interferes with the free movement of hazardous wastes across the state border or to other states.

In elaborating upon its legislative mandate, EPA has further divided and complicated the process by establishing in effect a pre-interim authorization stage (by dividing the interim stage into two phases), and by further dividing the second phase of interim authorization into at least three separate components.

Under EPA's regulations, to qualify for interim authorization state programs must: (1) control at least as nearly an identical universe of toxic, ignitable, corrosive, or reactive wastes as is controlled by the federal program or a larger state expanded universe; (2) cover all types of hazardous waste facilities in the state; (3) be based on standards that provide substantially the same degree of human health and environmental protection as do the federal standards; and (4) be administered through procedures that are substantially equivalent to those used in the federal program. Applications for interim authorization must include a program submission, authorization plan, certification letter from the state's governor and attorney general, state hazardous waste law, and the memorandum of agreement with the EPA regional administrator. This agreement should include the ability of a state to operate a manifest program in the future; an operating tracking system is not required at this initial stage.

In regard to final authorization requirements, EPA has defined consistency as preventing a state from imposing importation bans on hazardous wastes from other states. It interprets equivalency as requiring state programs to have the following elements: a manifest tracking system, a permit system, criteria standards, adequate resources and legal authority, a designated lead agency, and a public participation effort.

Because of its delay in issuing completed regulations, EPA decided that interim authorization would be divided into two phases. The first phase would be based on regulations promulgated in May 1980 which provided operating requirements for waste generators and transporters. The second phase would be further divided into several "components" corresponding to specific categories of hazardous waste facilities. States could apply for interim authorization to issue permits for those facilities as the federal standards are finalized. The first two components—A, storage and treatment facilities, and B, incinerators—were announced in January 1981, and the land disposal regulations in July 1982.

States had the option of waiting until all components in Phase II are promulgated before applying for interim authorization. Prior to such an authorization, EPA is supposed to regulate facilities within each component in the state.

EPA has also decided that if any state's hazardous waste program fails to qualify for interim authorization, EPA will enter into a "cooperative arrangement" with that state. In such an arrangement, the state agrees to perform tasks to implement the federal program and to work toward interim authorization in return for federal financial support for qualified activities. Until interim authorization, EPA maintains basic responsibility for the program.

EPA admitted that the establishment of two phases and the creation of components within Phase II "results in a more complex application process,"[23] but felt it had no choice since it could not promulgate all the regulations for the subtitle that it had planned to in time. The January 26, 1981, notice further states that all components of Phase II would be promulgated within a year, another "expectation" that it did not live up to. After much controversy and under court order the remaining land disposal regulations were hurriedly promulgated in July 1982.[24] They included provisions for land treatment facilities, landfills, injection wells, surface impoundments, and waste piles.

Other Intergovernmental Issues

There are several other intergovernmental problems in regulating hazardous wastes worth reviewing.

The first is the question of one state's refusing to allow hazardous waste from another state to be disposed of within its border. This issue has apparently been resolved by a recent Supreme Court decision, *Philadelphia* v. *New Jersey*,[25] which invalidated a New Jersey statute barring disposal within the state of solid waste originating or collected out of state. Regardless of whether the purpose of the law was to protect New Jersey's economy or environment, the court declared that such action interfered with interstate commerce. Although it concerns solid waste, this ruling should serve to prevent other states from discriminating against out-of-state hazardous waste.

On the other hand, there is the problem of "hazardous waste havens"— states with lax regulation and enforcement which inevitably attract out-of-state dumpers. Current regulations attempt to prevent this from occurring by maintaining federal requirements in the absence of acceptable state programs. However, cases have been reported of hazardous waste "migrating" to friendlier, dumping ground states. A 1979 congressional report, for example, included a complaint by Congressman Marc L. Marks (R-PA) that the state of New Jersey was dumping 65 to 70 percent of its wastes in Pennsylvania because New Jersey, but not Pennsylvania, had enacted strict hazardous waste laws: "Therefore, to avoid the expense and effort of complying with New Jersey's laws, hazardous waste generators in that state simply send their hazardous waste to Pennsylvania. Undoubtedly other states are following the same pattern." This report also claimed that the disposal industry shifted its shipments from New Jersey to Rhode Island where regulations were weak. However, when Rhode Island tightened its controls, the industry moved to Pennsylvania.[26]

Similarly, a Senate report mentioned the case of a Minnesota disposal company under court order to clean up its site. It then moved to Arkansas where it changed its corporate name. A public nuisance suit was filed there, too, and an Arkansas resident complained: "We feel we are a dumping ground for the entire country. Chemicals are hauled to our city from as far away as Canada." Finally, concern was also expressed at a public hearing in Massachusetts that states with environmentally acceptable sites may become dumping grounds for all the wastes in their region and that other states will not make a good faith effort to seek ultimate disposal.[27]

Ideally, regional solutions would be preferable and have indeed been called for by academics. However, given the failure of past attempts in both water and air pollution, such as the proliferation of 208 water quality management planning agencies that so far have failed to do the job,[28] it would appear to be fruitless to spend much time pursuing this chimera.

A more serious issue involves state-local conflicts over siting. Should a state override local zoning restrictions in order to locate a hazardous waste management facility in an area whose residents understandably oppose such a facility? In 1977 no state had the authority to override local exclusion of hazardous waste facilities; today eighteen states have passed such override legislation.[29] It will be essential to resolve this problem fully if shortages of hazardous waste disposal sites develop.

Another pending intergovernmental question is the compatibility of state cleanup funds with the federal Superfund. Five chemical and petroleum corporations, for example, have sued New Jersey claiming that its spill compensation tax is unconstitutional because CERCLA preempts states from establishing funds, liability schemes, or financial responsibility requirements which would duplicate the purpose of the federal legislation.[30] Until the courts conclusively settle this question, the New Jersey Spill Compensation fund and other similar state funds will be under a legal cloud and so state cleanup efforts will be retarded.

Perhaps the most crucial issue is the adequacy of state resources to deal with hazardous wastes. It has long been recognized that state and local funding and the amount of federal funds available to the states was the "key" and "determining factor" in meeting the goals of RCRA.[31] The General Accounting Office has been severely critical of state laws, staffing, and funding.[32] If the states are unable or unwilling to assume this responsibility, the entire thrust of EPA's approach—to devolve responsibility to the states—will be undermined.

The current cutback in federal expenditures clearly has negative implications for hazardous waste grants and assistance. While the FY 1983 administration budget proposed increases for Superfund activities, it also sought to reduce Subtitle C grants by more than 10 percent at the same time that federal funding and technical assistance for solid waste management planning had been completely eliminated. These proposed hazardous waste grant reductions have led at least ten states to consider withdrawing their Phase I interim authorization and returning these programs to EPA.[33]

Past national leadership can be faulted for giving a low priority to hazardous wastes, and so the program was plagued with insufficient staff and constant management turnover. The recent decision to eliminate (not even phase out in an orderly manner) federal nonhazardous waste activities may turn out to have the most detrimental consequences. State programs do not neatly separate hazardous from solid waste activities and municipal disposal sites often receive both types of waste. Further, improperly cared-for solid waste sites can develop into hazardous waste problems and possibly even into future abandoned sites under Superfund! For example, poor sludge waste disposal practices can lead to a serious problem of heavy metal contamination of groundwater sources.

Thus there is no sensible way to clearly demarcate solid waste from hazardous waste, especially in the case of small generators producing less than 1,000 kilograms of waste a month who do not fall within Subtitle C jurisdiction.

Changing federal regulations issued in increasingly complex stages, phases, and components also prevent orderly state program development. Most states need to adopt federal requirements through their own lengthy legal-administrative processes, which in some cases may take up to three years.

An EPA study noted that eighteen states which passed hazardous waste laws before RCRA were not anxious to proceed further until EPA regulations were issued. Laws in eight states even prohibited promulgation of state rules before federal rules are issued.[34] Thus any shifts or delays in issuing federal regulations have severe rippling effects on states and ultimately on the private sector, which remains unsure and confused as to whether it is complying with federal and state laws.

Conclusions: RCRA, EPA, and Implementation Delays

In conclusion, hazardous waste legislation like most other environmental laws was initiated and sponsored by Congress. In several stages these laws greatly enlarged the federal role to that of leadership while retaining a management and operation role for the states. After passage of RCRA, Congress maintained a strong oversight role, often criticizing Executive Branch delays and weaknesses in implementing the act. On the other hand, both Democratic and Republican administrations have accorded higher priority to budgetary and economic considerations than to aggressively controlling hazardous wastes. It is still uncertain whether EPA has been sufficiently galvanized by outside forces and crises to satisfactorily implement the well-meaning intentions of RCRA's authors.

RCRA mandated that regulations be finalized by April 1978. Now, more than four years later than the due date, the hazardous waste program is very much behind schedule. By September 1982 only thirty-five states had been granted Phase I interim authorizations, and only five states granted Phase II interim authorizations, A and B components. None are anywhere close to applying for final authorization, since EPA is just beginning to indicate how this may be reviewed. In the current period of limited interim authorization, it is sobering to note the GAO's finding that there was little assurance that interim status regulations were being met by hazardous waste facilities. GAO further concluded that these interim regulations do *not* provide for protection of public health and the environment.[35]

EPA had naively and optimistically predicted in 1980 that thirty-nine states would have full authorization by January 1982, and that all of them would be fully authorized by 1983.[36] However, the Council on Environmental Quality, recognizing the lack of EPA and state resources, recently indicated that a fully operating federal hazardous waste management program, especially one with full permitting of facilities, would be completed only "over the next 8–10 years."[37] Similarly, the General Accounting Office has criticized EPA's lack of progress and insufficient resources in implementing the Superfund program.[38] It will indeed be miraculous if RCRA is implemented and operating satisfactorily in 1986, a decade after its passage!

EPA's implementation of RCRA does not reflect much credit on the agency's administrative abilities. Some have charged its leaders with a deliberate conspiracy to delay implementation and enforcement of the law. Friedland attributes to EPA "a strategy of deferred regulation."[39] A slightly more charitable view is that mismanagement and incompetence rather than deliberate delay would better account for the extraordinary "slippages" that have occurred in promulgating regulations—the enforcement of these regulations, for all practical purposes, has yet to begin.

The pattern of delays, excuses, and apologies was set very early. The first study report on hazardous wastes was delayed close to a year and EPA officials had to concede in embarrassment to Congress that the initial planning for the study was "totally inadequate" and that the first stage of contracts that had been let were "not productive. Therefore we had to start over again from ground zero" and had to request the Senate to postpone the submission date.[40] EPA even ended up proposing hazardous waste legislation before their report was submitted. In the late 1970s a similar pattern emerged—regulations were promulgated before the justification documents explaining their rationale and rejecting alternatives were issued.

Given the administration's low priority and limited commitment to hazardous waste regulation, a basic reason for the implementation delays is of course that the Office of Solid Wastes lacked sufficient resources. It has been chronically understaffed and underfunded, even compared to other EPA program offices.

Second, the problems it confronted were relatively new and complex. As the 1979 report of the Council on Environmental Quality stated, lack of RCRA implementation was due in large measure "to limited data and availability of resources and to the complexity of the scientific and technical issues."[41]

The agency's commitment in the Carter administration to an extensive public participation program entailed holding numerous public meetings across the country explaining the act and then sifting through a considerable number of comments—there were about 1,100 written comments in a stack over six feet tall after 1,200 people attended five public hearings with more than 200 making oral presentations.[42] Also, the ever present specter of judicial scrutiny and prolonged legal appeals tended to slow the issuance of regulations. The agency's lawyers reviewed and rereviewed rules to make them as legally defensible as possible.

Finally, mismanagement played a role. The Office of Solid Wastes suffered from constant reorganizations and changing program leadership. Poor decisions, such as the Carter administration consolidated permit approach and the

Reagan administration initial landfill rule modifications, also accounted for delays of a year or more.

The dangers of these delays include disillusionment triggered by overly optimistic promises. For example, consider the misnomer of a Superfund, which clearly lacks funding adequate to clean up more than a fraction of the abandoned hazardous waste sites. States may not have sufficient incentives to develop their own programs to comply with federal regulations as they are issued in bits and pieces. And industry may continue to be frustrated by regulatory uncertainties and fearful of varying state requirements while it continues to dump hazardous wastes.

At best, the complicated and confusing implementation process may spur industry to avoid the government regulatory scene through increased emphasis on reuse of waste materials and in-house recycling and neutralizing of hazardous wastes.

We have come a long way from a 1976 estimate of fifty people in twenty-five state governments dealing with hazardous wastes.[43] Twenty states, for example, are now operating hazardous waste disposal programs with funds from user fee mechanisms.[44] Recent regulations also allow states more flexibility in obtaining authorization to operate their programs.[45]

Nevertheless, the cradle-to-grave system envisioned by RCRA's authors may either emerge stillborn or be aborted by impatient legislators frustrated by the scientific risk uncertainties and technical complexities of hazardous waste management and by the delayed intergovernmental implementation of RCRA.

III. Subnational Perspectives

5. Hazardous Waste Policy in Florida: Is Regulation Possible?

Bruce A. Williams and Albert R. Matheny

This chapter presents some of the results of our ongoing research into the regulation of hazardous waste in the state of Florida. Specifically, it aptly illustrates one of the most difficult problems of government regulation: the balancing of short-run economic considerations against longer term environmental, safety, and economic considerations. In addition, our analysis of hazardous waste regulation allows us to test the applicability of the two contrasting models of social regulation[1] that have guided previous research in this controversial area of policy making.

In this chapter's first section, we outline the assumptions of these two models. Most scholars currently engaged in debate over—and the study of—social regulation explicitly or implicitly adopt the first "conventional" approach. This approach assumes that successful social regulation is possible and that current problems in this area can be remedied by marginal policy adjustment. Its alternative, the "critical" approach to social regulation, is based primarily upon the work of Charles Lindblom, James O'Connor, William Ophuls, and K. William Kapp.[2] It questions the ability of existing political-economic institutions to successfully adopt and enforce social regulations. Although advocates of this approach have rarely looked at specific policy issues and have preferred to cast their argument in broader macrotheoretical terms, we believe that application of this approach to analysis of regulatory policy making is long overdue and exceedingly useful. In the second section of the chapter, we provide an overview of hazardous waste management in Florida. In the third section, we consider the explanatory utility of the conventional and critical approaches to hazardous waste management in Florida. We argue that, at least in the area of regulation, the critical model provides a more adequate explanation for the politics of regulation than does the conventional approach. The final section of the chapter discusses the theoretical and practical implications of our findings.

Models of Social Regulation

The Conventional Approach to Regulation

Despite the intensity of the debate over the proper level of government intervention in the economy, most scholars on both sides of it have adopted a very similar approach to analyzing regulations: we call this approach the "conventional" model. In this section we briefly summarize the debate and discuss the three basic assumptions of the conventional approach which unites the debaters.

In the late 1950s and early 1960s a body of scholarship emerged that was to serve as the intellectual underpinning for the social regulation of the 1970s. Scholars in this tradition observed the gradual "capture" of regulatory agencies by the very industries they were charged with regulating.[3] Rather than representing the public interest, regulatory agencies, as they aged, came to represent the narrow and specific private interests of the regulated industry. This view of regulation found widespread acceptance among both scholars and politicians, and led to attempts in the 1970s to restructure the regulatory process in a manner that would militate against the capture of regulatory agencies.

The causes of regulatory capture were typically identified as: the omnipresence of organized industry groups; the relative absence of groups representing the unorganized but affected public; and the bureaucratic ossification of regulatory agencies. In theory, the solution to these problems was the creation of regulatory agencies sufficiently powerful and resistant to influence to maintain their independence from the industries they regulated. In practice, the influence of the "capture" theorists is best exemplified in the enabling legislation of newer regulatory agencies such as the Environmental Protection Agency and the Occupational Safety and Health Administration. When such social regulatory legislation was written in the 1970s, it was built upon the implications of the capture theorists. First, the legislation itself often contained strict performance standards for the agency. Second, procedures were created for the promulgation of formal rules and regulations in open hearings. Such procedures were intended to eliminate agency discretion to bend rules in favor of industry interests. Third, agencies were to be protected in the appropriations process so that they would be resistant to pressures exerted by adversely affected private interests. Fourth, these agencies were to be staffed by trained professionals, not political appointees or those overly sympathetic to the interests of the regulated. Further, as James Q. Wilson points out, these professionals were themselves aware of the capture theorists and sensitive to the pitfalls discussed by them.[4]

In the 1970s and 1980s, the debate over social regulation intensified as a group of scholars reacted to and criticized the results of attempts to reform the regulatory process.[5] Just as the capture theorists provided the intellectual basis for the expansion of social regulation in the 1960s and 1970s, so too this new group of regulatory critics provided the basis for many of the regulatory—or deregulatory—reforms of the 1980s. They suggested that independent regulatory agencies had become too powerful; they asserted that the promulgation of many rules and regulations had raised the cost of doing business without any concomitant improvement in the quality of life. They argued, moreover, that such rules had serious adverse consequences for the efficiency of business and the economy in general. Finally, they suggested that regulatory agencies had become "anti-business" as a consequence of their capture by "public interest" interest groups.

The belief in regulatory excesses generally leads to two types of recommendations for regulatory reform. The first is to rely upon formal techniques for balancing the benefits of regulations against their costs. So-called "second generation" environmental regulations (e.g., the Toxic Substances Control Act and the Resource Conservation and Recovery Act) all incorporate requirements that regulations meet some type of cost/benefit standard. The second type of policy suggestion is that regulation decrease its use of punitive sanctions and increasingly rely upon market-oriented incentives to achieve regulatory goals: for example, regulations can provide positive incentives for desired behavior (e.g., tax credits for the installation of pollution abatement equipment) or, they can urge greater reliance on private legal action using restructured liability laws.[6]

In the 1980s debate continues between two schools of thought: one which is heir to the "capture" theorists and which suggests we need strengthened regulatory agencies and legislation to deal with failures of the private market; and one which has sprung from the recent neoconservative regulatory reform movement and which argues that we need less legislation and more market-oriented techniques for dealing with market failures. Despite their intense debate over the optimum level of regulation and the most efficient means to achieve that level, the debaters have some fundamental points of agreement. All but a few do not question the fundamental need for some form of social regulation by government in many areas, or the possibility that carefully designed strategies can achieve regulatory ends. Most of both the proponents and the ostensible opponents of regulation share a similar approach to social regulation. This approach is characterized by three general assumptions about social regulation in a democratic private market society. Its first assumption is that the failures of the market that create the need to regulate are not intrinsic to the economic system. That is, pollution, consumer fraud, and other objects of social regulation are treatable on a case-by-case basis and are not organically linked to the private market itself. They are consequences of individuals' stupidity, laziness, and venality. If we want to decrease their frequency, we must make reasonable laws and publicize them, explain to individuals how to change their behavior in such a manner as to obey the law, provide incentives for compliance, and increase the penalties for lawbreaking. Eugene Bardach and Robert A. Kagan, for example, suggest that firms with a moderate to strong predisposition to obey regulations ("good apples") comprise at least 80 percent of the firms covered by most regulatory programs.[7] Once a law is in place, these firms will obey it. Thus, one can solve the problems created by failures of the private market without fundamentally altering or eliminating the private market itself.

This first assumption leads directly to a second assumption of the conventional approach. Basically, the conventional approach assumes that it is theoretically possible to design regulations that simultaneously solve market failures and are compatible with a private market economy. More specifically, the conventional approach assumes the feasibility of regulatory policies that serve some public interest[8] without destroying the ability of regulated private interests to profitably continue in business. Strong disagreement exists over how to determine the optimal method for balancing public and private interests. As noted above, neoconservative regulatory reformers have urged the use of formal techniques such as risk-benefit and cost-benefit analysis to determine the optimal level of regulation. Critics of such techniques suggest that they tend to overemphasize the easily monetarized costs of private businesses at the expense of less easily quantified public considerations.[9] These critics advocate decision rules that give more weight to nonquantifiable benefits of regulation (e.g., clean air and water; undisturbed scenic vistas). The two sides to the debate disagree over the best manner of making regulatory decisions. Each side is convinced, however, that the adoption of its preferred decision rule will *adequately protect* the public and simultaneously allow private businesses to maintain adequate profits.

The third assumption of the conventional approach to social regulation is that once regulatory policies have been designed, it will be politically possible to pass and enforce them. Scholars of both conventional schools find fault with social regulation as it has been passed and enforced in the past, but, as in the case of market failures, attribute these failures to idiosyncratic rather than systematic causes. Neoconservative regulatory reformers see the failure of regulation stemming from the overreaction of legislators and regulatory agencies to the environmental and consumer groups of the 1960s and 1970s;[10] they argue that such failures can be overcome by adoption of more "rational" techniques for balancing the claims of countervailing forces. Regulatory critics from the "capture" school identify ineffective agency personnel or undue industry influence as the cause of regulatory failure;[11] they argue that the solution to this failure is the creation of more independent regulatory agencies staffed by better trained professionals. In general, both schools of the "conventional" approach to regulation have been optimistic about the system's perfectibility. Few critics of the regulatory system have questioned the fundamental ability of the American political system to pass and implement sound regulations. Most assume the existence of a reasonably well-functioning pluralist democracy that is capable of balancing the competing pressures that are brought to bear upon it. The two schools of the "conventional" approach argue heatedly about how this balance should be struck, but not about whether it is possible.

Due to the shared assumptions of most scholars writing about regulation in the "conventional" tradition, the debate over social regulation tends to be a rather narrow one; it centers around disagreement over the proper way to strike a balance between the need to protect the public from market failures and the need to avoid interfering unduly with the private market. The narrow bounds of the debate over regulation are well evidenced by the following quote from the editors of a volume written largely by conservative regulatory reformers:

> Perhaps understandably, in the late 1970s and early 1980s "regulatory reform" has taken on an entirely different meaning than it had in the 1960s and early 1970s. Today's reform usually implies moderating the "excesses," the "over-regulation" attributed to the effectiveness-enhancing regulatory reforms of the 1960s and 1970s. The regulatory pendulum, it is argued, has swung too far, has reached a point of diminishing (or negative) returns, and must be pushed back a bit, at least until the economy as a whole has regained its forward momentum.[12]

The Critical Approach to Regulation

While most scholars dealing with social regulation implicitly or explicitly adopt the conventional approach, an alternative model can be gleaned from the writings of environmentalists[13] and neo-Marxists.[14] This approach leads to the conclusion that, because of the nature of our political and economic systems, it is virtually impossible to deal adequately with many of the problems addressed by social regulation. The critical approach suggests that, in a private enterprise economy, the time frame of business is too short for the environment within which it operates. Current profit considerations tend to drive out any possible longer run concerns about the exhaustion of resources or the social effects of negative externalities.

Such attention to the market's defects is certainly not novel: the received wisdom of political science suggests that defects of private market economies provide the justification for government regulation. However, critical theorists part company with conventional theorists over the political system's ability to cope adequately with failures of the private market. They argue that so-called "market failures" generally increase profits and thereby benefit business: rather than extrinsic to the market in origin, they are intrinsic to our economic system. Moreover, because business dominates governmental decisions, such market failures cannot be rectified by political action.

The critical approach's assumptions contrast strikingly with the three assumptions that undergird the conventional approach to social regulation. First, in contrast to the conventional approach, critical theorists argue that market failures that create the need to regulate are systematic and inherent to a private market economy. O'Connor argues, for example, that the only way that private industries can remain profitable is by socializing many of their production costs: that is, by evading some of the actual costs of producing their products.[15] For example, one such method of socializing costs of production is to cut costs by improperly disposing of hazardous wastes as cheaply as possible. This forces the general public to bear the costs of such practices—either through the use of tax dollars to clean up waste sites, or, more likely, by bearing the costs of increased levels of disease and genetic mutation. Following a slightly different line of argument, Kapp argues that rapid depletion of resources and environmental degradation are inevitable consequences of private market economies due to the short-run profit considerations of businesses and their inability to take into account the long-run consequences of their actions.[16]

On the basis of this assumption about the nature of market failures, the critical approach concludes that it may not be even theoretically possible to design regulatory policies that allow the private market to function and at the same time adequately protect the public. Decision-making procedures designed to specify an optimal level of regulation—one that balances these two considerations—are inevitably doomed to failure.

According to the critical theorists, both conventional schools are astute in their criticism of one another: however, neither is sufficiently astute as to realize the validity of its critics' objections to its own proposals. According to the capture theorists, cost-benefit analysis and other similar techniques are inherently biased toward private profit makers because they miss an elementary point about the political economy of social regulation. Such regulation emerges in response to market failures, i.e., failure of the market to account for long-run social costs, or externalities, of unrestrained market operation. Many of these social costs are not and cannot be adequately valued by the application of the monetarization of cost-benefit or risk-benefit techniques precisely because they lie outside the bounds of market rationality. Thus, regulation is concerned with precisely those things that are improperly valued in the marketplace. And since costs and benefits must ultimately assume some economic valuation in cost-benefit analysis, advocating its use to determine optimal levels of regulation systematically minimizes the weight of inestimable values that may be politically or ethically important.

On the other hand, the neoconservative regulatory reformers argue that regulatory rulemaking that sets absolute standards for protecting the public (e.g., the Delaney Amendment, absolute ambient air quality standards) runs the risk of seriously reducing the profitability of many businesses and the general performance of the American economy. While conventional defenders of strict social regulation tend to view such arguments as mere bluff, advocates of the critical approach recognize that it may indeed be impossible to deal with many problems of market failure while simultaneously maintaining profitable industries.

Finally, even if it were possible to design regulatory policies that adequately balance private and public considerations, the critical approach assumes that the political system is incapable of actually implementing such policies. There are basically two explanations for this inability: structural and positional. The structural explanation is best developed in Lindblom's argument that market-oriented societies are inherently resistant to change and government interference in the private sector.[17] He argues that politicians' attempts to interfere in the market are systematically and automatically "punished" by economic sluggish-

Table 5.1. Alternative models of social regulation

		Conventional Model	Critical Model
Assumptions:	1.	Market failures are extrinsic to the basic operation of the private market system.	1. Market failures are systematic and inherently related to the private market system.
	2.	It is possible to design regulations that simultaneously solve market failures *and* are compatible with the private market system.	2. The conflict between private profit requirements and the public interest is so fundamental as to render the design of successful regulatory policy impossible.
	3.	The political system, with marginal adjustment, is capable of passing and implementing successful regulatory policy.	3. The political system is incapable of passing and implementing successful regulatory policy.
Predictions:		Continued tinkering with regulatory decision-making procedure will ultimately lead to adequately successful regulatory policies.	Market failures, unsuccessfully dealt with by regulatory policy, will ultimately lead to the fundamental restructuring of political and economic institutions.

Source: Compiled by the authors.

ness and higher unemployment. Since politicians are judged to a great extent on the way the economy performs, they are reluctant to—or even incapable of—seriously altering the outcomes produced by the private market.

The positional explanation is based on the direct political clout that business is thought to exert on government officials and the business-oriented ideology of government officials themselves.[18] In this explanation, the ability of business officials to gain access to and influence the decisions of government is seen as militating against any governmental decisions that would adversely affect business interests. In effect, business interests are able to veto the sustained enforcement of any regulations seriously inimical to their interests. Consistent with this critical approach, Edward S. Herman argues that the current anti-regulatory movement emerged after the passage of social regulation (what he calls "new regulation") that was seen as conflicting with business interests.[19] Moreover, he concludes that this movement's anti-regulatory fervor was far out of proportion to the quite minimal restraints that such regulation actually placed on industry. If this type of response follows even minimal constraints on business, critical analysts conclude that it is hard to imagine the passage and enforcement of truly effective social regulation.

It is apparent that, from the perspective of the critical approach to regulation, the parameters of debate over social regulation defined by the conventional approach are exceedingly narrow. Both sides in the conventional debate view the regulatory process as essentially perfectable and share the opinion that marginally adjusting regulation will result in an optimal set of sanctions and incentives: one group, influenced by the capture theorists, urges more and better constructed regulation; the other group of neoconservative regulatory reformers urges less direct government intervention into the market. The critical approach suggests, however, that this debate about the stringency and type of regulation is akin to arguing about the arrangement of deck chairs on the Titanic. Neither side can win this debate, because both sides share a false confidence in the ability of existing institutions to solve the problems posed by market failures. As one position (e.g., cost-benefit decision making) comes to dominate policy making and thereby to fail publicly, advocates of the other position (e.g., absolute standards for protecting the environment) gain strength. As the dominant position's failures push policy in the other direction, a new chorus of criticism is heard. Nothing ever seems to work—perhaps, as the critical approach suggests, because nothing *can* work.

In the next two sections we turn to an examination of a specific type of social regulation in a particular setting—hazardous waste regulation in Florida—and analyze the relative utility of the conventional and critical approaches to regulation that we have outlined.

Hazardous Waste Regulation in Florida

While we intend to study the problem in several states, Florida provides an excellent starting point for an inquiry into the questions we are raising in our research.[20] The problem of hazardous waste disposal is extremely severe in the state. According to EPA, Florida contains more potentially hazardous waste sites than any other state. This ranking is particularly disturbing because of the hydro-geology of the state. Most of Florida's drinking water comes from aquifers (water-bearing rock formations) that lie, in many areas, quite close to the surface. If these aquifers become polluted (and some may already be polluted), little can be done within the constraints of the current state of cleanup technology. Further, in 1982, Florida did not have even a single commercial hazardous waste disposal site. This means that any hazardous wastes that are not disposed of on-site (as is the usual practice for producers of large amounts of wastes) must be shipped at great expense to sites in Alabama and South Carolina. The high costs of shipping wastes to out-of-state sites creates incentives, especially for small operators, to dump the material wherever they can.

The structure of the Florida economy makes the development of hazardous waste regulation a particularly interesting issue. A large and politically powerful chemical industry may be expected to resist strict regulation. On the other hand, if improper disposal should seriously threaten the environment, large and politically powerful industries stand to lose a great deal. For example, one-third of the state's tax revenues are generated by the tourism industry. A "Love Canal" episode in Florida would severely damage this industry, especially if it occurred in the popular south or central portions of the state. Florida is also one of the largest producers of fruits, vegetables, and cattle in the United States. The impact of serious groundwater pollution on the citrus and cattle industries would be catastrophic. Both tourism and agriculture are widely acknowledged to be well-organized and politically potent lobbying forces. As such they might provide a countervailing force—crucial to the conventional approach discussed above—to the pressures of the chemical industry.

In addition, Florida environmental groups (e.g., Florida Defenders of the Environment, the Audubon Society, and the Sierra Club) have been a potent political force in the state.[21] Indeed, in the 1970s Florida was a pioneer in the passage of legislation designed to protect the environment. Insofar as environmental groups are still organized, the conventional model would suggest that they would act as another countervailing force to the chemical industry.

Hazardous waste disposal in Florida poses some of the most difficult regulatory problems possible. Over the long run, it is a potentially catastrophic problem; yet, there are strong market incentives that militate against any private market solution. Solving the problem requires the creation and enforcement, by government, of the regulatory policy that forces the sacrifice of short-run economic gain for the sake of longer-run considerations of public health and potential economic gain. An examination of federal and state government response to this problem provides an opportunity to compare the usefulness of the two models of regulation developed above. After we have sketched the history of waste regulation in the state, we will turn our attention back to the conventional and critical approaches.

Even though Florida has been a leader among states in its regulation of the environment, it has lagged behind other states in its attention to hazardous waste management.[22] In order to understand the regulation of hazardous waste in Florida, it is essential to begin with a consideration of its federal context as established by congressional legislation and EPA policy.

Legislative History

Subtitle C of the federal Resource Conservation and Recovery Act of 1976 (RCRA) directed the administrator of the EPA to implement a comprehensive "cradle-to-grave" program for managing hazardous wastes. This included the identification of wastes which are "hazardous," the monitoring of those wastes through a manifest system, the supervision of hazardous waste generators and transporters, and the permitting of hazardous waste management "facilities" (disposal sites). The legislation strongly encouraged the EPA to work with the states to develop their own hazardous waste management programs.

After four years of complex and controversial debate, EPA finally promulgated its regulations pursuant to Subtitle C of RCRA on May 19, 1980. The program provided for the complete supervision of hazardous wastes by November 19, 1980; it created a phased system for the permitting of hazardous waste management facilities, enabling the EPA and the states to spread out the anticipated overload of permitting activity in the months following announcement of the regulations. Phase I involves the application of nontechnical operating standards that can be quickly applied to the regulation of permitted facilities.

In the future, more technical Phase II regulations will be developed, and these, in turn, will be revised to simplify the permitting process under Phase III.

As of early summer 1982, Florida remained the only state in the Southeast not authorized to assume Phase I monitoring under Subtitle C of RCRA and EPA regulations. According to one source within the Department of Environmental Regulation (DER), the state's tardy response to hazardous waste management resulted from the belief among state officials that Florida was endowed with "clean" industries that posed no significant hazardous waste problem. This belief was held despite a 1977 study which found that the state produced over 580,000 metric tons of hazardous waste per year and still had no authorized facilities within the state to handle that waste.[23]

Florida's belated response came in the form of amendments to the state's 1974 Resource Recovery and Management Act (RRMA). These amendments were passed in 1980 and substantially mirrored the corresponding provisions in the federal RCRA. In addition, the 1980 amendments anticipated federal passage in December 1980 of the Comprehensive Environmental Response, Compensation, and Liability Act (Superfund), insofar as they established a Hazardous Waste Management Trust Fund designed to begin the accumulation of funds for use in cleaning up hazardous waste pollution and dealing with other related problems. The legislature supplied the Trust Fund with an initial $600,000 and provided for an initial 1 percent excise tax on the disposal of hazardous wastes. The tax is to escalate 1 percent per year to a maximum of 4 percent beginning in October 1981. Expenditures from the Trust Fund are to be repaid by those found responsible for hazardous waste problems remedied by the Trust Fund. In all, the Trust Fund is not to exceed $10,000,000.

The 1980 amendments are the basis for Florida's plans to assume control of hazardous waste management from EPA. Until that time, the DER has a "cooperative arrangement" with EPA: DER supervises hazardous waste management in Florida under EPA regulations and with the support of EPA funds ($1,555,000 for FY 1982).

As Florida attempts to move into Phase I compliance with EPA, it is important to examine the regulatory parameters provided by the 1980 amendments. These amendments will significantly shape the way in which Florida deals with its hazardous waste problems. Although they basically coincide with federal regulations, three important problems remain: siting procedures for hazardous waste facilities; the structure of the Trust Fund; limited state funding for DER's hazardous waste program.

Siting Procedures

The 1980 amendments provide a complicated scheme for locating authorized hazardous waste management facilities in the state. Anyone interested in establishing a hazardous waste management facility in Florida must first receive a permit from DER and then approach the county government that has juris-diction over the site of the proposed facility. If approved by the county, the site may be established subject to DER and local provisions. If denied by the county, an appeal may be made to the Regional Planning Council (RPC), an authority composed primarily of officials from several counties in a given area of the state. Even if the RPC overrules the county decision, final approval from the governor and cabinet is required before establishing the facility. If the RPC upholds the county decision, the site request is defeated.

The problem with the current siting provision is obvious. Local officials steadfastly resist approving the location of a "potential Love Canal" in their own backyards. The RPCs are largely dominated by local interests and are protective of county prerogatives. Not one hazardous waste management facility has been sited in Florida in the nearly two years since the passage of the 1980 amendments, despite the fact that scientists and engineers believe such facilities could be safely constructed and operated at a substantial profit.[24] All significant hazardous wastes that are disposed of legally are now shipped out of state, to facilities in Alabama and South Carolina.

The emphasis on local prerogatives in the siting provisions is partly due to the abiding fear of Florida "panhandle" residents that, if siting decisions were made on a statewide basis, the predominantly rural panhandle counties would become the dumping ground for urban south Florida's industrial excesses. These counties possess the only soil conditions (clay-based rather than sandy soils) which permit the construction of safe, inexpensive (unlined) landfills.

Our interviews suggest that many state legislators and DER staff members recognize the problem posed by existing siting procedures—a problem that would grow significantly worse if Alabama and/or South Carolina limit the amount of hazardous waste they accept. Despite this recognition, however, virtually no progress has been made in reforming the process. In the 1982 legislative session, new siting procedures were introduced that would have moved authority for reviewing local decisions from the RPCs to the statewide Environmental Regulatory Commission (ERC).[25] When local government and ERC disagreed over the siting of above-ground waste facilities, final decision-making authority would be given to the governor and the cabinet. In order to allay the fears of panhandle residents, these changes would only affect the siting of above-ground facilities: landfill siting would continue to be guided by the existing legislation. The authors of this new legislation hoped that it would allow the siting of state-of-the-art commercial facilities, most likely in central Florida, that would receive shipments of hazardous wastes, sell certain types of hazardous wastes that might be used by various industries (e.g., solvents), and either use chemical processes or high temperature furnaces to reduce the toxicity of the remaining wastes. This new legislation, however, never even reached the floor of the legislature. In both houses, it was killed in committee by a potent alliance of panhandle legislators, county government representatives, and en-vironmental lobbyists.

The Trust Fund

The Florida legislature apparently assumed that an annually escalating excise tax on hazardous waste disposal would soon fill the coffers of the Hazardous Waste Management Trust Fund. So far this optimism has been unfounded. During the first month (October 1981) of its collection, the tax generated $4.92 in revenue for the Trust Fund and, by mid-1982, revenues had totaled only $2,300. At this rate, in 1982 dollars, the Trust Fund will not reach its statutory maximum of $10 million for approximately 338 years! To put this dollar figure in perspective, expenditures on Love Canal alone have already exceeded $30 million for only partial cleanup.[26] Congress estimated that cleanups of abandoned sites would average $4 million per site.[27] Sources in DER informed us that engineering surveys to simply determine the severity of pollution at a single hazardous waste site and develop cleanup plans cost between $250,000 and $500,000. Even with federal Superfund monies paying 90 percent of the cleanup bill for private sites and 50 percent for municipal sites, the Trust Fund will not stretch very far.

An attempt was made in the 1982 legislative session to immediately raise the excise tax on hazardous waste disposal to 4 percent and add a 3 percent gross receipts tax on hazardous waste facilities themselves.[28] Legislators and DER officials, however, were handicapped when writing this legislation by the failure of the Department of Revenue to provide any estimate of the amount of money that would be raised by either existing or proposed taxes: this department had only assigned two people to duties relating to the excise tax's collection—which partially explained the small amount of revenue actually collected and which accounted for the department's unhelpfulness in drafting legislation. In any event, the new tax legislation was written into the siting legislation mentioned above and did not make it to the floor of either house of the legislature.

The shortage of Trust Funds has been greatly exacerbated by EPA's long delay in releasing any of the $1.6 billion in Superfund. As of this writing (August 1982), only $306,000 has been released to Florida for the cleanup of a single site (Whitehouse Oil Pits in Jacksonville).[29] According to sources within DER, the agency has been under pressure from the legislature to spend money from the state Trust Fund, even though it might forfeit its potential tenfold matching power when combined with federal dollars from Superfund. Indeed, an estimated $240,000 of unmatched funds have already been spent from the Trust Fund.

Funding for DER's Hazardous Waste Program

Within the confines of these amendments, DER, in 1982, managed hazardous wastes under a cooperative agreement with EPA. This agreement focuses primarily on permitting "on-site" hazardous waste facilities (since no authorized off-site facilities exist in Florida). Current and proposed (under Phase I "program recognition") permitting procedures are presented in table 5.2.

Under the cooperative agreement EPA provides approximately $1,500,000 for roughly thirty-five DER employees to monitor on-site hazardous waste facilities in Florida. This allows inspections of these facilities once every two years. In addition, the state itself provides $134,000 for all other activities in the DER's Hazardous Waste Management Program. These "other activities" have, of course, taken on great significance since the passage of federal Superfund legislation.

Superfund required EPA to undertake a nationwide survey of hazardous waste sites targeted for cleanup using Superfund monies. Each state was asked to submit a list of uncontrolled hazardous waste sites ranked for cleanup priority using a "site ranking model" designed expressly for the EPA's survey by the MITRE Corporation of McLean, Virginia. Florida's DER submitted twenty-seven sites for EPA consideration. Sixteen of these were ultimately included in the list of the 115 sites ranked nationwide as "most threatening," making Florida the state with the highest number of serious uncontrolled hazardous waste sites.

While we raise some serious questions below about the accuracy of the EPA's survey and the MITRE model, it is nevertheless clear that Florida has a severe problem. It is equally clear that the funds necessary for dealing with this problem have not been forthcoming. Among those we interviewed, estimates of the amount of money required to run a minimally adequate ongoing hazardous waste management program in the state run from a high of seven million to a low of three million dollars per year. As we noted above, however, in FY 1982 the state spent only $134,000 on the program. This is $16,000 less than the $150,000 DER officials estimate it will take to do a basic "needs analysis" surveying the status of all waste sites in Florida. The governor's budget for FY 1983 proposed an increase of only $200,000 in DER's hazardous waste budget subject to review by the legislature.

We now turn to an analysis of the reasons for the failure of the state of Florida to address adequately the problem of hazardous waste pollution. In this analysis we are particularly interested in applying the two models of regulation developed in the first section of this chapter.

Testing the Two Models of Regulation

In applying the assumptions of the conventional and critical models of regulation to the Florida experience, we hope to determine which model offers a better explanation of events in the state. We believe that this determination also provides us with the basis for making some predictions about the likely course of future events in this important area of regulatory policy making.

Table 5.2 Current and proposed permitting procedures for hazardous and solid wastes in Florida

On-site disposal by generator	Hazardous waste		Solid waste	
	Current[a]	Proposed	Current	Proposed
Domestic waste	Permit—EPA & DER (5-19-82)	Permit—EPA & DER (5-19-82)	No permit required	No permit required
Commercial waste	Permit—EPA & DER (5-19-82)	Permit—EPA & DER (5-19-82)	No permit required	Permit—DER
Mining waste (except phosphate)	Permit—EPA & DER (5-19-82)	Permit—EPA & DER (5-19-82)	No permit required	Permit—DER
Phosphate	EPA study to decide nature of waste (Same—DER)	Permit—DER (if designated hazardous waste)	No permit required	Permit—DER
Waste from government operations	Permit—EPA & DER (5-19-82)	Permit—EPA & DER (5-19-82)	No permit required	Permit—DER
Industrial waste	Permit—EPA & DER (5-19-82)	Permit—EPA & DER (5-19-82)	No permit required	Permit—DER

a. DER will take over EPA permitting authority when EPA awards program recognition to Florida DER.
Source: Florida Department of Environmental Regulation, "Handout," August 1982.

Market Assumptions

One of the fundamental differences between the two models was their assumptions regarding the nature of causes of the market failures that create the need for government regulatory intervention. The conventional model assumed that these failures were largely due to remedial factors that were extrinsic to the private enterprise system, while the critical approach assumed that such failures were intrinsically linked to the functioning of the private market itself. For reasons that we will present below, we find the hazardous waste issue conforms much more closely to the assumptions of the critical model than to the conventional model.

The governor's Hazardous Waste Policy Advisory Council divides producers of hazardous waste into three categories.[30] Small producers of waste are those who generate less than 1,000 kilograms per month; they constitute a large proportion of all generators, but account for only 1 to 2 percent of hazardous wastes produced in Florida. Medium-sized producers of hazardous wastes are those who generate more than 1,000 kilograms per month; this is a substantial amount of waste but not enough to support on-site disposal. They constitute 18 percent of all generators, but account for over 75 percent of all waste generated in the state. The balance of hazardous wastes is produced by large firms (e.g., Monsanto, Occidental) who process and dispose of hazardous wastes in their own facilities. For each of these categories of producer, the private market provides a set of incentives that make it virtually impossible to dispose of the wastes in a safe manner; further, it is unlikely that government—particularly, state government—can easily alter these incentives and either force or cajole firms into altering their practices.

Under RCRA, small producers of hazardous wastes are exempt from all state and federal regulations in this area. It has been reasoned that the vast numbers of such producers, coupled with the small contribution they make to the total amounts of hazardous wastes produced, make enforcement difficult and not very cost effective. Obviously, since they are unrestrained by regulation, such producers will dispose of hazardous wastes in the cheapest way possible: usually, this means simply throwing these substances in the municipal dump, or down the sewer. Safe disposal is exceedingly costly, especially in a state like Florida with no commercial hazardous waste storage facilities. Therefore, this is the only option available for businesses forced to minimize the costs of production in order to successfully compete. Unfortunately, while the individual contribution of these small producers is minor for any given year, their cumulative contribution to the hazardous waste problem in Florida may be quite large. Five of the twenty-seven sites modeled by EPA in Florida were municipal landfills—two of these sites were placed in the top five by the MITRE model. The accumulation, over many years, of small amounts of dumped hazardous wastes is clearly one of the chief reasons for these sites becoming dangerous hazardous waste sites. Thus, the inability of the state or federal government to regulate small-scale producers of hazardous wastes has contributed to the serious problem of municipal landfills becoming leaking hazardous waste sites. Indeed, municipal sites are a particularly severe problem for the state since Superfund covers only 50 percent of their cleanup. There are no current plans for addressing the problem of small producers and, given their large numbers, small immediate contribution, and the incentives operating on them, it is difficult to understand how they could be regulated at all.

Medium-sized producers of hazardous waste are generally acknowledged to pose the most serious threat to the state's fragile environment. Individually, too small to afford safe on-site disposal, these firms collectively produce three-fourths of the state's hazardous wastes. We believe that the problem of improper disposal by such firms is less a function of unscrupulous individuals with little or no regard for the public welfare—although there seem to be a number of them—than it is a problem of the competitive pressures faced by these firms. Most of these firms are electroplaters, electronics firms, wood preservers, battery and industrial drum salvagers, or other small companies operating in highly competitive markets. The cost of safe disposal of hazardous wastes, especially in Florida where there are no safe commercial storage facilities, would seem to be beyond the reach of such companies. In effect, if the costs of either proper disposal or the effects of improper disposal are included in the routine costs of doing business, many of these firms would simply be forced out of business. Indeed, many of the Florida sites on the EPA list are those left by businesses that have gone out of business, or businesses (in at least three cases) that have filed for bankruptcy *after* they were finally caught by the state and sued for cleanup costs.[31]

The state of Florida and the federal government are faced with a perhaps insoluble dilemma in the case of medium-sized waste producers. The costs of these companies' practices are exceedingly high. However, these costs will not be borne voluntarily by the companies themselves. Nor is there much hope that states will be able to force these firms to internalize their negative externalities. Due to interstate economic competition, one state's strict regulation of these practices will give a substantial advantage to firms operating in other less strict states. There is a good deal of evidence that states already are competing with one another by reducing the regulatory burden on companies and thus creating a "good business climate."[32] Further, it is exceedingly unlikely that the state or federal government will allocate the money for such cleanup out of general revenues. In effect, the solution that has been adopted is to clean up waste sites when the responsible company can be found and to determine the extent of the cleanup by the company's ability to pay. For example, in dealing with the Coleman Evans wood preserving firm near Jacksonville, whose dump site had polluted surrounding groundwater and scores the third highest score of the twenty-seven sites that were modeled, DER reported: "The City [of Jacksonville] is working out a cooperative arrangement with Coleman Evans that will attempt to solve the problems *within the company's* financial limitations. An initial plan to entirely remove the contaminated soils would have bankrupted the company."[33]

Two implications of this policy solution should be noted. First, the state of Florida's solution to the problem is not atypical, but rather is consistent with the direction of several Supreme Court decisions that have defined the feasibility of various environmental regulations in terms of their likely impact on the financial solvency of the regulated industries.[34] Second, this solution is consistent with the assumptions of the critical model. Clearly, the costs of hazardous waste pollution must be borne by someone. The solution adopted is that the state and industry will pay what they can, but that any remaining costs will be borne by the general population in the form of the ill-health effects of such pollution. Much as Kapp, Ophuls, and O'Connor would predict, what we cannot pay for now will be paid for by future generations as a temporal externality of current market activities.

Large producers of hazardous wastes who dispose of this waste in their own on-site facilities are generally not regarded as a severe problem in the state of Florida. Indeed, the report of the governor's Hazardous Waste Policy Advisory Council scarcely mentions such producers at all. When interviewed, DER officials, legislators, and committee staff members likewise seemed to believe that large firms were basically in compliance with the law. It would seem, then, that the regulations' application to large producers of hazardous wastes conforms to the assumptions of the conventional model. Hazardous wastes would be disposed of in an unsafe manner without government regulation but, once such regulations are adopted, the overwhelming majority of large firms ("good apples") will obey them with only a limited threat of punishment.

We believe that such an interpretation of the behavior of large generators of hazardous waste overlooks the degree to which such firms have been able to influence the legislation that regulates them. Where middle-sized producers, usually small firms, have been forced by the competitive pressures of the market into violating regulations, larger firms have been able to use their political power to make certain that these regulations do not interfere with their profitable operation. Two instances of this type of influence are particularly relevant to any analysis of hazardous waste regulation.

First, the congressional Superfund legislation was amended to limit severely the liability of chemical firms for pollution caused by hazardous waste.[35] As proposed by the Carter administration, the original bill called for a tax on the chemical industry that would have raised $4.1 billion. One-third of this money would have been earmarked to compensate individual victims of hazardous waste pollution. Further, the bill contained provisions for strict liability standards in determining fault in pollution cases as well as providing, under strict liability standards, the rights for individuals to sue chemical firms for damages in federal court. Compensation for individuals, strict liability, and the right of individuals to sue in federal courts were all eliminated from the final version of the bill. Further, the amount of money to be raised for the Hazardous Substances Response Trust Fund was reduced by 75 percent to $1.38 billion. We conclude that the Chemical Manufacturers' Association (CMA), the trade association and lobbying arm of the chemical industry, was able to get a bill through Congress with which they could readily comply. Indeed, consistent with the assumptions of the critical approach, the CMA claimed that whatever the benefits of the original Superfund proposals, the industry could not bear the costs it would impose on them.[36] The bill's severe limitation of liability and recourse for individuals, however, results in the costs of hazardous waste pollution being borne not by the firm but by the affected individuals. In effect, as O'Connor suggests, the costs of production have been socialized.

Second, at the state and federal level, several important producers of wastes that might be considered hazardous were specifically excluded from either state or federal regulations in this area. Both federal RCRA and state RRMA regulations exclude: "ashes and scrubber sludges generated from the burning of boiler fuel for generation of electricity or steam; agricultural and silvicultural by-product material and, agricultural and silvicultural process waste from normal farming or processing; and discarded material generated by the mining and beneficiation and chemical or thermal processing of phosphate rock, and precipitates resulting from neutralization of phosphate chemical plant process and non-process waters."[37]

Thus, whether or not scrubber sludge from electrical generating plants, agricultural runoffs, and phosphate mining by-products are hazardous wastes and cause a pollution problem, they are specifically exempted from such laws. In short, while medium-sized producers of hazardous wastes may be forced into the position of violating regulations to remain in business, other possible producers of wastes have managed to have themselves placed outside the authority of such regulations. Whatever costs are imposed by their practices will be borne by the general population now or in the future.

We will discuss the reasons for granting these exemptions when we evaluate the assumptions of the conventional and critical models about the ability of the political system to implement regulations. At this point it is sufficient to conclude that in the area of hazardous waste regulations, the critical approach provides a more realistic set of assumptions than the conventional approach about the nature of market failures and the ease with which undesirable behaviors can be affected by the alteration of market incentives by regulation.

Regulatory Design Assumptions

The second differing assumption of the two models concerned the theoretical possibility of designing regulations that could balance the private interest of industry against the public interest. Many scholars who adopt the conventional model assume, for example, that it is possible to design "objective" decision-making rules that will accurately balance the costs and benefits of regulation. The critical approach assumes that, due to the incompatibility of regulation's goals with the operation of the private market, such rules are bound to be seriously flawed. The search for such decision rules plays a prominent role in Florida's attempts to deal with hazardous waste pollution. Given the severe shortage of funds for cleanup and monitoring, decisions must be made to concentrate efforts in the areas that pose the gravest threat to the public and the environment. The severity must then be evaluated against the amounts of money available for dealing with hazardous waste. In order to evaluate the degree to which a successful balance has been struck, it is necessary to discuss the use of risk assessment techniques in hazardous waste management in Florida. Both the technical and political aspects of these techniques can be evaluated in terms of the assumptions made by the critical and conventional models of regulation.

As we noted above, the MITRE model was used by EPA to rank the relative danger of hazardous waste sites in Florida. The model is composed of three elements: (1) "routes or pathways" leading from the waste site, i.e., air, groundwater, surface water, combustion, or direct contact; (2) "impact categories," including the release potential of the site, the characteristics of its waste content, the target population or water supply, and the quantity of the hazardous waste; and (3) "rating factors," or scores from zero to three which are assigned to each category: the more threatening the site is in any category, the higher the rating.[38] The maximum score for any site is 100 percent. Once a series of sites are scored, they may be ranked according to clean-up priority or for other policy purposes.

The MITRE model's authors specify an elaborate set of assumptions, all of which must be met if "best engineering judgments" are to be converted into "policy prescriptions" through MITRE modeling. Three assumptions are prominent among these:

1. An absolute amount of data on each site must be available. (A maximum of 5 percent missing data for the entire site is allowed; beyond this maximum, the relative potential hazard of a site cannot be determined.)
2. The ranking scores are relative. (For example, a site score of 100 is higher than a site score of 50 but not twice as high; thus the scores may *not* be used in other risk assessment models as absolute scores.)
3. Political and economic factors are excluded and only engineering considerations are used to judge risk.

While the MITRE model is highly sophisticated, it is only as good as its assumptions: the modeling process used in Florida shows how poorly it performs as a risk assessment technique.

As mentioned above, twenty-seven hazardous waste sites in Florida were modeled by the EPA. This list did not represent the state's population of hazardous waste sites. At the time of the modeling process, over 160 hazardous waste sites had been identified by DER. This number was later raised to 200 such sites. The reason that only twenty-seven sites were modeled is simply that adequate information was available for only those twenty-seven sites. The remaining unmodeled sites may have been more or less hazardous, but the modeling process explicitly ignored them.

The failure of the information assumption in MITRE rankings of Florida's hazardous waste sites has important policy implications. Now that these sites have been identified and ranked, policy makers are likely to focus on them to the exclusion of potentially more hazardous sites which went unmodeled for lack of information.

An even more intriguing policy implication attends the model's exclusion of politico-economic factors in the ranking of Florida sites. Such considerations were to be accounted for separately by EPA in its nationwide ranking of hazardous waste sites. As table 5.3 indicates, EPA's original MITRE ranking of the twenty-seven sites compared with the agency's ranking of the final sixteen sites indicates little correspondence between the two. The dramatic difference between these two rankings indicates that EPA's decision rules include much more than the engineering considerations included in the MITRE model. The lack of correspondence between the two lists must cast doubt on either the MITRE model itself as an adequate risk assessment technique, or the ability of EPA to make decisions consistent with "objective" decision-making techniques.

Specifically, what sorts of factors do EPA and DER consider in their policy determinations based upon MITRE rankings? First, it is to the state's advantage to prefer the targeting of private hazardous waste sites as opposed to municipal sites, since the former receive a 90 percent Superfund subsidy in cleanup, while the latter receive only a 50 percent subsidy. Thus, the state may be tempted to distort the actual risk presented by a waste site as revealed in the MITRE model by factoring in the financial differences in federal compensation for different types of waste sites. Second, EPA has placed a special priority on waste sites which are "clean"—an unintendedly ironic term used to indicate sites which have either been abandoned or for which no legal entanglements are present.

As of summer 1982, five sites in Florida have advanced to the "cooperative agreement" stage of Superfund cleanup. Although money had been released for only one of these sites, these five are to be the sites cleaned up first by the state. The selected sites are overwhelmingly "private" and "clean"; this seems to indicate the importance of the two factors we have identified and the extent to which consideration of these factors tends to modify the results of the more formal MITRE model. The five sites are:

1. Biscayne Aquifer/Dade County (including the 58th Street Landfill, Miami Drum, and Miami International Airport);
2. Coleman Evans/Jacksonville;
3. Whitehouse Oil Pits/Jacksonville;
4. Sapp Battery/Panama City; and
5. American Creosote/Pensacola.

Table 5.3. Comparison of DER, MITRE, and EPA rankings of uncontrolled hazardous waste sites

DER ranking	Waste site / location	MITRE score	EPA ranking
(1)	Biscayne Aquifer / Dade County	79.85	(1) [a]
(2)	Schuylkill Metals / Plant City	59.16	—— [b]
(3)	Coleman Evans / Jacksonville	59.14	(10)
(4)	Davie Landfill / Broward County	59.14	(6)
(5)	U.S. Naval Air Station / Jacksonville	58.49	——
(6)	Gold Coast Oil / Miami	58.14	(14)
(7)	Miami Drum / Miami	58.04	(1) [a]
(8)	American Creosote / Pensacola	56.27	(4)
(9)	Reeves Galvanizing / Tampa	55.26	(3)
(10)	Whitehouse Oil Pits / Jacksonville	52.81	(9)
(11)	Solderless Terminal / Broward County	51.26	(11)
(12)	Pratt & Whitney / Palm Beach County	51.18	——
(13)	Tower Chemical / Lake County	49.60	(16)
(14)	Sapp Battery / Panama City	48.11	(15)
(15)	Zellwood / Orange County	47.19	(13)
(16)	Pickettsville Road Landfill / Jacksonville	44.00	(2)
(17)	Taylor Road Landfill / Tampa	42.40	(5)
(18)	Pioneer Sand / Pensacola	41.71	(7)
(19)	Timberlake Battery / Tampa	41.43	(8)
(20)	Chlorida Metals / Tampa	40.56	——
(21)	Alpha Chemical / Polk County	38.95	(12)
(22)	Cabot Carbon / Gainesville	35.67	——
(23)	Miami Intern'l Airport / Miami	34.67	(1) [a]
(24)	Vroom Site / Tampa	31.83	——
(25)	Regency Sludge / Jacksonville	20.73	——
(26)	Munisport Landfill / Miami Beach	18.28	——
(27)	Broward Industrial Plating / Ft. Lauderdale	11.75	——

a. EPA ranked Biscayne Aquifer / Dade County, Miami Drum / Miami, and Miami Intern'l Airport / Miami together as one site.
b. A dash (——) means these sites were ranked by DER but not ranked by EPA in the final list of 115 sites nationwide.
Source: Florida Department of Environmental Regulation, "News Release," October 23, 1981.

Clearly, the selection of these sites ignores the EPA's ranking of Florida's twenty-seven hazardous waste sites (see table 5.3). Four of the five sites (or five of the seven, if site 1 is broken down into three sites) are privately owned. According to DER officials we interviewed, these sites are either abandoned or operated by businesses far too poor to be held financially responsible for cleanup; indeed, American Creosote filed for bankruptcy in August 1982.

The inherent problems with the type of risk assessment embodied in the MITRE model process are illustrated by the jurisdiction used by officials for the choice of these five sites. When asked their basis for this judgment, they respond that they are selected on the basis of the purely technical MITRE modeling—despite the fact that the model produced a ranking that did not make those selected the five most serious sites in Florida, as well as the fact that the MITRE model itself was only applied to 27 out of a possible 200 sites. Policy makers in this case seem to have a tendency to fixate upon "technical desiderata," even when the assumptions of the risk assessment are violated or when its results are distorted in the policy process.

The thrust of our criticism of the use of the MITRE technique in Florida may be viewed as a strictly technical one. That is, the difficulties of the model are a function of: first, a lack of information that resulted in the inability to consider properly all sites in Florida; second, political "tampering" with the MITRE results, caused by the funding formula adopted in Superfund which leads to a tendency to place a higher priority on cleanup of private sites than of municipal sites even when strict risk assessment does not warrant such ranking. The implication of such a technical criticism is that the decision-making process could be perfected by gathering more information and by altering the funding formula to reduce distortion. Such a conclusion would be consistent with the conventional approach to regulation. It implies that even if the decision-making rules used in this area are flawed, they are subject to fairly straightforward correction: decision rules could be formulated that would in fact result in adequate regulation of this severe problem.

Such an assessment of our criticism of the MITRE model's use would be incorrect; in our research on Florida's hazardous waste management, we do not find support for the conventional approach to regulation. We do not believe that the lack of information and distortions in the funding formula are purely technical or easily remedied problems. Such a conclusion is too reminiscent of the misleading technical-political dichotomy that constantly reappears in the literature on the policy process. We cannot evade the consistent finding that overtly technical decisions have political components.[39]

First, the lack of information that flaws the results of the MITRE process is largely a result of lack of money allocated for the regulation of hazardous waste. This lack of money involves more than the underfunded hazardous waste program which we discussed above. One of the reasons hazardous waste pollution is such a severe problem in Florida has to do with its threat to underground aquifers. Yet, very little systematic information is available on the seriousness of extant aquifer pollution. Monitoring water quality is a joint responsibility of the Department of Health and Rehabilitation Services (HRS) and the Ground Water Division of DER; the former monitors privately owned wells and the latter has responsibility for all municipal wells. However, neither agency routinely monitors these wells for the presence of dissolved synthetic organics. Thus, no routine testing is done for the presence of the chemicals that are most likely to be released from hazardous waste sites (e.g., solvents, heavy metals, or chlorinated organics). Indeed, the HRS maintains only one laboratory for testing water and this lab does not have the capacity to test for anything other than the presence of bacteria. The only way the state ever finds out that dissolved organics are present in the water is when a complaint is made by users. At that point, if the complaint is about a municipal well, the water is checked by a series of tests. Such tests can be very expensive: a single sample tested by mass spectroscopy can cost up to $5,000. If dissolved organics are found, that particular well is simply closed.

Currently, the groundwater division of DER employs fourteen individuals. Seven of the fourteen are funded by the federal government and are involved full time in U.S. Geological Survey programs. Thus, only seven people are left to monitor groundwater problems in the entire state. While we were interviewing officials of the groundwater division, six of the seven staff members were out on a single well-testing operation. This meant that any new problems brought to DER's attention, no matter how serious, would have had to wait until the current problem was solved. In short, there is no systematic data collected on the degree to which hazardous waste pollution has already affected or will affect the groundwater in the state. Such a data gathering plan has been designed, but it would require at least doubling both the operating budget and staff size of the groundwater division. Such an increase is very unlikely, given current declines in state tax revenue. Indeed, as of August 1982, DER was faced with a 2 percent across the board funding reduction.

We believe that the lack of funding for the gathering of information is an indication of a wider problem in the application of quantitative cost-benefit and risk-benefit techniques. That is, the usefulness of such techniques depends upon the information that is gathered. This information is quite expensive to gather. Moreover, for a variety of reasons, it is against the interests of many policy makers and private groups that this information *ever* be gathered. In the case we examined, this information was largely unavailable due to appropriations decisions made in the legislature that were not subject to any sort of cost-benefit or risk-benefit analysis. Further, the amount of money that is available to use on the problems subjected to formal analysis is also arbitrarily limited by such a priori legislative decisions. It is at least possible that hazardous waste represents a much greater risk to the state of Florida than, for example, crime, poorly maintained roads, or inadequate education. However, the legislative decisions that allocate state revenues among these functions and hazardous waste are never informed by any type of cross-issue risk assessment techniques. Thus, before the first step is taken to apply technical decision rules, the most important decisions regarding funding have already been made. In the case we examined, these a priori funding decisions have resulted in a lack of adequate information upon which to base formal risk-benefit techniques as well as a severe shortage of funds allocated to dealing with the problem addressed by the risk-benefit technique.

We conclude, therefore, that the technical and political components of risk assessment in this area are inextricably tangled. Until the entire political process that surrounds hazardous waste management and its treatment relative to other problems can be reformed, any use of formal risk assessment techniques is likely to be seriously flawed. Further, these techniques come to be used as "objective" justifications for what are really highly political decisions. The problem, in short, is not the technical one that is assumed by the conventional approach but a political and economic one as is assumed by the critical approach. Until these broader political and economic issues are addressed, any formal techniques are bound to be used in a way that distorts the issue they are designed to clarify.[40] In our case, the MITRE rankings and their questionable use came to have a life of their own for policy makers despite the widespread realization that the results were seriously flawed.

Institutional Assumptions

We now turn to an examination of the assumptions made by the two models regarding the capabilities of the political system and these assumptions' application to hazardous waste management in Florida. The critical and conventional approaches differ in their evaluation of the ability of political systems to pass and implement regulatory legislation. The conventional approach assumes that, while regulations as currently passed and implemented often leave much to be desired, many of these problems can be addressed by either altering the decision rules used by regulators or alternatively by strengthening the regulatory agencies that implement legislation. In contrast, the critical approach assumes that effective regulation is made politically impossible by the political power of those economic interests that stand to be hurt by regulation. In order to understand the politics of hazardous waste regulation in Florida, it is necessary to understand the disjuncture between the passage of regulatory legislation and the appropriations process that allocates funds for the enforcement of the legislation. Politicians have been able to pass legislation that is widely perceived as addressing the public health hazards created by hazardous waste pollution, while simultaneously ensuring that regulations do not interfere with the functioning of well-organized and politically powerful economic interests.

The significance of the distinction between the passage and funding of regulatory legislation is indicated by the fact that, despite the scarcity of funds for enforcement of hazardous waste regulations in the state, Florida was ranked ninth among the states in the quality of its hazardous waste laws by the National Wildlife Foundation.[41] When it comes to passing regulations, the Florida legislature has been able to respond to the demands made to it by environmental groups; when it comes to funding their enforcement, however, legislators respond to interests other than those of environmental groups. However, the ranking by the National Wildlife Foundation ignores a second critical aspect of hazardous waste regulation: specific exemptions for certain classes of industries. While the laws may be comprehensive for those industries they affect, several large industries have been able to force specific exemptions. In the 1982 legislative session, for example, bills were proposed to revise the existing hazardous waste amendments; earlier, we discussed the siting and trust fund aspects of these revisions. Also included in them were clauses to eliminate the exemption for scrubber ashes and, as a prelude to eliminating the phosphate industry's exemption, the authorization of a state study of the possible "adverse" health impact of phosphate mining by-products. Both clauses were removed in committee hearings at the insistence of the representatives of the electric generating utilities and the phosphate industry respectively. In fact, we were informed by a member of the House Environmental Resources Committee, which held hearings on the bill, that the clauses were put in without any real hope of passage. The representative who offered the clause to end the exemption for scrubber ashes did so almost humorously: he wanted to force the utility lobbyists to do some work (i.e., show up at the committee hearings and use their influence to kill the clause). We were told by a lobbyist for the phosphate industry that once the phosphate clause was amended to suit industry interests, they simply lost interest in the bill. Thus, it would seem that it is only possible to pass legislation in this area with the tacit approval of the large industries that might be adversely affected.

Even more significant than the exemptions in regulatory authority is the lack of funding that hazardous waste regulation continues to receive in the state. The separation of regulatory passage from the appropriations enabled legislators to "have their cake and eat it too." That is, legislators can appeal to environmentally conscious constituents by voting for the passage of laws but, by later opposing funding for their enforcement, they can appeal to adversely affected economic interests and constituents who view themselves as hard-pressed taxpayers.

According to one state senator, who was a member of the appropriations committee, this ploy of supporting legislation while opposing funding for enforcement is a not uncommon occurrence. Further, representatives of two environmental interest groups that had been active in the passage of hazardous waste regulation admitted they did not have the personnel or expertise to monitor the appropriations process. Rather, they devoted most of their resources to the more visible legislative passage of regulations. So, the practice of supporting legislation but opposing funding is a relatively safe one.

The lack of adequate funding for hazardous waste management has also served to deflect criticism for the state's serious hazardous waste problem from the legislature to DER. One of the ironies of the current criticism of supposedly "bloated and inefficient" government bureaucracy is that many of the symptoms of this condition (i.e., unresponsive agency personnel, long delays in accomplishing assigned tasks, unreasonable and by-the-book enforcement of other responsibilities) are often indistinguishable from the symptoms of inadequate agency resources. This irony is fully realized in several legislators' criticism of DER. They argue that "DER is so bureaucratic and inefficient now that voting for increased appropriations will simply increase the problem." This line of argument provides the justification for voting for increased regulatory authority while voting against increased appropriations. When the regulations are not enforced, the fault lies with the bureaucrats at DER: "They could do it if they weren't so inefficient."

Summary

We suggested above that Florida was an interesting state in which to examine hazardous waste disposal because of the presence of strong environmental

groups and powerful industries that might be expected to favor strong hazardous waste legislation. Since the existence of organized groups and their
ability to bring pressure to bear upon policy makers is a crucial component of
pluralist theory and the assumptions of the conventional approach to regulation, it is important to understand why these groups were not more influential
in the policy process.

As is clear from our discussion, environmental groups have been moderately
successful at getting laws in this area on the books (although not in the 1982
legislative session). They have been unsuccessful at getting the legislature to
appropriate money for the laws' enforcement. We believe that these circumstances are explained by two factors: one has to do with the nature of hazardous
waste as an issue; the other, with the limitations of environmental groups as
political lobbyists. First, environmental groups have not made hazardous waste
a high priority item on their agendas. This is largely because hazardous wastes
have been much less visible as a threat to the environment than other environmental menaces in Florida—such as development's encroachment on wilderness
areas and the disturbance of the state's water supply. While hazardous waste
may become as visible as these other areas, until it does it will likely remain a
lower priority issue. Second, environmental groups generally do not seem to
possess the technical expertise or financial resources to follow or influence the
regulatory process.[42] While a representative of the Environmental Service
Center, an information clearing house and research organization serving a
variety of environmental groups, told us that there were plans to acquire such
expertise, it would be quite some time before environmental interest groups
were able to follow this crucial stage in the regulatory process. Further, the lack
of individual responsibility by legislators in the appropriations process makes it
difficult for environmental groups to bring pressure to bear on them.

The second potential source of interest group support for hazardous waste
regulation is the industries, like tourism and agriculture, that stand to lose a
great deal if hazardous waste pollution should become a highly visible problem
in the state. Despite their potential stake in this issue, everyone we interviewed
agreed that these groups had not been involved in the issue. Two explanations
were given for this inactivity. A member of the legislature suggested that the
issue of hazardous waste is one that probably would not affect these industries
for several years, and that these groups were much more oriented toward
defining their interest in the short run. Thus, hazardous waste management was
simply not an issue that they saw as affecting them at all. A second explanation
was offered by a representative of the environmental group, who believed that
in Tallahassee, an extremely isolated state capital, lobbyists for industrial trade
associations interacted quite frequently with each other. Because of this they
tended to focus upon issues that were common to industry in general and to
avoid, if at all possible, divisive issues. These two explanations lead to similar
implications for the activation of adversely affected industries: these groups are
not likely to support hazardous waste regulation until pollution has damaged
their members. Since this has not yet happened, they have not become involved.

At least in the area of hazardous waste regulation, our findings offer an important criticism of the work of conventional theorists like Wilson[43] and Jones.[44] Recall that these two authors suggested that a serious shortcoming of regulation was that the pressure of interest groups organized for protection of the interests of the public at large often resulted in an overresponse by vote conscious legislators. The result is extreme and unreasonable regulations. In our case, we find that the focus of such public interest groups on the passage of legislation makes it possible for legislators to avoid "extreme" regulation by offering symbolic responses that can be easily blunted in the appropriations process.

Once again, we find considerably more support for the assumption of the critical model than for the conventional model of regulation. Due to the low visibility of hazardous waste and the fact that its adverse effects are as yet poorly understood (largely due to lack of adequate information and monitoring) and not likely to be felt for many years, environmental groups have given it a low priority and many industrial groups have not recognized it as an issue at all. The structure of the policy-making process has funneled what interest group activity there is toward the passage of legislation. Little of this activity has been aimed at the less visible, but perhaps more important, appropriations process. Further, until this pattern changes, blame for governmental failures in this area of regulation is likely to be directed at DER, not the legislature.[45] This situation makes it likely that the political system will continue to respond to the issue in a largely symbolic manner, making it possible for the targets of regulation to avoid its effects.

Conclusion

If we are correct about the applicability of the critical approach to the politics of hazardous waste regulation in Florida, several profoundly disturbing conclusions follow. To the extent that our results are generalizable to other states, these conclusions are even more worrisome.

First, we expect that the state of Florida will continue to deal with the problem of hazardous waste disposal on a piecemeal basis: efforts will continue to focus only on the most dangerous of the sites that have already been identified. As long as hazardous waste sites periodically make the news, efforts will be made to clean up the most visible ones. However, given the limited resources allocated for environmental regulation in general, danger exists that some new threat will emerge and divert funds away from hazardous waste regulation. One DER official noted the tendency for funding to follow the "sexy issues" in environmental regulation. Funding for hazardous waste, for example, had come at the expense of funding for regulation of solid waste and for groundwater monitoring. If a new "sexy issue" emerges, we expect the state's already minimal financial commitment to hazardous waste regulation to erode.

A second disturbing implication of our analysis is that ignorance about the extent of hazardous waste pollution's threat will remain unabated. Given the tremendous expense of acquiring adequate data on this pollution and the difficulty of estimating its long-run effects, there is an incentive to simply limit the gathering of data. Already, we believe that a "see no evil" approach may be emerging with regard to the monitoring of groundwater quality in the state.

Third, recent trends of change in hazardous waste regulation are likely to decrease its effectiveness. Insofar as the structure of policy making in this area is changing, these changes are largely guided by those who subscribe to the conventional approach to regulation. In particular, recent changes have been influenced by neoconservative regulatory reformers. They recommend an increased reliance on state agencies and a lessened role for the federal government; however, this will increase the political leverage of those firms who resist regulation. Their threats to move or go out of business will be far more persuasive at the state level than the national level. Further, state regulators will be less able to use the threat of federal intervention as a "club" to gain compliance from recalcitrant state industries. Reductions in funding for EPA also bode ill for Florida's hazardous waste management program. Without current and past EPA grants to DER, the state would have, at best, a token program. Yet, there is little likelihood that the state legislature will increase funding to compensate adequately for any decline in federal funding. In the area of solid waste management, for example, federal grant cessation led to elimination of programs when the legislature refused to allocate funds for their continuance.

Finally, the critical approach to regulation implies that solutions to many regulatory problems require far-reaching structural reform of political and economic institutions. Kapp, Ophuls, and Schumacher all suggest that social decision-making mechanisms must be significantly restructured if they are to deal with environmental degradation.

A key failing of their writings, however, is that they devote little attention to the problem of achieving such changes. Indeed, designing *and implementing* fundamental change is something that is largely ignored by even the most extreme critics of the social order. Their absence of attention to this question is made even more troubling by the critical approach's conclusions: the longer such change is postponed, the more severe such change will have to be; the longer such change is postponed, the higher the likelihood of truly catastrophic ecological disaster.

6. Intergovernmental and Public-Private Sector Relations in Hazardous Waste Management: The New York Example

John A. Worthley and Richard Torkelson

Recent research efforts, notably Eckhardt,[1] Jorling,[2], and Worthley and Torkelson[3] have illuminated the nature of the public policy management problem regarding hazardous waste. Drawing from developments in the 1970s, these and other recent analyses presaged some significant federal and state policy developments of the early 1980s and suggested future directions. Our analysis of the Love Canal episode disclosed major problems of intergovernmental coordination and the relationship between the public and private sectors.[4]

Since these studies were completed much has happened that further clarifies the nature of this increasingly serious public policy problem and that more clearly suggests implications for policy management efforts. This chapter reviews these recent developments with an emphasis on the New York state experience and, in analyzing these together with the previously reported experiences, suggests areas for policy research, analysis, and management. Essentially, our findings are that intergovernmental tensions and public-private sector conflicts are now even greater barriers to policy development and implementation in New York state and that these are, therefore, critical issues for policy and managerial attention.

Developments in the Early 1980s

Several significant developments affecting intergovernmental and public-private sector relations have occurred thus far in the 1980s. They have considerable implications for public policy and management strategies concerning hazardous wastes. These developments revolve around regulatory, legal, organizational, political, industrial, and technological realities.

Regulatory Developments

Late in 1980 the federal Environmental Protection Agency (EPA) promulgated hazardous waste regulations. Based on the Resource Conservation and Recovery Act (RCRA) authority, they required three actions: (1) identification and listing of classified hazardous wastes; (2) annual reporting by all hazardous waste generators, transporters, and storage or disposal facility operators; and

(3) implementation of a manifest system providing documentation of the transportation of hazardous wastes from point of origin to place of ultimate treatment, storage, or disposal. While these rules clearly provide for the tracking and monitoring of toxic wastes, they fail to set standards for safe disposal or to address the problem of cleaning up existing dangerous dumps.

Since 1980, with the arrival of the Reagan administration, landfill disposal rules proposed by the previous administration have been delayed and the EPA has relaxed its enforcement of existing regulations. Specifically, the EPA has been granting more exemptions, allowing some industries to dispose of chemicals in small amounts, and exempting certain categories of chemicals from regulation; it has also cut its own budget by 30 percent and made significant cuts in its grants to states for hazardous waste management; it has proposed dropping rules requiring waste disposers to carry third-party insurance; it has permitted expansion of some disposal sites; it has dropped its requirement for annual reports by waste generators and transporters and substituted a sampling method; and it has substantially reduced its enforcement effort.[5] In sum, federal regulatory actions in the 1980s have tended to be more lenient than in the past, prompting both internal and external criticism. A career official at EPA itself, for example, recently lamented: "There are no regulations on the books to prevent chemical companies from dumping toxic wastes into landfills like Love Canal—it is still completely legal."[6] Both the congressional Office of Technology Assessment (OTA) and the General Accounting Office (GAO) have criticized EPA regulations as not protecting public health.

In New York state, however, home of the Love Canal and among the most active states in both hazardous waste related industries and in toxic waste control, regulatory measures have been more stringent. New York has not only enacted regulations that parallel the EPA rules but has also imposed more strict rules in requiring permits for transportation of hazardous wastes, in requiring annual reporting, and in ordering compliance with state criteria for treatment and disposal facilities. Furthermore, New York has increased its enforcement mechanism by establishing a Division of Hazardous Waste Enforcement within its Department of Environmental Conservation.[7] In effect, while federal regulation of hazardous waste has relaxed or retreated in the 1980s, New York state activity has increased.

Legal Developments

At the federal level, significant legislation was enacted in 1980. Not only did the Comprehensive Environmental Response, Compensation, and Liability Act, known as "Superfund," provide $1.6 billion over the next five years to clean up the worst hazardous waste dumpsites in the country (see chapter 2), but in 1981, New York enacted the "Felony Bill" that classifies many hazardous waste violations as felonies, provides for prison sentences for offenders, and empowers the courts to require violators to pay the cost of cleanup.[8] In brief, as New York has demonstrated, strong state statutes can be adopted.

Organizational Realities

Two organizational developments stand out. One is organizational restructuring, such as that in the New York Department of Environmental Conservation, to address the hazardous wastes problem. In 1981, a Division of Hazardous Waste Enforcement was established in New York with the specific charge to enforce regulations vigorously.[9] This development reflected two of the state's significant views: (1) that industry will cooperate and comply with policy efforts only if it is fearful of being caught; and (2) that federal enforcement activities are likely to be ineffective if they exist at all. This organizational development has provided a central focus for technical expertise and thus influenced policy deliberations in New York. Moreover, it enables more certain implementation of policy decisions in this area by providing staff and resources for monitoring and compliance.

The second development is the tension that has risen between EPA and state agencies. The federal "Superfund" was placed under the control of the EPA to finance the remediation, investigation, and maintenance of abandoned hazardous wastes sites. New York, and at least ten other states,[10] found the flow of money from EPA to be painfully slow, a situation that has created both antagonism and intergovernmental tension. The tension centers not only on the speed of transfer of authorized funds, but also on qualifying procedures employed by the EPA. The net result is that public sector hazardous waste management organizations are hardly acting in unison.

Political Phenomena

Two fairly remarkable sociopolitical developments regarding hazardous wastes have occurred in the 1980s. One is the formation of a national toxic waste citizen interest group headed by Lois Gibbs, the leader of the Love Canal Homeowners Association's efforts to get action. The Citizens Clearinghouse for Hazardous Wastes (founded in 1981) is committed to the notion that "you have to fight the hazardous wastes issue on a political level. . . . It cannot be fought on a scientific front because the science does not exist."[11] Gibbs's group has a small lobbying staff in the Washington, D.C., area and periodically it sends representatives around the country to help organize state and local political action. The significance of this formal organization stems from the ad hoc nature of past citizen groups concerned with hazardous wastes. In New York, citizen groups, aided by media coverage, have been influential but have lacked a long-term organizational infrastructure. Gibbs's efforts could help to fill that void in citizen participation capacity. One development already occurring is the mobilization of well-established neighborhood civic groups oriented toward the hazardous waste problem. The Old Bethpage landfill issue (in Oyster Bay, New York), for example, has been effectively pursued by such mobilized neighborhood civic associations.[12]

The second political phenomenon is an unstable public mood that evidently is apathetic to the problem in some regions and resistant to solutions in others. In November 1981 an EPA study found the area around Pitcher, Oklahoma, to be among the worst hazardous waste sites in the nation. Yet, the local population was apparently unconcerned or unaware. Because no immediate danger seemed to exist, local residents expressed actual disdain for the EPA study.[13] In New York, the visibility of the Love Canal episode has led to fairly large-scale control efforts. An example of such efforts was the proposed construction of a major toxic wastes disposal facility that would provide industry with a safe, controlled dump for its by-products. But, despite the unquestioned need for such a facility, the state has been unable to make progress toward construction because of intense community opposition to proposed locations for the site. Everyone recognizes the necessity of a facility, but no one will allow it "in their own backyard." Hence, the needs of the larger community (i.e., the state) clash with the land-use decisions of local interests. Political phenomena have thus emerged in the 1980s in new forms that significantly affect the nature of the hazardous wastes policy problem.

Industrial Actions

Similarly, there have been some significant, and in some instances countervailing, developments on the corporate front. On the one hand some notable cases of voluntary industry cleanup have occurred. Last year, for example, the General Electric Company agreed to undertake remedial action on several sites in New York on which the company had dumped PCBs and to pay a proportionate share of cleanup costs at those sites where it, together with other companies, had dumped toxic chemicals. In the Niagara Falls region of New York, Dupont and Olin Industries removed contaminated sediment from Gill Creek at a cost of $1.75 million. And recently, Hooker Chemical Corporation agreed to a $50 million cleanup and thirty-five-year monitoring program at a dump near Love Canal.[14] The chemical industry has established a hazardous waste response center in Washington called CHEMTREC where state and local officials who are concerned about hazardous waste spills can get advice on the seriousness of the threat and methods of cleanup.[15] There have, in other words, been some important corporate efforts to resolve the hazardous waste problem.

At the same time, however, there have been increased and concerted corporate efforts to avoid responsibility for hazardous wastes. The practice of "midnight dumping"—dispensing of hazardous wastes at deserted locations in the dark of night—has been found to be more extensive since the regulations went into effect.[16] There has, in brief, been evidence that business (together with organized crime) can, and is willing to, undermine efforts to resolve the toxic wastes problem. But there is also the clear evidence that business can be persuaded to help address the problem.

These conflicting, and perhaps confusing, phenomena can be explained in part by fiscal considerations: where industry has seen a fiscal advantage in cooperation it has participated; where it has seen a fiscal advantage in undermining policy initiatives it has tended to take that course. Indeed, cynical observers interpret the cause of business cleanup and remediation agreements as merely shrewd efforts to "buy off" future liabilities that might far exceed the current settlements. Less cynical observers see in these settlements industrial recognition of the advantage to business of social and political good will. In any case, industrial actions thus far in the 1980s have clearly emerged within a conflictual context vis-à-vis government. The settlements have been difficult to obtain and obstructionist business actions have been directed at neutralizing what business sees as government efforts to adversely affect business.

Technological Developments

Finally, the 1980s have witnessed intensified efforts to find environmentally sound and economical ways to dispose of the 80 billion pounds of toxic wastes generated each year in the United States. As Severo reports,"Despite an uncertain regulatory climate caused by the Reagan administration's pledge to free business from as many regulations as possible, there has been progress in two areas: the development of new technology, and the adaptation of old technology to new uses."[17] Most of the new technologies involve some form of incineration, such as "multiple-hearth" incineration in which waste is burned again and again; and "fluidized-bed burning" in which superheated wastes are passed through a bed of sand that becomes a catalyst speeding combustion. Some corporations have found ways to convert hazardous wastes into salable products through various "no-waste" technologies. These efforts, some of which are funded publicly, some privately, demonstrate that technological advances continue to offer some hope.

Related efforts have centered on "waste exchange" projects in which hazardous wastes of one company are matched with another company that uses such "wastes" in its production processes.[18] An exchange operates as a kind of broker linking potential users of wastes with producers of those wastes. Although they are commonly found in Europe, only twenty-four waste exchanges have recently been used in the United States, two of which are in New York. While most of these exchanges have been sponsored by private associations or private business, the New York State Assembly has considered establishment of a government-sponsored waste exchange in the state.[19]

Current Realities

Taken together, these several developments discussed above suggest two important characteristics of the hazardous waste management problem as it now exists. One is the increase during the 1980s of tension between government agencies. The other is intensified public-private sector interaction. The regulatory developments have tended to pit a slow-moving federal bureaucracy (especially EPA) against more aggressive state governments like New York. The slow implementation of the "Superfund" has similarly raised the tension level between Washington and state capitals over the flow of money; and state legislation like the "Felony Bill" has increased the tension between business and government.

More visible governmental agencies, like the New York Hazardous Waste Enforcement Division, have further complicated intergovernmental and industrial-governmental relations. The emergence of citizen interest groups has provided a mobilizing force demanding action by all parties that in turn generates tension between governments and between government and industry. Industrial actions, both in cooperating with government in cleanup efforts and in violating new regulatory programs, have clearly generated more public-private sector interaction. Moreover, new technological developments have intensified the level of public-private conflict in that they have produced new disputes over who should pay for the research and development of these technologies.

More subtle realities, however, have also contributed to the significance and complexity of these intergovernmental and public-private sector phenomena. As citizen awareness and concern over hazardous waste has grown, citizen efforts to confront the corporate purveyors have been frustrated. Often, abandoned waste sites, such as in the Love Canal case, are no longer owned by the original dumping corporation. In other states, the corporation may no longer be in business or is simply unknown; and in still other cases, known corporations simply turn a deaf ear to public complaints. As this has happened with the hazardous waste problem, citizens have turned to their elected representatives to intercede on their behalf. This has created a complicated and dynamic situation that is at the heart of intergovernmental and public-private sector tensions.

Our analysis of this dynamic interaction is based on the experience in New York state where concerned citizens, frustrated in their efforts at dealing with industry, have typically turned to their local governments which, owing to a lack of fiscal and technical resources, have turned to state elected and bureaucratic officials, who in turn have gone to federal officials only to discover a zealous caution about aiding a specific problem in one part of the country. At this point, the citizen is passed between levels of government, each of which is fearful that the total responsibility for a solution will rest on its already over-burdened shoulders. The ultimate outcome of these interactions has been an increase in tension and conflict among all levels of government, officials, and agencies.

The policy process in such a situation has developed—through consensus, disagreement, and controversy—a series of directions by which executive branch agencies have been functioning. In New York, the state accepted the responsibility for Love Canal and now, some four years later and forty million dollars poorer, is involved in massive litigation to recover the cost of the cleanup from the dumper—the Hooker Chemical Company. As a result of the protracted legal proceedings and further expenditures to litigate this matter, the state has now retreated and formulated a policy of withholding tax levy money for hazardous waste cleanup. New York state has determined that it simply cannot afford to pay the bill alone. It has, indeed, taken the dramatic step of declaring nonresponsibility for the financing of abandoned-site cleanup, a consequence of which is intensified effort by the state agencies to obtain financing from owners of sites for cleanup, remediation, and monitoring.

A recent example illustrates the dynamics involved. Several years ago, in response to intense citizen group action, the town of Oyster Bay in New York prompted the state Department of Environmental Conservation to investigate and eventually charge the Hooker Chemical Company with some 420 violations involving illegal dumping of toxic wastes at the Old Bethpage landfill. Since then the town has spent $8 million to cap the landfill and to test for seepage into groundwater. For three years the state has been negotiating with Hooker to reach an agreement on paying for the cleanup without undergoing extensive litigation proceedings. Frustrated and financially strapped by the delay, Oyster Bay's local government has been demanding that the state sue Hooker to recoup the town's costs.[20] But, because of the expense to the state that would be involved in bringing suit, the Department of Environmental Conservation policy has been to avoid court proceedings so long as other options exist. The state has placed this site on the "Superfund" list and is looking to the federal government to force the legal issue. The town of Oyster Bay, in the meantime, wants to avoid federal action and is trying to convince the state to remove the site from the federal list because of the treble damage provisions of the Superfund legislation. The state is refusing to do this until it is convinced that the town's proposed cleanup program is viable. The net result is high and still increasing intergovernmental conflict.

The conflict then extends to the public-private sector dimension. The state has now turned over a large number of its cases to its attorney general for litigation. These proceedings, costly and time-consuming, are a major aspect of hazardous waste policy making in the 1980s, and aggravate the conflict atmosphere between business and government. This conflict is also apparent in the response of state government to continued illegal dumping. The New York enforcement unit, for example, is using "covert investigation techniques and developing a network of informants" to enforce industry's compliance.[21] Clearly, a conflict situation exists in the 1980s as a major aspect of the toxic waste policy problem in New York.

Conclusion: Where Are We Going?

What do these early 1980s developments suggest for hazardous waste policy analysis and managment for the remainder of the decade? Where are we going and what can be done to hasten solution of this serious problem? Insight into these questions can be derived from clarification of the forces being generated by the developments discussed in the previous sections.

First, legal forces are likely to be considerable. Legislation on toxic waste is being formulated by state legislatures around the country with New York (and California) in the forefront. The New York legislature in 1982 did, indeed, enact legislation to develop a fund—financed by waste generators—for waste cleanup not covered by the federal Superfund monies and to cover the state share of Superfund costs where owners are unable to pay.[22] In addition, the legislature passed bills permitting the state to deny disposal and transportation permits to companies failing a "reputation check," and required waste disposers to carry insurance covering long-term liability. Addressing prevention, as well as remediation and enforcement, these vanguard actions will undoubtedly augur further strong statutory measures as the issue matures.

Concomitantly, as current litigation is resolved substantial case law will be produced and considerable legal expenses will accrue to both industry and government. If trends in court decisions continue, this case law is likely to require prevention and remedial action on the part of both industry and government.

Second, fiscal forces will clearly be a major determinant of future developments. The cost of cleanup and long-term maintenance of sites, of compensation to toxic waste victims, and of control of waste disposal is already enormous and will surely escalate, barring the discovery of cost containment measures. Indicative of this factor is the case of the Life Science Products Company, a chemical plant in Hopewell, Virginia, that was found to be contaminating the James River in 1975. The source of pollution could have been cleaned up for about $250,000 at that time, but the firm delayed action and since has paid out $13 million in damage claims. According to experts, the firm faces a $2 billion price tag for cleanup operations.[23]

Hints of the direction of these fiscal forces can be found in New Jersey's tax measure on industry (challenged in the courts on a claim of unconstitutionality) and the governor of New York's directive to the Commissioner of Environmental Conservation to "explore alternative funding mechanisms for toxic waste remediation since the state does not have the fiscal resources needed and Superfund is proving to be inadequate and inapplicable in many cases."[24] The governor's concern over costs can be traced from the fact that Love Canal has already cost more than $60 million and that 335 similar danger spots have thus far been identified in the state.

Third, political forces are likely to gain in strength. As noted earlier, Lois Gibbs, experienced and successful in mobilizing political action at Love Canal, now heads the national interest group effort discussed above. Her efforts, and those of similar state and local groups that seem likely to mobilize, will contribute to local political demands for policy action in this area.

Fourth, the development of technological options will receive attention. Research continues in both government and industry and is likely to produce new capabilities for hazardous waste management. Two areas of great promise are modification of manufacturing operations to reduce waste, and the development of waste exchange programs whereby one company's waste can be used in another company's manufacturing operations.

Fifth, industrial forces seem to be developing that augur action by business itself to address the problem. Speaking to the American Petroleum Institute, Harvard business professor Robert Reich drew from a historical analysis of government regulation of business, and warned corporate executives that they must act now to avoid a new, post-Reagan wave of stringent government regulation. "The irony is that unless American business actively seeks to anticipate and respond to emerging public concerns in advance of the next wave [of regulations] business will have already been seriously eroded by the time it breaks."[25] Professor Reich's remarks were soberly received and reprinted on national business pages. Thus the desire to escape future regulation may spur waste generating industry into preventative action.

What is apt to result from the exercise of these combined forces? In the near term, intergovernmental tension will undoubtedly increase as the political forces that demand a rapid response from all levels collide with the fiscal forces that prompt a literal passing of the buck. Public-private sector conflict is likely to intensify further in a kind of "game of chicken" as each side attempts to resist increasingly complicated legal, political, and fiscal forces until the other side succumbs. But, in the long term it can be argued that we are heading toward intergovernmental and public-private sector collaboration simply because there is no alternative and because the forces for action will not dissipate. It has already become clear that neither business nor government alone can afford the bill for remediation and cleanup, that neither business nor government alone will be able to cope with the political and social demands arising, and that neither business nor government alone stands to "win" in the courts.

The great need is, therefore, to discover and develop processes and mechanisms that enable, facilitate, and hasten the collaboration and cooperation that seems to be the only feasible strategy. It is in this area that policy and management researchers and analysts can make major contributions to the solution of our hazardous waste problem.

There is recent evidence that cooperation can be realized, that effective processes and mechanisms can be developed. The federal government has established Superfund to help the states; large corporations have agreed to finance cleanup efforts and may be developing a conducive perspective of practical self-interest; technological developments are proceeding and may indeed offer assistance; tax incentive proposals are on the legislative agenda and regulatory enforcement programs are developing. Fiscal, industrial, legal, technological, and political incentives are indeed emerging.

If one conclusion is clear in New York state, it is that government has demonstrated a willingness and an ability to get tough and to take action that can cost business dearly. It is also apparent that business has the will and ingenuity to undermine and frustrate government efforts. Where are we going in toxic waste management policy? While the future is uncertain, it does appear that positive interaction of industry and government is the only route that offers hope and, therefore, policy research and analysis efforts should focus on ways of producing such a partnership. Failing that, the "game of chicken" will certainly continue until a series of Love Canal-like emergencies prompts additional political, social, and corporate crises.

7. Institutional Fragmentation and Hazardous Waste Policy: The Case of Texas

Kenneth W. Kramer

This chapter presents an in-depth analysis of the context within which hazardous waste regulatory policy is made in Texas. In Texas, the policy response to the hazardous waste issue is characterized as fragmented, technical, and plagued by discussion over its effectiveness. Interconnecting factors which have shaped this policy response include: (1) federal initiatives, (2) the ascendance of state-level bureaucrats, (3) the pervasive influence of interest groups (especially industrial interest groups), and (4) diffused media attention. The interaction of these elements has resulted in the policy configuration described in the following pages.

The chapter itself is divided into three parts. The first part details the severity of the hazardous waste problem in the state of Texas. Next, we focus upon the state policy response to the problem. Finally, we discuss in detail the factors which have shaped the Texas response.

Severity: The Hazardous Waste Problem in Texas

In the twentieth century, especially in the last fifty years, Texas has been undergoing a dramatic transformation from a predominantly agricultural economy and frontier-rural society to a highly industrialized and heavily urbanized part of the country. Although cattle ranches, cotton farms, and wide, open spaces still characterize much of Texas, so do drilling rigs, petrochemical refineries and other industrial plants, and large cities. The state is now not only first among the fifty states in cattle and cotton production but also a leader in the production of energy and chemicals, third in population, and first in the number of Standard Metropolitan Statistical Areas (SMSAs).[1]

Unfortunately this industrialization and urbanization has created some negative by-products: for example, air pollution from industries and automobiles in such cities as Houston and El Paso, water quality problems in both rural and urban areas of the state, and increased generation of solid waste,[2] much of it hazardous. In fact, Texas ranks sixth among the states in its generation of hazardous waste.[3] Although exact totals are hard to determine (partly because of definitional problems), an estimated two and a half million metric tons of hazardous wastes are generated in Texas annually.[4] As might be expected, the petrochemical industry is the largest single generator of those wastes.[5] Chemical plants account for an estimated 60 percent of the hazardous wastes produced in the state (not including wastes from oil and gas production and certain other activities). The closest "competitor" is the metals industry, with only 12 percent. By one estimate Texas generates more than half of the hazardous petrochemical wastes in the nation.[6]

Hazardous wastes in Texas are managed in a variety of ways. Most of the hazardous wastes which are classified as "municipal solid waste"[7] are buried in landfills. The "industrial solid waste"[8] produced in Texas, which includes but is not limited to hazardous wastes, is usually routed to underground injection wells (35 percent of the industrial wastes), landfills (30 percent), surface impoundments (15 percent), or "temporary" storage areas (15 percent).[9] The remainder is "land-farmed," incinerated, chemically treated, or managed in some other way. Most of these wastes are actually stored or disposed of "on-site"—in other words, at the plant or other site where they are generated. Approximately six hundred on-site disposal facilities currently exist.[10] Estimates vary, but 75 to 90 percent of the industrial hazardous wastes in the state are probably disposed of on-site.[11]

The rest of these wastes are stored or disposed of at "off-site" commercial facilities—in other words, facilities located away from the site of generation and usually operated by a commercial waste disposal company. These facilities handle wastes coming from a number of generators. As of 1982, twenty-three authorized commercial industrial solid waste sites were operating in Texas[12] (one other authorized site, owned and operated by a governmental entity— Gulf Coast Waste Disposal Authority—was also in operation). Three-fourths of those sites are located in counties along the Texas Gulf Coast; approximately one-half of those sites are in the Houston-Galveston area. (The majority of on-site facilities are also in the Texas coastal area.)[13]

The location of so many facilities in the coastal region is not surprising since the petrochemical industry and many other industrial waste generators are concentrated in that region. The proximity of industry has created a market for commercial waste management facilities in the coastal area and industries preferring not to retain wastes on-site have used the nearby facilities because of convenience and economics (lower transportation charges, for example). However, these commercial facilities along the coast also have become a magnet attracting industrial wastes generated elsewhere in the state and even elsewhere in the nation. For example, in the north central Texas area (primarily Dallas–Fort Worth) in the one-year period from March 1979 to February 1980 approximately eight million gallons of hazardous wastes (primarily generated from the manufacture of machinery and production of some chemicals) were shipped away from the site of generation.[14] During that period in that area, there were no authorized commercial hazardous waste disposal facilities, only one authorized storage facility for such wastes, and only a few nonhazardous industrial waste disposal facilities. Of the total industrial wastes generated in the area in 1979–80 and shipped off-site, two-thirds were shipped directly to sites elsewhere in Texas, most of them in the coastal region.[15] This situation has led some citizen groups to charge that the Texas Gulf Coast is becoming the "dumping ground" for the state's industrial wastes.

Not only are there many active hazardous waste management facilities in the coastal region but also many abandoned dumps. For example, approximately two-thirds of the thirty-one inactive sites identified as "problem sites" in a study by a state legislative subcommittee in 1980 are located in coastal counties.[16] Of the four abandoned disposal sites initially selected by the U.S. Environmental Protection Agency (EPA) for cleanup work as part of its "Superfund" implementation, three are in the Houston-Galveston area.[17]

Overall, the problem posed by abandoned and, in some cases, improperly managed active hazardous waste disposal sites in Texas appears to be extensive —in terms of number of sites, specific concerns about public health and the environment, and the costs of cleanup and correction. Although the legislative subcommittee which studies problem sites concluded " . . . Texas apparently does not have any industrial waste facilities which pose environmental or health hazards the magnitude of Love Canal . . . ," the total number of active and inactive problem sites which the subcommittee described in its study was fifty-seven.[18] Moreover, the subcommittee indicated that it had not intended to provide a complete inventory of problem sites but rather a sampling of such sites. Indeed, a study by the U.S. House Commerce Committee's Subcommittee on Oversight and Investigation identified over 300 potential problem sites.[19] Some of these sites were removed from the list following more thorough evaluation, and while the full extent of the problem of abandoned or improper disposal sites remains to be determined, it is potentially immense.

Certainly several major problem sites which have been identified have already been evaluated as being costly to clean up. Before a state legislative committee hearing in 1981 the deputy director of the agency responsible for regulating industrial waste disposal estimated that the cost for cleaning up just seven major abandoned sites would be approximately $56 million.[20] Of that total, $31 million alone was estimated as necessary to tackle the problems at what appears to be the worst abandoned site in the state: the Motco dump near Texas City, just off the interstate highway running between Houston and Galveston.

The Motco site, also known as the Texas City Wye, was abandoned in 1976 when its last owner went bankrupt. It consists of seven open pits, which are unlined and contain a morass of toxic substances including benzene, polychlorinated biphenyls (PCBs), styrene tars, and vinyl chloride. Prior to the start of EPA cleanup efforts, the eleven-acre site even lacked a fence to keep people and animals out and a dike around the site (which is located in a floodplain) to keep wastes from washing off the site during rainstorms.[21] According to a Texas Senate committee report, from 1973 to 1978 approximately 600 people lived in a trailer park directly adjacent to the waste pits at the Motco site but were never informed of potential dangers from the site.[22] Results of a preliminary environmental study of the site in late 1981 were encouraging in that air, soil, and water samples showed lower levels of contamination than had been expected,[23] but the long-term effects of the existence of such problem sites are not certain.

In addition to concern about abandoned sites such as the Motco dump, official and public concern has also focused on several active hazardous waste facilities in the state. Two of the commercial disposal facilities which have received the most adverse publicity in recent years are the Malone Service Company's Texas City disposal site and the ironically named Texas Ecologists, Inc. (TECO), site in the rural community of Robstown near Corpus Christi. Although the Malone Co. has sued a Dallas TV station and its reporter for allegedly false accusations, the site has been criticized in the media for supposed seepage of toxic chemical wastes through its dikes and into Galveston Bay, among other charges.[24] Moreover, the company's technical director was discredited by revelations that he had faked his academic degrees, including a claimed Ph.D. in public health.[25] The TECO site has been alleged by the Texas Department of Water Resources to have contaminated 14 million gallons of groundwater at the site, has been cited as poisoning a nearby maize field through an accidental release of powdered herbicide into the air, and has been charged by neighboring farmers with causing health problems for their farm animals and their families.[26] Less extensive charges of mismanagement have been made against other operating industrial waste sites run by commercial operators, and concern among some public groups exists as to what problems might be revealed by the more stringent monitoring requirements only now being placed on on-site facilities as a result of the implementation of the Resource Conservation and Recovery Act (RCRA), adopted by the U.S. Congress in 1976.

In summary, the hazardous waste problem in Texas is characterized by massive generation of wastes, the existence of a large petrochemical industry responsible for the generation of the majority of the "industrial solid waste," primary reliance on injection wells and landfills (and related practices) for the management of such wastes, a heavy predominance of on-site disposal, a concentration of disposal facilities in the coastal areas of the state (including many in floodplain areas), the consequent "importation" of hazardous wastes into the coastal region, an extensive number of abandoned sites which will be costly to clean up, allegations of mismanagement at some currently active waste facilities, and concerns about adverse public health and environmental impacts ascribed to both abandoned and existing sites. All of these characteristics indicate a hazardous waste problem which may not be as dramatic as that described in other states such as New Jersey and New York but which is nevertheless significant, complex, and, perhaps above all, disturbing because of the uncertainty about exactly how extensive the problem is and how it might best be addressed. This is the setting in which public policy makers and concerned groups in Texas who have responsibility for or interest in public health and environmental protection have had to work, and this is the nature of the problem which they are attempting to address.

The Policy Response to the Problem

The public policy response in Texas to the hazardous waste problem is perhaps best characterized as fragmented, technical, and uncertain (in terms of its appropriateness and effectiveness in addressing the problem). A review of these characteristics illustrates how the policy does or does not address the particular characteristics of the problem itself.

The Fragmented Nature of the Response

Although it is a truism that the response to many problems is fragmented in a federal system of government (because of the respective roles played by national, state, and local governments), the fragmentation discussed here is primarily at the state level itself. The state level policy response to the hazardous waste problem in Texas is fragmented both in the sense of legislative enactments and of administrative responsibilities. Statutory authority for regulating and otherwise addressing the management of hazardous wastes stems from the Texas Solid Waste Disposal Act (initially passed in 1969 and most recently amended in 1981),[27] the Texas Well Disposal Act,[28] the Texas Oil and Hazardous Substances Spill Prevention and Control Act,[29] the 1981 Amendment to the Water Code,[30] those parts of the state's Natural Resources Code which provide the general authority for the Texas Railroad Commission,[31] and several other statutes dealing primarily with air and water quality. Responsibility for the implementation of these laws is divided among several major state agencies, primarily the Texas Department of Water Resources (TDWR), the Texas Department of Health (TDH), the Texas Railroad Commission (RRC), and the Texas Air Control Board (TACB). Unlike several states, Texas has no overall state environmental protection agency, nor does it have a principal agency for regulating toxic substances and/or hazardous wastes.[32]

The specific breakdown of authority among the state agencies responsible for hazardous waste is as follows: TDWR regulates the management of industrial solid waste; the Department of Health regulates the management of municipal solid waste (for example, commercial and institutional activities may produce hazardous wastes which are disposed through municipal garbage collection, and these wastes thus come under the regulation of the Health Department); the Railroad Commission (which regulates energy production in the state as well as intrastate commercial transportation) is responsible for the regulation of the wastes produced by oil and gas development and other energy production activities; and the Texas Air Control Board regulates air emissions from hazardous waste facilities (for example, stack emissions from incinerators used to burn hazardous waste materials). Additional responsibility is given to TDWR in that it is the administering agency for the fund which provides the necessary state match for federal Superfund monies; it is also the lead agency for the oil and hazardous substances spill program.

The Water Resources Department and the Health Department are the primary state agencies which implement the federal Resource Conservation and Recovery Act (RCRA) in Texas. Oil and gas production wastes are exempted from coverage by RCRA; thus the Railroad Commission is not involved in the implementation of this federal law. The programs of these two agencies, authorized by the Solid Waste Disposal Act, have been judged as meeting the test of being "substantially equivalent" to the EPA hazardous waste regulatory program operated under RCRA, and the Texas agencies initially have been delegated the authority to operate that program in the state in lieu of EPA.[33] Final delegation has been slowed by EPA's delay in issuing all the regulations necessary to implement RCRA.

Because of the federal requirements under RCRA, the general characteristics of the TDWR and Department of Health regulatory programs are not only similar to each other but also to basic programs in other states. Generators of hazardous wastes must record the amounts and types of wastes they generate on a "manifest" or "trip ticket," which must then accompany specific waste shipments to their ultimate disposal site. Both generators and waste disposers must submit to the respective agencies monthly reports detailing the amounts and types of wastes shipped or disposed of. These requirements are intended to provide a "cradle-to-grave" monitoring system, allowing the tracking of wastes to see that they are properly managed. Facilities which dispose of or otherwise manage hazardous wastes, whether on-site or off-site, are required to obtain a permit from the appropriate regulatory agency to authorize them to manage the wastes. (Some facilities are required to obtain not just waste disposal permits under RCRA but also wastewater discharge permits and air emission permits under other statutory requirements.) The civil and criminal penalties for violation of these and other requirements of the state regulatory program are the same as those specified by RCRA.[34]

The impact of RCRA, however, especially considering its exemption of oil and gas wastes and the EPA's sanctioning of both the TDWR and Health Department programs through its delegation of RCRA authority, has not significantly affected the fragmentation which characterizes the policy response of Texas to the hazardous waste problem. This fragmentation has led to agency jurisdictional questions which some groups and officials have characterized as negatively affecting the efforts to deal effectively with hazardous waste problems. An example of this situation, as it relates to oil and gas wastes, was provided by a former state assistant attorney general and reported as follows in a position paper developed by the state Sierra Club chapter:

A problem that has developed in the Legislature has left to the RRC [the Texas Railroad Commission] the responsibility for determining what waste materials result from activities associated with . . . oil or gas. . . . The RRC has included such materials as drilling muds in that categorization and thus has claimed regulatory authority over them. However, neither the RRC nor the other state agencies involved in the regulation of hazardous waste materials have determined whether *certain wastes resulting from oil and gas drilling operations*—for example, waste transformers and transformer oil, used engine and other machine oils, and sludges from storage tanks—fall under the general jurisdiction of the RRC or under the regulatory authority of the Solid Waste Disposal Act which is administered by the other agencies.

Thus, these hazardous or potentially hazardous waste materials are simply *not being regulated* at present [1981] . . . .[35]

This general type of jurisdictional issue has had some specific adverse effects. An example of such an effect was recounted in an earlier work by this author:

. . . in one recent situation where pollution resulted from the spill of diesel fuel that was stored and used at an oil-drilling site, *both* TDWR and RRC denied jurisdiction over the problem. TDWR claimed that the pollution resulted from oil production activity and thus fell under the responsibility of the Railroad Commission. But the Railroad Commission asserted that the spill was not a direct result of oil production activity and thus was not its responsibility. The dispute was finally resolved with the aid of the attorney general's office but not before the landowner whose property was affected by the spill experienced considerable expense and inconvenience trying to get the state to address the problem.[36]

Such situations and similar ones resulting from the disposal of industrial wastes at municipal waste sites may be avoided through a memorandum of understanding (MOU) approved in January 1982 by the three major state regulatory agencies with hazardous waste responsibilities. This MOU was required by a 1981 amendment to the Solid Waste Disposal Act in response to concerns over jurisdictional questions. Some of the questions about division of waste regulation in Texas have been resolved by this MOU; but the guidelines in the agreement are not precise enough to answer all such questions, and some problems apparently still remain about the definition of oil and gas wastes exempt from RCRA regulation.[37]

Moreover, MOUs are not as definitive as divisions of responsibility which might be set forth in the form of statutes and rules, and they are certainly subject to change and interpretation themselves, even more easily than statutes or rules. The contribution of this MOU among the three primary hazardous waste regulatory agencies in the state to alleviating the jurisdictional questions resulting from administrative fragmentation will depend upon its implementation over the long term. The development of an MOU to alleviate jurisdictional questions among the agencies does not address, of course, other issues raised by fragmented administration, such as the question of efficient use of resources.

Texas is certainly not alone in the fragmentation of its hazardous waste programs, as is illustrated by the finding in the National Wildlife Federation's 1979 study that at least thirty states did not have one "state agency or official with overall responsibility for coordinating or regulating toxic substances."[38] While the extent to which fragmentation of administration truly influences the effectiveness of regulatory programs has never been satisfactorily answered, it is interesting to note that of the ten states ranked highest in terms of the strength of their toxics regulatory program by the National Wildlife Federation study, eight had a single agency or official responsible for the program.[39] Whatever the effect, however, fragmentation is definitely a characteristic of the Texas approach to regulation of hazardous wastes.

The Technical Nature of the Response

The management of hazardous wastes is in many ways, at least when done properly, a rather technical process in the scientific and engineering senses. Thus, a policy which seeks to regulate such management would logically address those technical aspects of the process and the problem. But the use of the term "technical" here to characterize the Texas state policy response to the hazardous waste management problem is perhaps more restrictive than its common usage. "Technical" here refers to a policy response which is highly formalized and based primarily on engineering considerations—as opposed to a policy response which might allow more informal participation in its implementation and which would be based on economic, land-use, scientific, and social considerations as well as engineering ones.

The characterization of state hazardous waste regulatory policy as highly formalized relates primarily to the procedural aspects of the policy, from the "trip ticket" system of monitoring waste shipments to the very routinized process for the permitting of waste disposal facilities. The latter is necessitated not only by the specific administrative rules related to waste management but also by the well-intended but cumbersome Administrative Procedure and Texas Register Act,[40] a statute which specifies procedures that must be followed by Texas state agencies in issuing permits and conducting other administrative activities. These various requirements result in a permit application process which, especially in the case of contested permits, is more akin to a judicial rather than an administrative process. (As a matter of fact, in the case of permits for industrial solid waste management and other activities which the Texas Department of Water Resources regulates, there is a virtually independent and basically judicial body of three people, appointed directly by the governor and called the Texas Water Commission, whose specific responsibility is to decide on the issuance of permits.) Thus, citizens wishing to contest a permit application are virtually required to retain an attorney to represent them in the public hearing on the application—a hearing which is conducted much like a court case, with rules of evidence, cross-examination of witnesses, and the like. The major difference between such a formal hearing process and a court trial is that the authoritative decision will be made not by the hearing examiner (the person who presides at the hearing) but by a higher agency official(s). The hearing examiner's recommendation usually carries considerable weight with the final decision maker,

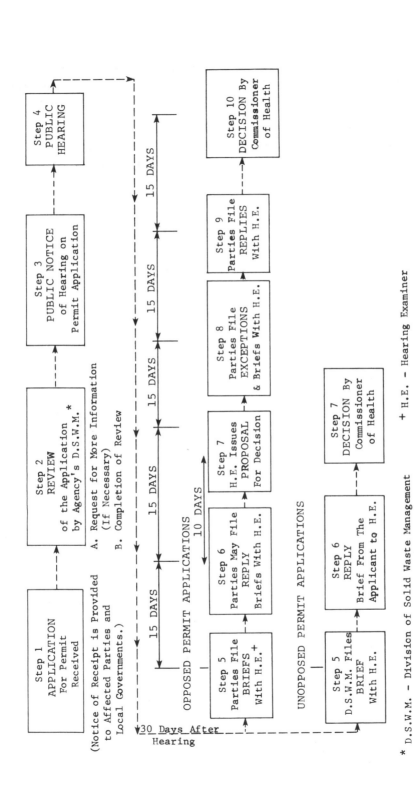

Figure 7.1. Permit application process for municipal solid waste facilities in Texas, 1981

* D.S.W.M. – Division of Solid Waste Management   + H.E. – Hearing Examiner

Source: Compiled by the authors.

however. Figure 7.1, which diagrams the permit application process for municipal solid waste facilities, illustrates the formality of these procedures.

Although such a formalized process has its advantages—in terms of such aspects as documentation, assured responses to questions posed by the participating public, and perhaps consistency of decisions because of the effect of precedent—it also tends to limit public participation more than a less formal process would. Moreover, because the statutes and the administrative rules governing hazardous waste management primarily outline the procedures to be followed by permit applicants and the engineering, financial, and similar legal criteria to be met, the issues which may legitimately be raised through the hearing process and which may be appropriately considered in the final decision on a permit are relatively narrow. Neither the statutes nor the rules governing waste management specify land-use or socioeconomic criteria for the location of facilities. Thus, those concerns which most often are voiced by citizens who oppose waste facilities—that such facilities would be incompatible with existing land uses, that values of nearby property would be adversely affected, that certain areas or neighborhoods are bearing the brunt of the establishment of such facilities—are not directly addressed by law or regulations. If the procedures for permit applications are followed correctly, and if the engineering and similar technical specifications are met, then usually the permit for a facility will be granted.

Some qualifications to this general observation must be made. Organized, determined, and well-financed opposition to the location of a facility at a specific site might lead to the withdrawal of a permit application, if the applicant did not wish to endure the adverse publicity or the financial costs of potential delay incurred by legal hassles.[41] Also, decisions about the permit might be made on other than technical grounds but couched in technical reasoning.

Finally, although the statutes or rules do not specify siting criteria, the Health Department regulations governing municipal solid waste facilities do require submission of certain land-use data with the permit application and do assert a rather vague authority of the department to consider land use and the public health in deciding whether or not to grant a permit (an authority which recently withstood a court challenge).[42] TDWR has issued technical guidelines on "site selection and evaluation" for landfills and surface impoundments which list geologic, topographic, hydrologic, soil, and climatologic "factors that should be given consideration when trying to select a disposal site that will pose the least amount of threat to the environment."[43] Agency personnel have argued that permit applicants try to follow these guidelines as much as practical in order to assure a favorable review by the permitting agency's technical staff. While this is a logical argument, there is no empirical evidence to support it. Indeed, a review of the enumerated siting factors indicates that several, if followed, would probably preclude the location of facilities in the coastal areas where they are now concentrated.[44] However, most of the existing landfills and surface impoundments in that area predate the technical guidelines, and most permits for hazardous waste disposal requested and issued since the guidelines have been for disposal wells. So, to date, there have been no major tests of the guidelines with respect to siting criteria. But several facilities are being considered for coastal counties, and some of the possible sites are in areas which the technical guidelines suggest be avoided—calling into question their influence.[45] Even if they might have some informal influence, they have no legal effect.

In general, then, the state policy response to hazardous waste problems is highly technical, characterized by formal procedures and engineering and similar criteria. The prevalent view of official policy seems to be that hazardous waste disposal will be properly accomplished through state-of-the-art techniques (injection wells, landfills, and surface impoundments) if appropriate facility design standards are employed. The policy generally does not extend beyond this rather narrow approach to the problem.

The Uncertainty of the Response

Whether the fragmented and technical policy response described is appropriate and effective in addressing the state's hazardous waste problem is a matter of dispute. The state legislative committee studies undertaken in 1980 and 1982 and the debate over proposed hazardous waste regulatory legislation during the 1981 legislative session provided opportunities for defenders and critics of the state's policy to present their views. A review of public statements and materials generated by various officials and groups during this period indicates the nature of the dispute over state hazardous waste policy in Texas.

In general, the regulatory agencies and relevant industrial groups (the generating industries and the waste management industry) publicly defend the existing policy as appropriate and effective. Conservation groups and property owner groups (associations of people living near existing or proposed waste facilities) have often been critical of the policy. Other groups, such as state legislators and academic researchers analyzing the policy, have tended to be much more split in their views on policy effectiveness.

Criticisms of the policy have generally centered on six major areas: (1) the jurisdictional questions noted earlier; (2) the lack of specific land-use and similar siting criteria in the law or regulations; (3) public participation mechanisms; (4) the monitoring of waste management activities, and other aspects of enforcement of regulatory policy; (5) financial requirements for facility operators; and (6) lack of strong encouragement of alternatives to landfilling and related waste management practices. The first three criticisms have generally been covered explicitly or implicitly in the earlier discussion of the fragmented and technical nature of the policy. A diverse set of groups, including the state League of Women Voters, the Sierra Club, Texas Farm Bureau, and numerous local citizens' groups such as People Against a Contaminated Environment (P.A.C.E.) in Liberty County, have leveled these criticisms against state hazardous waste regulatory policy and have made recommendations to deal with jurisdictional, siting, and public participation issues.[46] A study group at the

Lyndon Baines Johnson School of Public Affairs, which prepared a report on the siting of hazardous waste disposal facilities for a special subcommittee of the Texas House Committee on Environmental Affairs, also reached many of the same conclusions as the aforementioned groups.[47] The LBJ School study especially recommended the formulation and application of formal land-use criteria in the site permitting process and suggested a number of ways in which public participation opportunities in the decisions about facility siting might be increased.[48]

In light of the Texas legislature's 1981 changes to state hazardous waste regulatory law to meet some of the "substantial equivalence" requirements of RCRA, much of the concern about financial assurances that waste facility operators will be able to cover any necessary cleanup or closure (and post-closure) costs at their facilities has subsided. However, substantial concern does still exist about monitoring of waste management facilities and other enforcement activities by the regulatory agencies. The concern may be summarized as focusing on both the resources and the will or commitment of the agencies to adequate enforcement. The extensive number of off-site and especially on-site industrial waste facilities, municipal waste sites, and oil and gas drilling operations in the state, when compared to the number of enforcement personnel available to or employed by the agencies, have led many groups to conclude that personnel resources are not adequate to address the hazardous waste problem. For example, although EPA reportedly estimated that Texas would need a minimum of 148 people *additional* to those authorized in the 1980–81 biennium to carry out RCRA enforcement responsibilities regarding industrial waste facilities,[49] the Texas Department of Water Resources currently has only about seventy-five field inspectors to monitor activities under all of their regulatory programs (including wastewater discharge permits, which are much more numerous than waste disposal permits).[50] Although regulatory agencies probably never have enough enforcement personnel, this divergence between the recommended and the current numbers of personnel has led to widespread calls for additional inspectors.[51]

Also of concern with regard to enforcement has been the lack of interest or commitment which some groups charge as being characteristic of the regulatory agencies. Evidence to prove this assertion is, almost of necessity, circumstantial. The major criticism leveled against the Health Department and TDWR in this regard appears to be the contention that they are reluctant to take violators of their regulations to court.[52] The officials of these agencies do not deny this reluctance. In fact, they assert that trying to work out an arrangement with violators to allow them to alleviate the problems to the satisfaction of the regulatory authorities is the best use of the agencies' limited resources and may avoid unnecessary delays in cleanup actions which might result from attempts to deal with the problems through law suits against violators.[53] However, critics of this approach argue that it sends the wrong signal to waste managers—a sign that violations will not be prosecuted—perhaps leading to inadequate incentives to force compliance. Unfortunately, the empirical evidence has not been gathered to determine whether the viewpoint of the agencies or their critics is correct, whether the enforcement glass is half-empty or half-full.

Although the Health Department and TDWR have been criticized for their enforcement philosophy, the most extensive criticism has been directed at the Railroad Commission. Numerous landowners who have charged that their land and/or water supplies have been contaminated by improper disposal of oil and gas operations have severely criticized the commission for what they view as a lack of concern about oil and gas waste disposal.[54] Also officials of cities dependent on water supplies in areas with heavy oil and gas production activity have charged that the commission is not enforcing its own rules on waste disposal.[55] Perhaps this perspective on the commission's attitude is best captured in the implicit criticism contained in the LBJ School study group's recommendation on oil and gas waste regulation:

> The Texas Railroad Commission is responsible for wastes from oil and gas production up to the point when the oil or gas reaches the first container at a refinery. Since the location of wells cannot be subject to siting criteria, it is especially important that the resultant hazardous wastes be closely monitored. The wastes in question include spent acids (which contain metal salts), fracturing fluids, and some drilling muds. The Railroad Commission has allocated the part-time services of one attorney and no funds to investigate this aspect of oil and gas production. The Legislature should direct the Railroad Commission to take an active role in ensuring that hazardous wastes under its jurisdiction are disposed of in an environmentally safe manner.[56]

The fact that oil and gas wastes are exempt from RCRA coverage and that EPA has apparently been hesitant to step into the situation under its Clean Water Act authorities have added to the concerns of many groups about this particular aspect of hazardous waste regulation.

A final criticism which has been made concerning hazardous waste policy in Texas is that its emphasis on engineering criteria for injection wells, landfills, and surface impoundments implies that these methods are appropriate and safe if those criteria are met. Consequently, the argument is made that there is little or no incentive for industry to pursue waste reduction and/or more technologically sophisticated methods of waste management (especially if the latter is much more costly than the traditional methods of waste management).

This lack of incentive is especially disturbing, the critics charge, because of studies and experiences indicating that the traditional methods, even if "properly" implemented, do not assure permanent containment of wastes nor protection of the environment and public health. For example, studies by Texas A&M soil scientists indicate that the clay liners which have been widely used at landfills in the Texas coastal counties (where Beaumont clay is the primary soil type) and elsewhere are not as "impermeable" as once thought.[57] Also, synthetic liners used in disposal operations have not always proved to be effective for containment of wastes.[58] Even EPA in its new "landfill" regulations recognizes that there is no such thing as a "secure landfill"—that liners will eventually leak—and that consequently landfills may only be viewed as collection facilities or "controlled releases" into the environment.[59] TDWR has acknowledged this in its part of the state's *Solid Waste Management Plan* but has not moved away from its traditional approach to landfills.[60] On the other hand, the governing board of the Texas Department of Health in late 1981 adopted a policy which states that "resource recovery" (defined as any alternative to landfilling, including recycling and systems which convert waste to energy through incineration) is the "preferred method" of solid waste management, and is initiating efforts to encourage municipal solid waste managers to move away from the traditional reliance on landfills.[61] This move has been applauded by several conservation and other citizens' groups. However, concern still exists that in the area of *industrial* solid waste Texas is not moving forcefully, unlike states such as California, to promote alternatives (as indicated by the breakdown of waste management practices for industrial waste in Texas). While increased engineering requirements for landfills and similar methods will detract from the economic advantage enjoyed by this approach, the contention is that such requirements are inadequate to promote other methods.

All of these criticisms are generally dismissed by the generating and waste disposal industries, at least in public statements. For example, in a hearing held by a special state legislative subcommittee on hazardous waste facility siting, the uniform message of a waste disposal company (Browning-Ferris Industries or BFI), a petrochemical company (Diamond Shamrock), and the trade association of chemical manufacturers, the Texas Chemical Council, was that no new legislation on hazardous waste regulation is needed in Texas.[62] The trade association for the disposal industry, the National Solid Waste Management Association, did tentatively suggest that the regulations might be "fine-tuned" following release of the EPA landfill regulations and made more specific in order to aid waste facility developers.[63] This could be interpreted as a willingness to accept at least certain siting criteria in the form of agency rules, if this would assist the process for locating waste facilities. But the industry perspective has generally reflected the view that all of the major disposal techniques are safe, if proper steps are followed, and that monitoring requirements are adequate. They point to the fact that some aspects of the federal RCRA requirements— such as the trip ticket or manifest system—are modeled after requirements earlier initiated by TDWR in its state regulatory program as evidence that Texas is a leader in the regulation of hazardous wastes.

In summary, the perspectives on the appropriateness and effectiveness of the state hazardous waste regulatory program in Texas are diverse. This diversity may result in part from the different philosophical outlooks of the groups who are interested in hazardous waste management, but it probably also reflects the existence of strengths and weaknesses in the program and the lingering public and scientific uncertainty as to what constitutes proper waste management.

Overall, then, Texas hazardous waste regulatory policy is complex in terms of its fragmented and technical nature and uncertain in terms of its appropriateness and effectiveness. The complexity and uncertainty of the policy response is not surprising in that the hazardous waste problem itself has been characterized by these elements. But these general elements of the problem do not explain the specific policy response which has just been outlined, at least not by themselves. What mediating factors have shaped the particular approaches of Texas state regulatory policy toward hazardous waste management?

Factors Which Have Shaped the Texas Response

In a cosmic view of the effects of the public policy environment on the development and implementation of a specific policy, an almost infinite number of factors shaping that policy might be identified. In the case of state hazardous waste management regulatory policy in Texas, four factors seem to be most relevant in explaining how policy has developed and why it has taken the form described here. These factors are (1) the impact of federal solid waste and hazardous waste policies, (2) the dominance of bureaucratic actors in this policy-making area, (3) the strong influence of interest groups (especially industrial groups) on policy making in Texas, and (4) the lack of a central focus for media attention on the hazardous waste issue. The first factor has been a mechanism for facilitating or forcing action on solid waste and hazardous waste management in the state. The other three factors have operated to assure that the actions taken have been based on decisions made generally within a political subsystem rather than within the macropolitical arena. That, in turn, has resulted in a hazardous waste regulatory policy of the particular nature described.

The Impact of Federal Solid Waste Policies

Although much has been made of several elements of the Texas hazardous waste regulatory program having been used as a "model" for the RCRA requirements developed by EPA, the initiation of the state's solid waste program actually was spurred by a federal effort. The Texas Solid Waste Disposal Act was first approved in 1969, and the first solid waste plan for the state was published in 1971.[64] The legislation and the plan were directly or indirectly the result of the Solid Waste Disposal Act adopted by the U.S. Congress in 1965.[65] This federal legislation was the first move by the national government into the solid waste issue-area, and it provided financial and technical assistance from the federal government to state and local government solid waste programs. Funds from that 1965 act financed a 1968 state of Texas survey of solid waste management in political subdivisions around the state and a 1969 survey of regional solid waste planning agencies in the state.[66] The findings of these surveys—perhaps especially the large number of substandard landfills identified (835 out of 875 landfills) and the rather phenomenal growth in the volume of solid waste generated in the state in the 1960s (an 81 percent increase, compared to a 17 percent population increase)[67] led to the adoption of the legislation and the solid waste plan (whose development was also financed by the federal grant money made available under the 1965 act). Efforts related to hazardous waste issues were subsumed under these efforts to deal with solid waste in general. "Prior to the enactment of the Texas Solid Waste Disposal Act . . . in 1969 there were no effective state regulations governing the disposal of solid wastes."[68]

The 1969 legislation authorized the two agencies designated as solid waste regulatory agencies by the act (the Health Department and the Texas Water Quality Board, which was later merged into TDWR) to establish design criteria and a permitting system for solid waste management facilities. Regulations implementing the legislation were subsequently adopted in 1970.[69]

That same year, Congress passed the Resource Recovery Act of 1970, which emphasized the development of new technologies for solid waste management and encouraged comprehensive studies on such topics as hazardous wastes and waste storage sites. However, appropriations and federal administrative action were not forthcoming to achieve the aims of the legislation.[70]

Therefore, many specific steps taken by the state agencies between 1970 and 1976 (the date of passage of RCRA) to address solid waste and especially hazardous waste issues were *not* directly the result of federal legislative prodding. This observation must be qualified in several ways, however. First, this period was one of intense interest in environmental protection, and a number of major pieces of federal environmental law were enacted during this time, including the dramatic new air and water pollution control statutes. Many of these dealt with hazardous waste disposal issues as part of the general context of combating air and water pollution, including the establishment of an administrative framework of air emission and water effluent controls intended to address hazardous as well as more "conventional" discharges into the environment.[71] One of the most relevant statutes here was the Safe Drinking Water Act of 1974, which established the federal Underground Injection Control (UIC) program and authorized its delegation to the state agencies with the resources and the authority to implement it.[72] This is the program which directly addresses the injection well method of hazardous waste disposal (because of its relationship to possible contamination of underground water supplies). However, the implementation of the "hazardous components" of the air and water pollution control statutes has been consistently behind schedule, compared to other components of the laws (the UIC regulations from EPA, necessary to provide the criteria for delegation of the program to the states, were not finally promulgated until 1980).[73] Moreover, none of the federal statutes in the 1970 to 1976 period (prior to RCRA enactment) established requirements regarding land disposal of hazardous and other solid wastes.

A second qualification, though, to the observation that many steps taken by states during this period to address solid waste programs were not prompted by federal directives, is that the trend toward eventual federal action was clear.

Many articles and official statements indicated that although air and water pollution problems were being addressed through statutes enacted by Congress in the early 1970s, the problems of land pollution had not, and the federal government needed to take action to respond to these problems.[74] The annual reports of the Council on Environmental Quality documented many of these concerns and suggested specific steps that the federal government might take to address the concerns.[75] Thus, some of the steps taken by states to combat land pollution problems may have been done in anticipation of federal action, perhaps indeed in an attempt to help shape the nature of the federal response and/or to retain some independent responsibility in this area.

Texas was one of the states which took such action on the solid waste problem. In 1975, the Texas Water Quality Board, for example, established performance standards for waste disposal operations, prohibited discharge of industrial wastes into ground or surface water (perhaps an instance of direct response to certain federal initiatives), prohibited disposal of such wastes at "unauthorized locations," authorized staff development of technical guidelines for different disposal methods, and established a manifest system for tracking shipments of certain industrial wastes (making Texas one of the first states to do so).[76] The adoption of these rules has been cited by state officials as having helped to shape eventual RCRA requirements set by EPA for regulating hazardous waste disposal. Indeed, some of the requirements in EPA's 1980 RCRA regulations appear to be patterned after some of the 1975 Texas regulations.

However, the adoption of RCRA in 1976 established additional requirements for hazardous waste regulation for those states wishing to receive delegation of the new EPA regulatory program authorized by that act. Among these, for example, were more stringent penalties for violations of hazardous waste statutes and regulations, stricter financial liability requirements for waste facility owners and operators, permitting requirements for on-site facilities as well as off-site ones, and higher standards for hazardous wastes than for other so-called Class I industrial wastes (a term used by TDWR for that category of wastes receiving the most regulatory attention). These requirements have been specified by RCRA itself and in the regulations which have been issued by EPA to implement the law. They led to TDWR and Health Department adoption of new rules for hazardous waste in 1980 and 1981 and to the adoption of amendments to the Texas Solid Waste Disposal Act in 1981, most of which were characterized by legislators and agency officials as necessary to qualify Texas for delegation of the EPA regulatory program.[77] Indeed, although delegation is being accomplished by EPA in steps, Texas has been one of the first states to obtain delegation at each step because the state had an existing regulatory program and because it is making the changes necessary to receive delegation of the EPA program.

Thus, although Texas has preceded the federal government in some specific areas of hazardous waste regulations, the impetus for other regulatory requirements has come from RCRA and substantial amounts of funding for the state regulatory program has come from EPA. The impact in Texas of federal efforts on hazardous waste management is also seen in the area of cleaning up abandoned hazardous waste sites that pose problems to public health and the environment. An attempt was made by some Texas legislators prior to the establishment of the federal Superfund to create that type of fund in the state to finance cleanup efforts regardless of the fate of the concept at the national level.[78] Indeed some of the state regulatory officials publicly urged that such a fund be established.[79] But the legislature did not adopt those recommendations. Instead, after the creation of the federal fund in late 1980, the Texas legislature in 1981 created a fund financed only by a specific appropriation of money and to be used *only* as the necessary 10 percent state match to receive cleanup money from the Superfund.[80] A competing bill which would have established a state fund with a variety of revenue sources and which would have financed *both* state and federal cleanup efforts did not even get out of committee.[81] Thus, in terms of hazardous waste cleanup, the state has been primarily reacting to the federal thrust rather than adopting an innovative role.

In general, then, state efforts in Texas have been prompted by federal initiatives, although the state stepped ahead of the federal government in certain hazardous waste regulatory requirements. Certainly the fact that Texas now has a state solid waste regulatory program in general stems initially from the findings produced by the federally financed solid waste surveys. And the desire for the right to implement the later EPA RCRA requirements has led to state changes in policy.

Bureaucratic Dominance in this Policy Area

Among Texas state officials (be they legislative, executive, or administrative) the desire for delegation of RCRA authority is certainly in keeping with what are perceived to be prevailing political attitudes in Texas against "federal intrusion." But the fact that the desire for delegation has been strong among Health Department and TDWR officials is especially important since state bureaucrats in Texas, in conjunction with industry interest groups (to be discussed later) tend to dominate policy making in the hazardous waste regulatory area, as well as in a number of other policy areas.

The reasons for this bureaucratic dominance are varied. They include the technical nature of the policy area, the relative lack of legislative professionalism and other constraints on legislative effectiveness, and the lack of effective gubernatorial leadership in this area. As has been noted, in many respects hazardous waste management is a very technical field, requiring knowledge of chemistry, engineering, geology, hydrology, and similar disciplines in any efforts to properly manage wastes or effectively regulate their management. The agencies, of course, have generally had such expertise available to them. Others without that available expertise or without easy access to it are thus at a disadvantage in many respects in trying to evaluate waste management and regulation of management practices.

This observation is especially important in Texas, vis-à-vis the legislature, since the state ranks very low in terms of legislative professionalism.[82] The Texas legislature is characterized by biennial sessions, limited duration of session (no more than 140 days for the biennial regular session and no more than 30 days for each called special session), consequent poorly paid, part-time legislators, high turnover of legislators, inadequate staff, and inadequate research support.[83] Legislators seldom have the opportunity to develop the necessary expertise which facilitates evaluation of policy issues and responses. Consequently, agency expertise is rarely challenged by the understaffed and underfunded legislative offices. For instance, studies by interim legislative committees, including some done for committees dealing with hazardous waste issues, have been primarily literature reviews coupled with input from public hearings and from interviews and other more informal contacts with relevant people. But seldom is there a systematic examination of similar policy issues and responses in other states, any sophisticated data gathering and analysis using modern social science research methods, or much use of outside academic or other institutions for research. (One qualification to that is the increasing use of the LBJ School of Public Affairs in Austin to research certain policy issues, including the study of hazardous waste siting.) Thus, the agencies come to be relied upon for information in making policy decisions and in actually drafting proposed legislation. For example, in 1981 in making amendments to the Texas Solid Waste Disposal Act, the legislature adopted essentially the draft bill developed by TDWR staff. Only minor modifications were made to the bill before passage (attempts to make significant changes in the bill were successful in the House, with the private concurrence of the agency, but were scuttled in the Senate and not included in the conference report which eventually passed). Perhaps this reliance on the agencies is one reason why the Texas legislature has been ranked so low in evaluation of "effective lawmaking" (forty-fifth out of fifty states in a 1971 study).[84]

Neither is the governor generally an effective counterforce to the bureaucracy in such policy areas as hazardous waste. In addition to a lack of sufficient expertise available to the governor except from the agencies, he lacks formal power to force agencies to take action. For example, in Texas agencies such as the Health Department and TDWR are headed by multimember governing boards, which although appointed by the governor are not subject to easy control. The board members are appointed for six-year staggered terms, so no governor in one four-year term is able to appoint all of the members of a governing board (such boards have a minimum of three members, with one appointed every two years). Moreover the governor's power to remove members (only granted in 1980) is very constrained. The governor can only remove board members which he or she *personally* appointed, and then only with the concurrence of two-thirds of the Texas Senate. These procedures leave the governor with little effective formal control over the agencies. Indeed, below the governing board levels and chief administrative levels within the agency, the employees of agencies such as the Health Department and TDWR are protected by civil service guarantees just as federal employees are. (These agencies are among only a dozen Texas state agencies with merit systems, all the result of "strings attached" to federal grants, which constitute a substantial amount of the funding of those agencies.)[85] Thus, from top to bottom, these state agencies are rather strongly insulated from formal gubernatorial leadership. In formal terms, the Texas governor is weak.[86]

This does not mean, of course, that such agencies might not be influenced by the exercise of the governor's informal powers (such as his access to the media and consequent ability to use the media to focus attention on specific issues and on bureaucratic actions or inaction). However, no governor in the "environmental decade" ever made hazardous waste regulation a priority issue and used the media to push for action on the issue. The only qualification to that has been the position taken on low-level radioactive waste management taken by Governor William Clements. Interestingly, in that situation, due to a variety of factors, Texas in 1981 adopted strong new regulatory legislation. The governor did not take a similar interest in nonradioactive, hazardous waste legislation that same session.

What is the effect of this relative lack of legislative and gubernatorial control and consequent bureaucratic domination in this policy area? One major result has been that hazardous waste management policy in Texas has tended to be based on the dominant perspective among the bureaucrats responsible for the administration of that policy: the engineering perspective. For example, the respective divisons within the Health Department and TDWR which operate the waste management regulatory program are headed by professional engineers, and most of the professional people in those divisions also have engineering backgrounds.

Previous research on the orientations of policy makers involved in environmental issues has identified two competing orientations (health-environmental versus business-economic) affecting the evaluation of policy proposals and the determination of policy decisions.[87] The suggestion here is that, at least in the case of bureaucrats involved in environmental policy administration, there is actually a third competing orientation: technical-engineering.

In the case of hazardous waste management, the three orientations might be described as follows:

A health-environmental orientation focuses on the impacts of hazardous waste on public health and the environment. This orientation perhaps views hazardous wastes as so dangerous that practically no release into the environment is an acceptable risk. A policy stemming from such an orientation might ban land disposal and injection well disposal of hazardous wastes, require the reprocessing and recycling-reuse of whatever hazardous materials are generated as industrial and commercial by-products, and require destruction or neutralization of whatever hazardous materials could not be recycled or reused.

A business-economic orientation focuses on hazardous wastes as necessary by-products of a number of industrial processes and is concerned about the costs of waste management practices and the economic impacts of regulatory requirements. Such a view would probably hold that some negative environmental impact is inevitable and acceptable where hazardous materials are concerned. A policy based on such an orientation would encourage the use of benefit-cost analysis in determining regulatory requirements and allow market forces to determine the level and type of recycling and reuse of hazardous materials that would take place.

A technical-engineering orientation presumes that there is an engineering solution to any problem of a technical nature. Thus, the hazardous waste management issue is a question of "proper" versus "improper" management. Proper management occurs when a facility is "well-engineered." For example, in the case of a landfill, that would require a liner for the bottom and sides of the pit to prevent migration of wastes. The mechanism for assuring that such engineering requirements would be implemented might well be a permit system, which may be obtained by the facility operator only after a technical review of the engineering of the facility. Cost effectiveness of the techniques required might be considered but would not be the determining factor in choosing required practices. Environmental damage would be presumed to be avoided or minimized if such practices were followed.

If the descriptions of these orientations are valid, then Texas state hazardous waste regulatory policy fits most closely the type of policy stemming from the technical-engineering orientation. The characteristics of the policy stemming from the first orientation are not part of the state's policy. Benefit-cost analysis, arising from the second orientation, is not required in the Texas regulatory program, although concerns about costs are frequently raised and recycling-reuse is dependent on economics. But the development and implementation of state policy definitely assumes that there are engineering solutions to the hazardous waste problem. The dominance of engineers within those agency divisions responsible for hazardous waste management, and the dominance of those agencies upon policy making in this area apparently has had a considerable impact on the nature of the policy.

Influence of Industrial Interest Groups

Although the bureaucratic actors are the most influential within the state government in shaping hazardous waste policy, groups outside the state government also attempt to influence the policy making as well. In the hazardous waste area, these attempts in Texas have come primarily from ad hoc citizens' organizations (formed to oppose particular waste facilities), civic and environmental groups, local elected officials, and industrial interest groups.

The ad hoc citizens' organizations have been the most creative in their efforts, choosing such names as P.A.C.E. (People Against a Contaminated

Environment) and C.A.P.O.N.E. (Citizens Against Polluting Our Neighborhood Environment). The civic environmental groups, such as the League of Women Voters and the Sierra Club, have staked out a claim to the "public interest" in hazardous waste. Local elected officials have generated a considerable volume of public statements and resolutions, usually in opposition to hazardous waste facilities in their areas. However, industrial interest groups seem to have had the most influence over state policy in this regard.

The general influence of business and industry in regulatory policy making in Texas has been widely discussed by students of the state's government and politics.[88] Due to the traditional weaknesses of political parties in the state, interest groups have become particularly important in Texas, and business and industry groups have been the most significant in influencing policies of an economic and regulatory nature. Their power is based on close connections with policy makers, economic and other resources, and the strategic importance of several industries to the economy of the state. Regulated industries and their representatives have developed close relationships with the regulatory agencies in a number of instances in Texas.[89]

Perhaps the most relevant example of this to the discussion of hazardous waste management is the close tie between the oil and gas industry and the Texas Railroad Commission. The commission members are elected, and most of the campaign contributions come from the regulated interests, primarily oil and gas production. No one antagonistic to oil and gas interests has served on the commission.[90] The publication of an advertisement in the 1972/73 *Texas Almanac*, paid for by two oil industry associations and bannered with "Since 1891 the Texas Railroad Commission Has Served the Oil Industry," has often been cited as illustrative of a cozy relationship.[91] Some people who question the commission's attitudes toward oil and gas waste disposal have blamed the ties between the commission and the industry for those attitudes. In a related issue, critics of the oil and gas industry's waste disposal activities point to the fact that a U.S. senator from Texas, Lloyd Bentsen, who also has close ties to the industry, was the member of Congress who successfully pushed for 1980 amendments to RCRA which exempted oil and gas wastes from coverage under the law. That move eliminated the potential for EPA to become involved more heavily in oil and gas waste disposal and/or to exercise more oversight of the Railroad Commission.

Another example of industry influence in the hazardous waste area was the role played in 1981 by the chemical industry through its trade association, the Texas Chemical Council, in limiting the scope of amendments to the Solid Waste Disposal Act by the state legislature. The chemical industry and waste disposal industry were in the posture during the 1981 session of wanting the legislature to amend the state law so that Texas would qualify for the delegation of RCRA authority from EPA. One of the reasons why the industries wanted Texas to have RCRA delegation was to eventually end the dual regulatory scheme. Since both Texas and EPA had hazardous waste regulatory programs, and Texas had not yet been delegated authority to operate the EPA program (effectively merging the two programs), industrial companies were forced to comply with two sets of requirements: for example, filling out two manifests (one state, one federal) and, in the case of industries disposing of their wastes on-site or the operators of off-site facilities, having to get both a state and an EPA permit for a facility. This kind of situation is, of course, costly and might be plagued with conflicting requirements. Therefore, industry was interested in seeing that changes were made in state law and regulations to make the state program "substantially equivalent" to EPA's program, thus allowing for delegation.

Industry was not interested, however, in seeing the state go beyond the minimal RCRA requirements. Therefore, in the 1981 legislative session the affected industries supported House Bill 1407 to make the limited changes required by RCRA but opposed (with varying degrees of intensity) another more comprehensive bill which would have established siting criteria, set stringent requirements on facilities sited in floodplains, created a specific role for county governments in siting decisions, facilitated public participation in the permitting process, and made other changes.[92] The latter bill was supported by several ad hoc citizens' groups, the Sierra Club, League of Women Voters, some local elected officials, certain labor unions, and (regarding some provisions) the Texas Farm Bureau.[93] By means of a compromise between the authors of the two bills in the House, some provisions of the stronger bill were incorporated into HB 1407 as floor amendments. The compromise bill passed the House, after the industry for strategic reasons decided to forego a battle in the House and put their efforts into removing the strong amendments in the Senate. With the help of a prominent senator from a district with several petrochemical companies, those amendments were stricken in Senate committee after the senator noted that they "were not very palatable . . . to the industrial interests in the state" and could "disrupt industry."[94] The conference committee approved the Senate version of the bill on the last day of the legislative session, and the legislation making minimum changes to conform to RCRA was adopted.

Qualifications must be made to these examples of industrial influence in this regulatory policy area. The most important qualification is that industry is not monolithic in its views. For instance, ironically, some of the commercial waste disposal companies have privately indicated their willingness to accept certain land-use siting criteria in the law or regulations if all facilities, on-site and off-site, are forced to "play by the same rules." The companies who generate hazardous wastes, and their umbrella Texas Chemical Council, have opposed such siting criteria. Essentially, there are different economic interests at stake. If siting criteria were adopted and applied to all facilities, chemical plants relying on on-site disposal but not "sited" in proper disposal areas might have to send their wastes to off-site commercial facilities (perhaps at a higher cost). Thus, the commercial waste disposal companies might gain, but the chemical companies might lose—in an economic sense. This dichotomy of interests in industry may dilute its influence in the future, especially if other groups exploit these differences of industrial opinion successfully.

The second qualification to the strong influence of industrial groups in this regulatory area is that some of their adversaries on the hazardous waste issue are gaining strength and resources. For example, environmental groups in Texas are growing in membership, economic resources, and expertise.[95] One group, the Sierra Club, has grown by approximately 100 percent in membership in the early 1980s, has increased its lobbying budget and now retains three lobbyists, and is currently helping to coordinate legislative efforts with other environmental groups, some of whom have not been active in this area before. Moreover in metropolitan areas at least, public opinion among Texans seems to support environmental protection efforts.[96] Thus, industrial influence, although continuing to be strong, may be countered more effectively in the future.

Absence of a Central Focus for Media Attention

One reason why the state regulatory agencies and industrial groups have been the dominant actors in the shaping of state hazardous waste management policy is the fact that such policy has been made in Texas within a traditional political subsystem: agencies, clientele groups, and, to some extent, legislative committees. This issue of hazardous wastes has never really entered the highest political arena in the state, where major elected officials, such as the governor, might be able to exercise more influence and which would prevent the traditional actors from working largely outside the view of even the informed public.

One factor in the failure of the issue to enter this highest arena has been the absence of sustained, extensive media coverage of the hazardous waste problem in the state. Although a number of problem waste sites have been the subject of many news reports, and controversies over proposed facilities have received considerable local publicity, hazardous waste has not been as widely covered a topic in Texas as in several other states, such as New York and New Jersey. One reason for that perhaps is the absence in Texas of a central focus for media attention on hazardous waste problems such as the focus which has been provided elsewhere by problem sites such as Love Canal in New York and the "Valley of the Drums" in Kentucky. Both of these sites, for example, have been viewed as major threats to public health and the environment either because of proximity to population and identified adverse health effects or because of the extensiveness of the hazardous wastes at the site (four sites in the "Valley" together comprising tens of thousands of rusting and leaking drums containing discarded chemicals).[97]

Official statements and media stories in Texas have generally contended that the state has no problem site of the severity of New York's Love Canal.[98] Whether or not this is true based on objective measures—the Texas Ecologists, Inc., site near Robstown has been mentioned as a contender for the "Love Canal" title in Texas,[99] but has received less attention because of its location in an agricultural rather than an urban area—the prevalent *perception* in the state has been that no known "Love Canal" exists in Texas. Statewide media coverage has generally reflected that perception, despite the investigative reporting of several television and newspaper reporters with an interest in the hazardous waste issue. Consequently, most people in the United States as a whole and most people in Texas have heard of Love Canal, but few have heard of the Texas City Wye, the Sikes pit, or the French Limited disposal site, all of which are ranked worse than Love Canal (see Appendix A).

Of course, problem sites or proposed sites have received extensive local coverage, and that public awareness on the local level affects subsequent decisions made regarding those areas. Moreover, national publicity about Love Canal has undoubtedly affected many people in Texas as far as their attitudes toward hazardous wastes are concerned. But the absence of more extensive statewide media coverage on the waste problem thus far has meant that major state policy decisions about hazardous waste (as opposed to decisions about permitting a particular waste site) have taken place without intensive public scrutiny. The situation with nonradioactive hazardous waste may be contrasted with the rather extensive publicity on the radioactive waste issue generated by such problem sites as the Todd Shipyards storage site in hurricane-prone Galveston, which was found in violation of its permit and charged with massive mismanagement.[100] That publicity helped to spur the interest of major state legislators and the governor in addressing the problem, and the legislature responded with a new, stringent regulatory policy for low-level radioactive waste. That legislation included many of the same provisions which were knocked out of HB 1407 on nonradioactive wastes, eliminated quietly in a Senate committee session not covered by the media. That type of action would have been more difficult in a state where the hazardous waste issue commanded more attention from major officials, the media, and the people.

In summary, these four factors do not constitute all of the influences which have shaped state hazardous waste regulatory policy, but they are among the most important. Federal policy actions prompted the development of a state waste policy in the first place and probably has helped to produce a stronger policy than otherwise would be the case. Texas has been in the forefront in setting certain regulatory requirements on shipping and land disposal, but the nature of these requirements has been technical-engineering, reflecting the dominance of that orientation within the respective offices of the state regulatory agencies and the dominance of agency officials in this type of policy area. The nature of these regulatory requirements has consequently not calmed the concerns of those with a health-environmental orientation who may view the requirements (and their enforcement) as inadequate to address the problem. Moreover, the current policy has been influenced by the lobbying activities of interest groups representing various segments of industry. Given the generally favorable climate for business and industry in Texas, and the strength of industrial lobbying groups, state hazardous waste regulatory policy does not present as strong a regulatory approach as otherwise might be the case, or as some environmental groups would desire. The nature of media coverage of the issue has contributed to that situation. However, the National Wildlife Federation in its 1979 study ranked Texas thirteenth among the fifty states in terms of the strength of its toxics regulatory program (although only twenty-ninth when the program response was compared to the "need" indicated by the state's ranking on hazardous waste generation).[101] That is a higher ranking than many political observers would have expected. To some extent, the Texas response reflects a "paper program," characterized by extensive legislation but without adequate monitoring and enforcement capabilities.

Prognosis: Future Policy Trends in Texas

This complex interplay of factors will continue to shape state hazardous waste regulation policy in Texas as it inevitably evolves to meet changing circumstances. Further actions by EPA on the hazardous waste issue are anticipated in the early 1980s as the federal government completes the initial implementation of RCRA and evaluates that policy and its implementation. Given its consistent desire to obtain delegation of federal environmental programs, Texas can be expected to modify its regulatory program to conform to federal requirements.

However, dramatic changes in state hazardous waste regulatory policy in the early 1980s are not likely, barring something akin to the revelation of a "Love Canal" in Texas—which might prompt a "crisis" situation requiring some major policy response. Otherwise, the factors which have helped to keep the policy response to the hazardous waste problem within a relatively narrow context—bureaucratic dominance in the policy area, the technical-engineering orientation of the relevant bureaucratic actors, and the continuing influence of industrial interest groups—will probably continue to be the operative ones shaping policy.

This is not to say that no significant changes will take place in the near future. Strong support for land-use or similar siting criteria does exist among a variety of groups, and gradually they may have some success in achieving this goal. Their efforts may be aided by the beliefs of some people in the regulatory agencies and affected industries that the adoption of such criteria (if not so stringent as to preclude disposal sites anywhere) may be a necessary concession for restoring public confidence in hazardous waste management. If the public can be convinced that waste facilities will be sited so as to avoid presenting threats to public health and the environment, the theory goes, then the process of siting waste operations may be facilitated—avoiding costly legal and other delays brought on by public opposition.

Other changes may result from certain attempts at policy dialogue on the hazardous waste issue which have been undertaken in Texas. These are efforts by some organizations to conduct a process whereby a diversity of interests concerned about the hazardous waste issue (academics, ad hoc citizen groups, industry, elected officials, environmental groups, state regulatory officials) are brought together to discuss the issue and to reach some consensus about addressing the issue—or, at least, a better specification of points of disagreement. One such effort is being conducted by the Keystone Conference Center in Colorado, initiated by a request from the Gulf Coast Waste Disposal Authority in Texas, and involves a group of approximately thirty-five people (mainly from the Houston-Galveston area). The other effort is the Texas Roundtable on Hazardous Waste, established by the Texas Environmental Coalition, which is operating at the state and regional level and seeks to create a greater public awareness of the hazardous waste problem and a resolution of conflicts about the problem. These policy dialogue efforts may produce changes in legislation, regulation, or private actions.

Thus, hazardous waste policy in Texas is still evolving. The waste problem, however characterized, still exists. Addressing the problem has been as complex as the problem itself. As more is learned about the problem and about the effectiveness—or ineffectiveness—of the policy response, some of the uncertainty about hazardous waste will be alleviated. As that happens, state policy will change, but it will evolve not only on the basis of changing perceptions about the problem but also within the context of the political and other factors in Texas identified here.

8. Technological Policies and Hazardous Waste Politics in California

David L. Morell

The country's most populous state has responded to its version of the national hazardous waste management crisis through a complex blend of innovative technology-forcing policies and traditional state-local political compromises. In the process, California is leading the nation away from reliance on land disposal as the principal method for dealing with hazardous wastes. The wastes are to be treated instead. Scientific and technological innovations comprise the central focus of this state's toxic substances control program. This effort even includes the initial framework for an overall policy to regulate carcinogens—the first policy of this kind in the United States, at any level of government.

At the same time, however, California's search for the proper balance of state, regional, and local authority over siting has been tentative and traditionalist. For better or worse, the state has not assumed a national leadership role concerning the politically explosive siting issue. And while important modifications were made in 1980–82 in the administrative structure for hazardous waste management in California, concerns remain over the institutions' capabilities to implement effectively the state's new science-based policies.

Contemporary hazardous waste politics in California can be analyzed from the perspective of several different issues, each of which forms the central focus of one section of this chapter. The first set of issues revolves around technological innovation and science policy. What political pressures led to California's decision to use the regulatory process to ban the land disposal of selected categories of toxic wastes, requiring that they undergo treatment instead? While the legislature and certain interest groups certainly participated in the decision, these events are essentially an example of executive branch initiatives in a complex arena of scientific policy.

The second cluster of issues involves the use of political authority over hazardous waste management. Which institutions are to make what decisions? Who wields power over hazardous waste policy, a topic of growing significance and increasing political salience? Who will hold the authority to approve—or veto—proposed new hazardous waste management facilities? Here legislative politics and the pursuit of consensus have been the dominant characteristics of California's experience. Innovation has been overwhelmed by caution, and at times one has seen near paralysis in the face of powerful political forces seeking incompatible objectives.

Debates over political authority quickly spilled over the boundaries of the Sacramento policy-formulation arena into state-local politics in Southern California. While this ought not to be surprising given the magnitude of this region—Southern California alone includes over thirteen million people—a case study of the complexities of the Southern California Hazardous Waste Management Project offers additional insights into the political tensions associated with responsible hazardous waste management.

Severity of California's Hazardous Waste Problem

Without question, California faces a serious hazardous waste management challenge. Nationally, the state ranks fourth among all the states in the extent of hazardous waste generation.[1] The Los Angeles County Department of Health Services has estimated that such wastes are generated by at least 17,000 businesses in that county alone.[2] Often these wastes are managed poorly, and at times they are dumped illegally. Like so many other states, California also has had to cope with numerous legacies of past improper waste disposal practices. Cleanup activities supported by federal and state Superfund monies are underway at several of the most dangerous old sites, including three on the list of EPA's top Superfund sites.[3]

Patterns of Hazardous Waste Generation

The state's Department of Health Services (DHS) in August 1982 prepared a summary of current hazardous waste generation patterns.[4] While recognizing the inadequacy of much of the available information, the department presented the estimates on a statewide basis, with totals[5] shown in table 8.1, and off-site disposal amounts[6] displayed in table 8.2. Los Angeles County in Southern California, Kern County in the San Joaquin Valley, and Contra Costa, Santa Clara, and Sonoma Counties in the San Francisco Bay area are the top five areas using off-site disposal. Together they account for 59 percent of the total off-site waste stream.[7] These wastes all go to one of seven licensed Class I hazardous waste disposal sites in the state. The BKK Corporation's landfill in Los Angeles County is the state's largest, handling over 450,000 tons per year; Chemical Waste Management's landfill at Kettleman Hills in Kings County, in the Central Valley, accounts for another 200,000 tons per year.[8]

Data about on-site management of hazardous wastes in California are far less reliable. Three categories of wastes shown in table 8.3 comprise almost 92 percent of the 4.8 million tons per year identified as being disposed of on-site.[9] Almost all of these wastes are disposed of in injection wells, primarily in Kern County.[10] These data do not include the many on-site industrial surface impoundments commonly used for on-site storage or treatment in many manufacturing areas around the state.

Additional details on patterns of hazardous waste generation have been compiled by the Southern California Hazardous Waste Management Project. The best estimate is that waste generators in this region are responsible for some

Table 8.1. Estimated statewide amounts of hazardous wastes disposed of annually in California

Hazardous wastes disposed of off-site	1.4 million tons / year
Hazardous wastes going to on-site disposal	4.8 million tons / year
Total	6.2 million tons / year

Source: University of California–Davis, *Hazardous Wastes in California: On-site Storage, Treatment and Disposal.* (Davis: Chemical Engineering Department, draft of May 20, 1982).

Table 8.2. Estimated statewide amounts of hazardous wastes disposed of off-site annually in California, shown by category

Categories	Amounts (tons/year)
Acid solutions	145,000
Drilling muds	99,000
Alkaline solutions	89,000
Oil and water mixtures	84,000
Alkaline sludges	77,000
Pesticides	76,000
Tank bottom sediments	68,000

Source: California Department of Health Services, *Report to the Hazardous Waste Council on Current Hazardous Waste Generation in California* (Sacramento: Toxic Substances Control Division, Hazardous Waste Management Branch, August 31, 1982).

3.4 million tons/year of hazardous wastes (this excludes the large figure for rinse water and wash water). This amounts to some 1.4 pounds of hazardous waste being generated per person per day throughout the region. Of this amount, some 2.9 million tons (85 percent) are managed on-site; the remainder are sent to off-site disposal facilities.[11] The vast majority of these wastes are being generated in the Los Angeles/Long Beach industrial area.

Table 8.4 presents these data, dividing the Southern California region into ten areas. As shown, the upper coastal Los Angeles area alone accounts for 1.3 million tons per year, nearly all of it dealt with on-site. The north Los Angeles area and Orange County account for another 500,000 tons each, 94 percent of which is managed on-site. San Diego accounts for an additional 316,000 tons, almost all managed on-site.[12] The Long Beach area ranks next in waste generation, with 315,000 tons/year. Management of this waste stream is divided almost equally between on-site and off-site facilities.

Corrosive wastes are by far the most significant category of wastes being managed on-site (still excluding rinse waters and wash waters, which are predominantly sewered). Some 62 percent of the on-site wastes are termed corrosive, followed by acidic solutions with heavy metals (9 percent), halogenated solutions with heavy metals (6 percent), aqueous solutions containing dissolved cyanide (4 percent), and mixtures of oil and/or gas with water (4 percent).[13]

Table 8.3. Major categories of hazardous wastes disposed of on-site annually in California

Category	Amount (tons/year)
Rinse water and waste water	2,924,000
Aqueous solutions with organic residues less than 10 percent	1,300,000
Corrosives	153,000

Source: California Department of Health Services, *Report to the Hazardous Waste Management Council on Current Hazardous Waste Generation in California* (Sacramento: Toxic Substances Control Division, Hazardous Waste Management Branch, August 31, 1982), 19.

Distinctions in the types of wastes occur from one part of Southern California to another. Corrosive wastes dominate in upper coastal Los Angeles (84 percent), northeast Los Angeles (58 percent), San Diego County (75 percent), northern Los Angeles (50 percent), and Orange County (49 percent). Paint sludges are the leading waste category in central Los Angeles (50 percent). Oily tank bottoms are the largest category in Long Beach (31 percent). In southeast Los Angeles oily sludge is in the top category (only 15 percent, with the patterns of waste generation very divided). Aqueous solutions with heavy metals are the highest category in San Bernardino-Riverside (47 percent), while non-halogenated solvents account for 44 percent of the waste flow in Ventura County.[14]

The BKK landfill in West Covina (Los Angeles County) is the dominant facility in the region for off-site waste management, receiving nearly 80 percent of the region's total off-site waste flow. This landfill receives between 75 and 80 percent of the hazardous wastes from Los Angeles, Orange, San Diego, San Bernardino, and Riverside counties destined for off-site disposal. Most wastes generated in Ventura go to the Casmalia Resources Company's Class I disposal site in Santa Barbara County, while Imperial County's generators normally use the Imperial Valley site operated by the IT Corporation.[15]

Problems of Past Disposal

Toxic wastes, carcinogens, corrosive chemicals, and other dangerous materials dumped on the ground in various California communities have found their way back into the air, into underground supplies of drinking water, and into surface waters. For example, pesticide wastes placed in unlined land disposal facilities near Lathrop, in the Sacramento area, were discovered in 1979 to have been percolating down to the groundwater since 1969. Presence of these wastes, including the dangerous DBCP (dibromochloropropane), forced closure of several wells. In 1977 male workers at the Occidental Chemical Company plant in Lathrop found that they had become sterile after workplace exposure to

Table 8.4. Patterns of hazardous waste generation in Southern California (tons/year)

Region	Generation of wastes sent off-site	Generation of wastes managed on-site	Total on-site & off-site waste generation
Upper coastal Los Angeles	148,206	1,185,597	1,333,803
North Los Angeles	24,846	499,332	524,178
Central Los Angeles	31,782	132,726	164,508
Northeast Los Angeles	37,110	83,509	120,619
Southeast Los Angeles	26,274	7,860	34,134
Long Beach	156,606	158,779	315,385
Orange County	40,242	445,950	486,192
San Diego County	25,614	290,688	316,302
San Bernardino–Riverside counties	11,610	44,001	55,611
Ventura County	15,486	11,025	26,511
All	517,776	2,859,467	3,377,243

Source: Southern California Hazardous Waste Management Project, *Hazardous Waste Generation and Facility Development Needs in Southern California* (Los Angeles: Los Angeles Sanitation Districts, draft, November 1982), table II-2.1.

DBCP at levels considered to be safe. DBCP is now thought to be present in 25 percent of the water wells in the state.[16]

Local groundwater pollution in Sacramento County has been attributed to the suspected carcinogen TCE (trichloroethylene). This material has also been found contaminating numerous wells used for drinking water in Southern California.

In 1975, 5,000 gallons of liquid hazardous wastes from Los Angeles deposited at an evaporation pond in the San Francisco Bay area reacted with other contents of the pond to produce a large, odiferous cloud visible over ten miles from the disposal site. Residents of the city of Richmond filed hundreds of complaints, several claiming illness from the odors. Wastes had ended up in the Bay area because Class I disposal site operators in Southern California had rejected these wastes due to their odors.[17]

Underground storage tanks for toxic waste solvents were found in 1982 to be leaking into the groundwater in the Santa Clara–San Jose area, home of the "Silicon Valley" electronics industry.[18] In September 1982, two workers at a chemical plant near Bakersfield died after cleaning a tank that had contained the agricultural fumigant EDB (ethylene dibromide).[19]

Because of severe groundwater contamination, the Stringfellow Quarry, a commercially owned Class I hazardous waste disposal site near Riverside east of Los Angeles, was closed in 1972. Constructed in 1955–56 as a series of evaporation ponds used to concentrate liquid wastes, Stringfellow received some 32 million gallons of wastes, mostly spent acids and caustics. The facility overflowed its containment dikes in heavy rains in 1969. Reopened after repairs, it had finally closed in 1972. The state began an active abatement program here in 1977, the total costs of which are expected to exceed $6.5 million.[20]

Extensive site cleanup analyses are underway at the McColl dumpsite in Fullerton, also in Southern California. Local residents and elected officials from this area in 1982 were pressing the state to begin its Superfund cleanup immediately. The Department of Health Services insisted on detailed scientific analysis as the precursor to an environmentally sound mitigation effort. The legislature appropriated $1.5 million from the state Superfund for remedial action at McColl, and specified that none of this money could be used for "studies."[21]

These are only a selected sample of the many incidents of mismanagement of hazardous wastes in California. These incidents continue to require allocation of scarce resources of money and personnel, and increase local fears of hazardous wastes and thereby constrain the state's ability to implement its new siting policies.

The Politics of Technological Innovation: Restrictions on Land Disposal of Hazardous Wastes

In 1981 California embarked on an innovative effort to phase out the land disposal of highly toxic wastes. This technology-forcing intervention into the private marketplace was designed to improve public health by ending the traditional practice of simply placing these extremely hazardous materials onto or into the ground. This technique is typically less expensive than alternative methods of treatment (ignoring the high costs of remedial health care and environmental cleanup). Epstein, Brown, and Pope termed this effort "a radical shift away from the state's precarious dependence on land disposal in favor of an aggressive program to direct the most hazardous wastes to new recycling, treatment, and destruction facilities."[22] Analysis of the evolution of this regulatory program, the first of its kind in the United States, sheds light on several aspects of hazardous waste politics.

After careful study of the problem, and revision of the early statements, several categories of wastes were identified. A tough schedule, shown in table 8.5, was promulgated in late 1982 to prohibit these wastes from land disposal over a two-year period.[23] These wastes were all identified as highly toxic, persistent in the environment or bioaccumulative, and mobile after land disposal. They present the greatest long-term risk to public health and the environment

Table 8.5. State of California schedule for prohibiting land disposal of specific hazardous wastes

Waste category	Deadline for land-disposal prohibition
Cyanide wastes	June 1, 1983
Toxic metal wastes, acid wastes, polychlorinated biphenyls (PCBs)	January 1, 1984
Liquid wastes containing halogenated organic compounds	January 1, 1985
Organic sludges and solids containing halogenated organic compounds	July 1, 1985

Source: California Department of Health Services, *Notice of Proposed Changes in Regulations in the Department of Health Services Regarding Hazardous Waste Land Disposal Restrictions* (Sacramento: Toxic Substances Control Division, August 18, 1982) (R-32-82). Attachment: "Case Histories of Pollution Incidents from Land Disposal of Hazardous Wastes in California," pp. A-1, A-2.

when placed on the land. The new regulations carry a maximum penalty of $25,000 per day of violation.

Compliance and enforcement, however, depend on construction of treatment facilities. That is, the wastes prohibited from land disposal under the revised regulations must instead be safely recycled, treated, or destroyed. If technologies to perform this kind of treatment are not available, the phaseout deadline will be extended. Therefore, it became imperative in 1981–82 for state agency proponents of the land-disposal phaseout program to encourage the siting of new waste treatment facilities in order to implement their program.

Development of the Landfill Restrictions

Work on this regulatory program commenced in the fall of 1981, as an extension of an analysis of hazardous waste policy options begun the year before by the governor's Office of Appropriate Technology (OAT). In a real sense, therefore, this is a case in which an innovative public policy emerged as a direct result of the creation of an unusual government agency—OAT—and that agency's willingness to pursue nontraditional approaches to the resolution of a critical public policy problem. Explicit recognition must also be given to Governor Edmund G. Brown, Jr., whose personal prestige and considerable intellectual and political energies went into the establishment of OAT and into support of OAT's decision to explore new ways to deal with toxic wastes. Without the governor's personal support, it seems unlikely that this controversial program would have taken form so rapidly.

OAT's 1981 report—*Alternatives to the Land Disposal of Hazardous Wastes*[24]—assessed the types, volumes, and sources of hazardous wastes being generated throughout California for off-site treatment and disposal, and identified a variety of treatment technologies which could be used to reduce, recycle, treat, or destroy specific portions of this waste stream. The analysis concluded that about 75 percent of all hazardous wastes sent off-site for disposal were amenable instead to treatment by proven, available technologies.

On the basis of the OAT report, Governor Brown in October 1981 issued an executive order directing the Department of Health Services—the state agency with lead responsibility for overseeing the management of hazardous wastes—to prohibit land disposal of highly toxic wastes. The governor also directed DHS to:

(b) exercise its existing regulatory authority . . . to impose higher fees on the land disposal of high priority, extremely hazardous toxic wastes, until such time as land disposal is banned for these wastes.

(c) increase monitoring and enforcement inspections at all hazardous waste disposal sites.[25]

The governor also directed DHS to establish a new effort to encourage private corporations to construct and operate new waste treatment facilities in order to implement the land-disposal phaseout program, and to involve the public actively in its overall hazardous waste management effort.[26]

In response to these directives, the Office of Permit Assistance in the governor's Office of Planning and Research established a special program to assist applicants in obtaining permits for new waste treatment facilities; DHS created a new Office of Public Education and Liaison; and the department organized an interagency task force composed of agencies with direct regulatory authority or expertise in waste management. This task force included representatives of the Air Resources Board, the Water Resources Control Board, the Office of Appropriate Technology, as well as the Department of Health Services. This group published a discussion paper in January 1982, allowing initial public comment on the proposed new policy direction. This discussion paper was circulated to over 1,500 persons, organizations, and companies; and some 340 people participated in two public workshops held in Berkeley and Los Angeles.[27]

When first announced in October 1981, the executive order caused great consternation in industry circles. OAT had had little contact with industry, which was shocked to be faced suddenly with an aggressive schedule to prohibit cheap land disposal of its wastes. Oil companies, chemical firms, electroplaters, and others argued that the governor's office had been captured by environmental zealots who had no comprehension of the realities of corporate economics. As later reported in the *Hazardous Materials Intelligence Report*:

Joan Berkowitz of Arthur D. Little (ADL) told HMIR that California officials are overly optimistic about the availability of treatment technologies, and that California's estimate that the annual cost to industry of using alternative treatment technologies of $30 million is probably too low. . . .

Berkowitz said that the planned rapid implementation of the ban might encourage illegal disposal, cause industries to ship wastes out of state or relocate to other states, and force small- and medium-sized generators out of business. She added that it was ironic that California should be the first state to phase out landfilling, because evaporation exceeds precipitation in California, making the climate ideal for landfills.

John Cupps of the California Council on Environmental and Economic Balance (CEEB), a consortium of business, labor, and public-interest organizations that includes representatives of Dow, Monsanto, du Pont, ARCO, Union Oil, Getty Oil, Hewlett-Packard, and Bechtel Engineering, disputed OAT's predictions that adequate treatment facilities would be available in time for the ban and said that the state should have waited to institute the ban until the facilities had been sited and were operating.[28]

In sum, critiques of the OAT study and the executive order typically encompassed several of the following arguments:

1. The state's regulatory and legal framework could not possibly be altered quickly enough to implement the ban on land disposal set forth in the executive order.
2. The OAT report did not provide sufficient proof that California's existing Class I landfills pose a significant threat to public health or the environment.
3. Data presented on waste generation patterns in California were insufficient as the basis for rapid implementation of an entirely new program. The OAT evaluation was based on an extrapolation of only two months of data, and excluded the 70 percent of all wastes that are managed on-site.
4. The state's analyses did not examine the residuals or sludges to be generated by use of the recommended treatment technologies and the resulting requirements for ultimate disposal of these wastes on the land.
5. Future facility siting problems could render the entire program inoperable.[29]

As one specialist testified at a hearing of the Assembly's Committee on Consumer Protection and Toxic Materials in February 1982, "Through review of California's new program . . . we have found that a good start has been made—. . . [but] a siting and permitting process must be developed—not just streamlined—to bring *any* alternative or conventional type of 'priority' wastes management program into reality."[30]

As time progressed, the level and content of communication increased between industry representatives and government proponents of the new policy. The conclusions of the OAT report received wider distribution and provoked extensive discussion, both in small group meetings and in public sessions. A second paper from the interagency task force, summarizing the major issues, was circulated to a wide audience in May 1982. This discussion paper included substantial revisions to the initial proposals. In particular, the state clarified its position with respect to prohibiting land disposal: the scheduled ban would be enforced *only* to the extent that treatment technologies actually were available.[31] If the technologies were not available, the deadlines would be extended.

It was difficult for industry to attack such a state posture as "unreasonable," and criticism of the phaseout policy dwindled considerably. (Conceivably, the real battle was simply being deferred until 1983 and later, when decisions on actual availability of treatment technologies would be made. How would "availability" be defined, not only in terms of engineering capacity to treat certain waste categories in certain volumes, but also in terms of cost and geographic accessibility? Industry's cooperative stance in 1982 could well shift toward recalcitrance once again when these detailed certification decisions have to be made.)

In August 1982, the interagency task force completed revisions to the proposed regulations.[32] The phaseout schedule, provisions for waivers if the necessary technologies were not available, and other aspects of the program were all scrutinized at an October 1982 public hearing held in Sacramento. The overall public response was quite favorable, criticism being restricted to confusion over some of the details of program implementation (especially the waiver provisions). After the public hearing the regulations were formally promulgated through the Office of Administrative Law. They were set to take legal effect in January 1983.

Permits for New Treatment Facilities

Central to the success of this effort to phase out the land disposal of certain toxic wastes will be the state's ability to work closely with waste generators and with the waste management industry to issue permits for the necessary new treatment facilities. In this respect, California's early progress was encouraging, although a number of formidable political obstacles still lie ahead.

The 1981 executive order, and subsequent one-on-one contacts between state agencies and officials of the already sizable waste management industry, triggered an impressive response. By the time the regulations were formally proposed for adoption in late 1982, several large waste producers and waste processors had anticipated the changing regulatory climate and had begun to reduce somewhat the volume of wastes going to land disposal. A number of firms had expressed interest in constructing new treatment and recycling facilities in order to take advantage of the large existing market once the regulations start to come into effect in June 1983.[33] The following pages describe several of the projects initiated during this period.[34] These projects illustrate varying levels of citizen support or opposition to new hazardous waste treatment facilities.

Cement kiln incineration, Lebec. General Portland, Inc., has received permission from the Department of Health Services to burn up to 1 million gallons of hazardous waste solvents as a supplemental fuel in its cement kiln located in

Lebec, a small community on Interstate Highway 5 in Kern County just north of the Los Angeles metropolitan area. The solvents substitute for expensive fuel, and are rendered nontoxic through chemical changes induced by the intense heat. This one-year demonstration is being closely monitored to provide air emissions information applicable to future waste incineration projects. No significant public opposition emerged to this incineration test project, perhaps partly because Lebec lies outside the highly polluted Los Angeles air basin and perhaps partly because of its remoteness from population.

Wet oxidation, Casmalia. The Casmalia Resources Company has received permission from DHS and the local air pollution control district to install a wet oxidation system to detoxify cyanides and aqueous organic wastes at its existing Class I hazardous waste landfill facility in Santa Barbara County. This is a portable treatment unit which alters the chemical composition of the liquid wastes through adding heat under pressure. Extensive air pollution controls are to be employed, and a special air quality sampling and analysis program will be instituted during the six-month trial period due to begin in early 1983 when the unit first becomes operational.

At the public meetings on this proposal, local residents' concerns were focused almost entirely on problems they identified with the existing landfill, rather than with the proposed new treatment facility. In this sense, their criticisms reinforced several prevailing notions about the politics of siting hazardous waste facilities in California. First, popular opposition to new treatment facility siting proposals may be significantly less than to new landfills. Yet open discussion of the proposed treatment facility may offer an opportunity for the public to raise sharp questions about the adequacy of existing landfills. Such questions add to the pressure to implement the program of land disposal restrictions. Second, siting of treatment facilities at existing landfills—on-site and off-site—may be easier than opening a new facility at a completely new location. People are accustomed to (if not always pleased with) the presence of hazardous wastes at these locations, and thus will be relatively more accepting of new activities there. A sacrifice-area syndrome in siting does exist (at least to some extent). In all these cases the developer already owns the land, easing his site-approval anxiety.

Moreover, treatment at Casmalia, in Santa Barbara County, requires that the wastes generated in the Los Angeles area be brought in by truck along congested U.S. 101 from the Los Angeles area through Ventura County. In contrast, criteria for siting waste treatment technologies (as opposed to landfills) devised by DHS call for locating them in urban-industrial areas.[35] Despite such limitations, the proposed treatment process at Casmalia will test the wet oxidation system and thus contribute to the state's capacity to move away from land disposal.

Industrial waste treatment, Wilmington (Los Angeles). The BKK Corporation has identified a site in the industrialized Wilmington area of Los Angeles. As of late 1982 BKK was preparing full permit applications to construct a large facility to treat industrial wastes. In November 1982 the company delivered to the Los Angeles City Planning Department the draft environmental impact report (EIR) for this facility required under the California Environmental Quality Act (CEQA).

This facility is designed to detoxify liquid inorganic wastes and acids, producing dewatered residuals for subsequent land disposal. BKK estimates that its new treatment plant could handle 80 percent or more of all the hazardous wastes that have, until now, been hauled to the corporation's large Class I landfill in West Covina, a city of some 80,000 people on the eastern edge of Los Angeles County. The corporation hopes to have its regional treatment center operational by October 1983.

Response to this proposal from the public and from involved local officials in the city of Los Angeles has been supportive and relatively noncontroversial—a tribute both to the general perception that treatment is far preferable to land disposal, and that in the proposed location a treatment facility would be fully compatible with its industrial neighbors. At present BKK operates a municipal solid waste transfer station on the site. Nearby neighbors include several automobile junkyards and oil storage tanks. The site is near a freeway entrance. The nearest residences are some 1,500 to 2,000 feet away. The city of Los Angeles has committed to provide the required sewage treatment capacity to receive BKK's treated effluent discharge, and city and state permit processing was well underway in late 1982. While strong local opposition might still materialize, it was certainly not apparent as of that date.

Transfer and treatment facility, Otay. The BKK Corporation has also come forward with a proposal for a hazardous waste transfer station and small waste treatment facility at Otay, in San Diego County. This site formerly was certified as a Class I hazardous waste landfill, but was closed voluntarily by its operators in 1980 rather than meet the financial security provisions of EPA's more stringent new standards under the Resource Conservation and Recovery Act.

The BKK transfer station at Otay is designed to support the company's larger treatment facility in Wilmington, as well as to funnel appropriate wastes to BKK's landfill in West Covina. The treatment facility was tested at the West Covina site while permit processing was underway for the Otay location. It can handle a range of liquid wastes.

This siting proposal was in the final stages of the required local and state permit process at the end of 1982. Federal approvals were also pending. The facility was built and was ready to operate. Relatively little public opposition had emerged, although some residents expressed concern at a public meeting in November 1982 that their area was becoming a dumping ground for all of Southern California.[36] The Otay site, like Casmalia, was already known for past hazardous waste activities. Thus, public opposition was somewhat muted.

This proposal is illustrative of a second siting principle, land-use planning. The Otay location, from a hydrogeologic perspective, seems suitable for a transfer station and for a small treatment facility. Some sources, however, indicate that a large planned residential community is contemplated for a tract of land just to its east.[37] While the proposed facility conforms to existing land uses, if this ·housing were actually constructed, the facility's proximity to large numbers of nearby residents would not be desirable.

This phenomenon of residential housing encroaching upon licensed hazardous waste facilities has been a problem already at several California locations, as well as elsewhere in the country: at BKK's own landfill in West Covina, at the McColl dumpsite in Fullerton, at the Palos Verdes landfill in Los Angeles County, and at the Calabasas landfill in western Los Angeles County. As yet no effective mechanisms exist in this state to ensure that existing waste management facilities are kept relatively remote. Institution of low-density buffer zone regulations around waste management facilities, and the enforcement of such regulations in perpetuity, would ensure needed separation between residential and industrial land uses. New hazardous waste landfills in California are required under A.B. 2370 (passed in 1981) to have a 2,000-foot buffer zone separating them from existing residential housing.[38]

Demonstration-scale waste treatment unit, West Covina. The BKK Corporation has been responsible for still another treatment facility siting project in Southern California, this one at its large landfill in West Covina. The company in 1981 received a DHS permit to test and demonstrate two portable waste treatment systems. This demonstration project was designed to produce results that would contribute to the company's engineering decisions and permit processing for the much larger Wilmington facility. It enhanced BKK's understanding of the treatability of the various wastes now coming to the landfill. This equipment was subsequently moved to Otay, where it will be used with larger storage tank capacity to treat wastes in the San Diego area.

Here again, locating a treatment facility at a licensed hazardous waste landfill ran quite smoothly. State permit processing was routine and reasonably speedy, and citizen opposition was minimal. As at Casmalia, local residents concerned about the Class I landfill used this opportunity to complain about its continued operation, but could not find much to criticize in the new treatment system itself.

Defeat of Siting Proposals

Two other recent siting attempts in Southern California offer a less optimistic view of the politics of siting new treatment facilities, however. These involved a treatment facility and transfer station in Long Beach, a large industrial area, and a comprehensive hazardous waste treatment plant and disposal facility in the Sand Canyon area in the northern portion of Los Angeles County.[39] Both cases preceded development of the land-disposal phaseout regulations, and both triggered concern over the adequacy of the state's siting process.

Comprehensive treatment and land disposal facility, Sand Canyon (Los Angeles). The facility at Sand Canyon was proposed by the IT Corporation to handle industrial liquid wastes through a variety of treatment and disposal techniques, including chemical and physical treatment, incineration, storage ponds for biodegradation, solar evaporation ponds and spray evaporation areas, neutralization ponds, and a chemical waste landfill.

Prior to 1970 this site had been a hog ranch. In 1971, Operating Industries, Inc., applied for a permit for a Class I (solid waste) landfill here, but was rejected by the county. The 1977 version of Los Angeles County's Solid Waste Management Plan later did show this area as a potential solid waste (non-hazardous) facility location. In 1977, IT selected the Sand Canyon site after searching for an area suitable for a new hazardous waste facility. After investigating the geology of the site and completing detailed design work for the facility, IT submitted an application for a conditional use permit to Los Angeles County's Department of Regional Planning in mid-1978. IT hired a consultant to prepare additional environmental information for use by the county. In early 1979, the consultant submitted a preliminary environmental impact report to the county. Extensive review of the document followed, with additional information being requested by the county and supplied by IT. During this period, considerable controversy developed over the environmental feasibility and land-use suitability of the project. A citizen group calling itself "Dump the Dump" organized in the area, sparking intense opposition to the proposed facility. In 1979, when IT approached the community to present its siting proposal formally and to discuss ways to mitigate the facility's potential impacts on the community, the company encountered the strong local opposition. At one public meeting some 1,400 angry residents overflowed the high school gymnasium, protesting against the facility. Eventually, in 1981, finding the public opposition to the project to be insurmountable, IT withdrew its permit application.

The Sand Canyon siting attempt illustrates a number of issues regarding the siting process. First, a clear decision was never reached. IT found itself in the position of expending considerable effort and money to answer a seemingly endless series of questions. Yet the company never received a decision on the adequacy of the EIR, nor did it ever receive definitive government approval or denial. After approximately four years of work on this facility siting proposal, and expenditure of some $1.5 million, IT became convinced that the Sand Canyon project would never be allowed to succeed, and withdrew its proposal.

From the perspective of public involvement, this case illustrates how a determined core of public opposition can defeat a proposal, even in a county as large as Los Angeles (over 7 million residents). The IT Corporation expended considerable effort attempting to answer questions which arose, but found it difficult to satisfy residents already convinced that such a facility did not belong in their community. Local opponents saw the company's offers to compromise as too late, self-serving, and not credible.

The proposal was eighteen months old before the local public was aware of it. This delay may have contributed to the opposition which arose. Some observers argue that the proponent should have involved the public prior to submittal of the permit application. IT's failure to involve the public early in the planning stages exacerbated the community's outrage. Initially, public opposition arose because the Sand Canyon site had previously been designated as a potential landfill in two major county planning documents: the solid waste management plan and the general plan. Neither plan adequately considered the compatibility of a landfill with existing neighboring land uses. The public was either uninvolved in or unaware of the possible significance of the county's decision to include this site in its planning documents. Consequently, the public became suspicious of the county's overall planning process.

Waste treatment and transfer station, Long Beach. The siting controversy in Long Beach further illustrates aspects of hazardous waste politics in California. Chemical Waste Management, Inc. (CWM) proposed in 1981 to upgrade an existing waste treatment facility (an oil-water separation plant for oil field wastes), and to construct a new hazardous waste transfer station at the same location. The transfer station would temporarily store and consolidate non-sewerable residues and a variety of hazardous wastes prior to their bulk transport to the company's Kettleman Hills Class I disposal site in Kings County, some 200 miles north of Los Angeles. No treatment or disposal of hazardous wastes was to occur at the proposed Long Beach facility. This project was designed to serve industrial liquid waste generators in the Long Beach, Carson, Torrance, and Wilmington industrial areas, and to provide a more efficient and cost-effective method to transfer hazardous wastes from this region to the remote Kettleman site.

This proposal, too, engendered a great deal of opposition in the local community—and it, too, was eventually defeated by local resistance. Many Long Beach residents believed that the location selected by CWM was totally unsuitable for a hazardous waste facility. The area is zoned for single-family residential homes (the oily wastes separation facility is a pre-existing use, having operated there for over fifty years).[40] Thus the proposed new hazardous waste transfer facility would require a general plan amendment, a zoning change, and a conditional use permit. The immediate area is primarily residential, with approximately 8,000 people living within a twenty-block radius of the proposed site. Some houses reportedly were as close as sixty feet to the boundary of the proposed site.

The Long Beach City Planning Department, using a risk assessment model, concluded that no suitable areas for a waste transfer facility of this kind existed anywhere within the city's boundaries. In February 1982, the city formally adopted an ordinance which effectively prohibits (through restrictive conditions on proximity to existing residences) the siting of any hazardous waste storage, treatment, transfer, or disposal facilities within the city's boundaries (with the possible exception of locations within large existing oil refineries).[41] This local ordinance thus runs directly counter to the state's attempt to locate new waste management facilities in urban-industrial areas like Long Beach (though not in immediate proximity to residential housing as had been proposed by CWM).

Both of these siting rejections took place before the state's new policies on waste treatment and landfill phaseout had taken their final form, and before state agencies had taken an aggressive stance favoring the siting of new treatment facilities. Both locations may have been inappropriate for facilities of the kind proposed, and as such these proposals deserved to be defeated by local political opposition. Nevertheless, the two cases do illustrate that facility siting can enounter grave difficulties in California, and that easy implementation of the treatment policy may not be a foregone conclusion. The Hazardous Waste Management Council's members are very much aware of the Sand Canyon and Long Beach cases, and are attempting to balance these experiences against the apparent successes in Otay, Wilmington, Casmalia, and Lebec in deciding what changes to recommend in the state's overall facility siting process.

Science Politics

A State Cancer Policy

California's ability to promulgate new policies calling for treatment rather than land disposal of hazardous wastes indicates the potent combination of political forces effectively orchestrated by Governor Brown and a few of his principal aides, who acted cooperatively with several senior members of the state's public health bureaucracy and the legislature. The state's cancer policy was formulated in 1982 by the large and impressive group of public health scientists—physicians, epidemiologists, chemists, and others—based in the Epidemiology Studies Section and the Hazard Evaluation System and Information Service of the Toxic Substances Control Division in the Department of Health Services. No other state has attempted such an ambitious scientific and regulatory effort to prevent cancer.

In July 1982, this group published the first of a proposed set of three documents dealing with cancer: *Carcinogen Identification Policy: A Statement of Science as a Basis of Policy.*[42] The next document covers risk assessment. Finally, the third document sets forth a framework for policies to control exposure to identified carcinogens.

The first document summarized the level of scientific knowledge about carcinogenesis, as the basis for eventual action to limit cancer risks to individuals and the general population.[43] Causes of the major forms of cancer were identified: diet, use of tobacco, and exposure to cancer-causing substances. The uses, and potential misuses, of various methods to predict the potential of one of the 50,000-odd chemicals and other substances present in commerce in California received attention. The report covered epidemiological studies, cancer bioassays in animals, and short-term tests for mutagenecity, DNA damage, and cell transformation. Its authors paid particular attention to the issues involved in interpreting test results, thereby laying the basis for subsequent regulatory requirements.

The report concluded that, despite the many scientific complexities in carcinogenicity, a working definition of a "carcinogen" is an essential basis for protecting public health in California. Two potentially controversial assertions are that "a chemical that is carcinogenic in animals should be considered potentially carcinogenic to humans unless there is clear and sufficient scientific evidence to the contrary," and that "because of the risks involved, the long latency period, and the requirements to act in a health-protective manner, policy decisions to regulate exposure to carcinogens cannot require that the identification of risk be without necessity."[44] The Department's degree of regulation of chemicals identified as carcinogenic remained to be determined, based on the scientific results of tests and on the overall risk to the general population.

The DHS policy document represented a step toward shifting the burden of proof in environmental health regulation. While the study received a great deal of scientific scrutiny in its preparation—twenty-nine scientists outside of the state's Health Services Department receive acknowledgment in its preamble—industry groups immediately began to press for further scientific review of the report by an outside scientific advisory panel. Such demands were voiced in public workshops on this document, and on its October 1982 successor: *Carcinogen Identification Policy: Methods for Estimating Cancer Risks from Exposures to Carcinogens.*[45] This document describes various risk assessment models, and identifies a set of policy issues involved in estimating human risks from animal data and epidemiological studies.

Toward the end of 1982, DHS issued in draft its third document: *Carcinogen Policy: A Policy for Reducing the Risk of Cancer,*[46] and scheduled a public workshop for mid-February 1983. While this document would have no direct regulatory impact, it "will serve as the basis for future initiatives by DHS, and may to some extent guide the actions of other state agencies concerned with carcinogen control."[47] The overall goal was to reduce human exposure to carcinogens.

Debates over science policies also took place in the legislature. A bill proposed in 1982, S.B. 1777 (Paul Carpenter), would have required major alterations in the scientific basis of the state's regulatory decisions. The bill would have created a Hazardous Waste Science Advisory Board and restructured the process by which wastes are classified as hazardous or nonhazardous. This bill was strongly supported by industry, but was defeated after extensive and occasionally acrimonious debate.

The Department of Health Services did, however, agree with key legislators to review its California Assessment Manual (CAM)[48] with an eye toward necessary revisions. This document and associated regulations set forth the technical procedures to identify a waste as nonhazardous, hazardous, or extremely hazardous. For each of four categories—toxicity, flammability, reactivity, and corrosivity—the revised CAM will detail threshold criteria, definitions, and acceptable test procedures.[49] The basic CAM document has been under development by DHS for several years. Public workshops on revisions to CAM were held in November 1982.[50] According to a report in *Chemical and Engineering News*: "A number of spokesmen for businesses or business associations are critical of the draft regulations. Their comments focus mainly on the criteria for establishing toxicity."[51]

Control of Toxic Air Pollutants

At the same time, the Air Resources Board (ARB) proposed a series of procedures designed to control previously unregulated release of suspected toxic compounds into the air.[52] After a tempestuous public hearing in Sacramento on October 27 and 28, 1982, the ARB adopted procedures and criteria which qualify a suspected compound for regulation. These criteria closely resemble those set forth by the Department of Health Services in its cancer policy. The Air Resources Board amended existing air quality regulations to grant local air pollution control districts authority to enforce the new standards, and approved formation of an independent panel of scientists and medical experts to evaluate various chemical compounds to determine whether any are candidates for regulation as airborne toxics.[53] Implementation of this new ARB initiative remained uncertain as of late 1982, dependent on decisions by the autonomous local air pollution control districts, on actions in the legislature, and on decisions of the new Deukmejian administration.

Cancer Politics in California

Several overall political factors are noteworthy in California's science policy efforts. The scientific group in the Health Services Department chose an aggressive and innovative stance in this controversial area, far in advance of health agencies elsewhere in the nation. California has the size and the relative autonomy to envision implementing such a policy on its own. The state's cancer policy initiatives reflect a high level of scientific competence within the state's regulatory and public health agencies. In addition, senior state officials showed a willingness to place a priority on control of cancer in California, even in the face of substantial industrial pressure for "greater scientific validity." An environmental toxicologist for PPG Industries, for example, told the ARB in its October 1982 hearing that the proposed amendments to the Health and Safety Code were "based on bad science."[54]

Cancer politics has the unique feature of placing industrial advocates at a comparative disadvantage—no one wants to be identified as "pro-cancer." As a result, controversy frequently involves definitions and standards, and focuses on acceptable measurement protocols. Debates over the reliability of animal bioassays or the appropriate threshold limit value or the correct ambient air quality standard, for example, frequently become a surrogate for deeper controversy over political values or economic profitability.[55] In this sense, the disagreements over S.B. 1777, the cancer policy documents, and the ARB's toxics regulations may be simply the opening furor in a continuing struggle over the regulation of cancer-causing substances in California.

Reorganization for Toxic Substances Control

Initiatives from the governor's office to reorganize the government's regulatory apparatus fared less well than did the landfill phaseout concept. On this issue, Governor Brown and his chief toxics aides were forced to compromise with key figures in the legislature. However, some significant institutional improvements were made. In the wake of the October 1981 executive order phasing out the land disposal of certain categories of wastes, Brown's Special Assistant for Toxic Substances devised an innovative institutional proposal. A separate executive branch department would regulate toxic materials and implement the new treatment policy.

California has a highly fragmented overall environmental regulatory structure, with separate statewide boards for air quality regulation, water quality control, solid waste management, and coastal zone management. Air and water quality regulations are implemented by semiautonomous regional or county boards: The South Coast Air Quality Management District in the Los Angeles area, for example.

The California Hazardous Waste Control Act[56] gives the Department of Health Services institutional responsibility for hazardous waste management. The program first created by the legislature in 1972 called for DHS to establish and enforce regulations to ensure the safe handling, storage, use, processing, and disposal of hazardous wastes. The State Water Resources Control Board has important responsibilities to protect groundwater and surface waters under the Porter-Cologne Water Quality Control Act.[57]

As the state's principal public health organization, DHS responsibilities include a wide range of health functions: hospital oversight, maternal and child health, county health departments and programs, and Medi-Cal insurance, to name a few. This has posed problems of effective administration of hazardous waste management programs; administrative, personnel, computer, and other support functions have had to come from an overburdened DHS structure whose priorities lay elsewhere.

During the early 1970s, regulation of land disposal facilities by DHS, along with the Water Resources Control Board, was among the most advanced in the country.[58] California's abundant land, much of it in arid areas, allowed licensing of eleven Class I landfills in the state for disposal of hazardous wastes, all under strict regulations for soil permeability. The state also devised the country's first cradle-to-grave hazardous waste manifest system, which later served as a model under RCRA for mandatory use nationally.

The 1972 state act has been amended several times since, broadening DHS' regulatory authority, strengthening enforcement provisions, adding increased penalties for violations, and requiring DHS to encourage waste recycling and source reduction. With passage of the federal RCRA in 1976 and gradual expansion of California's waste management program, many aspects of the state's program became "substantially equivalent" to the federal requirements. Despite remaining problems of program implementation, California was authorized by EPA to operate on an interim basis in lieu of the federal program.[59]

In 1978 DHS established a Hazardous Materials Management Section to implement its program. This section in Sacramento, assisted by regional offices in Berkeley, Los Angeles, and Sacramento, issued operational permits for waste management facilities (Class I landfills and other surface impoundments), conducted periodic inspections, and enforced the law. Its FY 1980 budget was $5.6 million.[60] The DHS Hazardous Materials Laboratory and the Epidemiological Studies Section in Berkeley conducted scientific assessments of hazardous waste issues.

The department's overall hazardous waste management program is supported by a combination of disposal fees and waste-generation fees. Operators of disposal sites pay a fee of $4 per ton, set by statute.[61] The monies from this fund cover the costs of administration of the state's hazardous waste control program. However, revenues from these fees are expected to fall about 30 percent short in FY 1983 with an anticipated deficit of $1.2 million.[62] Pressures may therefore grow to increase these fees, perhaps in line with the potential hazards of disposing of different wastes. A second account, the Hazardous Substance Account, is maintained at $10 million annually.[63] Fees assessed under this account are levied by a formula which heavily taxes extremely hazardous wastes, along with lower rates for wastes going into evaporation ponds, for example. Revenues from this fund are used for a variety of functions: cleanup of old sites, restoration of natural resources (such as groundwater), and health effects studies, to name a few.

By the early 1980s, California's hazardous waste program had grown to significant dimensions. The scientific labs' capabilities, in particular, had advanced to a position of national prominence. Yet the Department of Health Services' components were still dwarfed in significance by the staffs of the Air Resources Board, the Water Resources Control Board, or the Energy Commission. The legislature explicitly recognized Health Services' dominant role in hazardous waste management. Total control over approving the hazardous waste elements of required county solid waste management plans, for example, was vested in Health Services rather than in the state Solid Waste Management Board. Yet the program had relatively little public visibility.

This began to change in 1980 and 1981, after Love Canal gained national prominence. Citizens for a Better Environment (CBE), concerned about inadequate implementation of hazardous waste programs by DHS, sued the agency in 1980 to gain access to previously unavailable data on the status of permit issuance. With these documents in hand, CBE published a report sharply critical of the DHS permit program.[64] The state's auditor general then pursued the matter further, issuing in October 1981 a report which argued that DHS as then organized had largely failed to regulate hazardous waste disposal adequately to protect the public health. Permits had not been issued in sufficient numbers; inspections of operating facilities had been inadequate; transportation of wastes presented unacceptable hazards.[65]

By late 1981, in response to such public criticism, the legislature and the governor's office agreed that major programmatic and institutional expansion were imperative for the state's hazardous waste regulatory apparatus. The governor proposed to create a separate executive branch department for toxic substances control, equal in bureaucratic status with the Air, Water, and Solid Waste Boards and with the Health Services Department itself. This was beyond the threshold of the legislature's political tolerance, however. The legislators were reluctant to enhance the power of a new competitor to the traditional environmental regulatory boards. As a result, the proposed reorganization failed to gain the necessary support in the Assembly and Senate.

Some changes had to be made, though less than the governor's proposal for a whole new agency. A compromise was reached in 1982. A new Division of Toxic Substances Control was created within Health Services, combining the Hazardous Waste Management Section (now a branch) with several scientific units. The lab's director was promoted to division chief, reporting directly to the Health Services director. While still a part of the public health agency, and thus using its personnel, contractual, and administrative procedures, hazardous waste management in California had a new and more visible institutional home. Presumably, this reorganization has enhanced the state's structural ability to implement its new hazardous waste policies through the actions of the new division. In late 1982, once the landfill phaseout regulations had been completed, OAT's Toxics Group was shifted into the DHS Toxics Division, reconstituted as a new program development branch.

Election of George Deukmejian as governor in November 1982 added an important new variable to the organizational equation. The state's Republican attorney general under Brown, with years of experience in the legislature, the new governor can be expected to espouse principles of efficient government. While California's overall structure for environmental management remained fragmented, the new division represented increased institutional coherence regarding the difficult topic of hazardous waste management. However, the type of leadership, resource commitments, and policy goals of the new administration remained to be seen as it assumed responsibility for implementing the new controls on land disposal of hazardous wastes, including the siting of new waste treatment facilities.

Siting, Local Authority, and the Controversy over State Preemption

As in every other state, some of the proposals to site new hazardous waste facilities in California have encountered severe opposition. Rejection of the facilities proposed at Sand Canyon and in Long Beach, as described above, raised concerns over the possible ability of other facilities to receive local approval. Nationally, over twenty states have responded to intense local opposition by opting for some form of state preemption: eliminating local control over siting decisions.[66] Should California, too, take this route?

In contrast to its striking innovations in hazardous waste policy, especially the phaseout of land disposal of highly toxic wastes, California has approached the issues of siting authority and preemption very cautiously. In this arena a rather different set of political forces and political actors have been at work. Here the scientists in hazardous materials labs and the lawyers with access to top state officials were eclipsed by state legislators, and more so by local elected officials who rebelled against any suggestion that they lose their authority over facility siting.

Preemption Versus Home Rule

While the state legislature did consider a proposed preemption bill in 1981, it decided to defer formal action to revise existing siting authority. Instead, the legislature created a Hazardous Waste Management Council to study the overall siting issue and report back with its recommendations for necessary changes—if any—in the balance of state and local authority. The Department of Health Services in conjunction with OAT's Toxics Group accelerated efforts to site new treatment facilities under existing siting mechanisms (that is, with local consent formally required, either from the city council in an incorporated area or from the county board of supervisors in an unincorporated area). The Southern California Hazardous Waste Management Project, a DHS-sponsored effort involving state and local officials, pursued planning and siting of new treatment and land disposal facilities in this seven-county region, working within the constraints of existing local and state permitting authority. (This effort is discussed in detail below.) The experience gained in Southern California added realism to the council's deliberations. Success in achieving local approval to site new facilities would lessen pressures to recommend preemption; failure would lead in the opposite direction.

In 1981, in response to the vehement local rejection of the new facilities proposed at Sand Canyon and Long Beach and to industry concerns that the state's hazardous waste management capacity would soon be inadequate, state Senators Montoya and Roberti introduced S.B. 1049,[67] a preemption statute drafted in large measure by experts from large waste-generating industries.[68]

S.B. 1049 called for a new state Hazardous Waste Facility Site Evaluation Council to assume the dominant role in siting previously played by local governments. This council would conduct public hearings on siting proposals, determine whether a proposed site was inconsistent with local or regional land-use plans or zoning ordinances, and if necessary preempt local land-use plans or zoning ordinances to allow the hazardous waste facility to be sited in the proposed location.

Support for S.B. 1049 was expressed by such powerful industry groups as the California Council for Economic and Environmental Balance (CCEEB) and the Western Oil and Gas Association (WOGA). Strong resistance to the bill immediately became apparent in the legislature, however. Counties and cities throughout the state—especially in Southern California where controversy over siting hazardous waste management facilities had been most evident— spoke out against preemption. The County Supervisors Association of California (CSAC) and the League of California Cities took the lead in opposing S.B. 1049. Their representatives argued that preemption would not be appropriate in California since good facility-development proposals could gain sufficient local acceptance to receive the necessary permits from local governments. To these groups, the Sand Canyon and Long Beach sites were poor ones which had deserved to be defeated in the local siting process.

Representatives from Southern California then proposed a compromise measure, S.B. 878, sponsored by Senator Diane Watson (D-Los Angeles). This bill avoided state preemption but also eliminated total local veto authority over proposed hazardous waste facilities. Under S.B. 878, municipalities (and counties in the unincorporated areas of the state) would have exercised initial decision-making authority over proposed new facilities, as at present. Local rejection of a facility, however, would be susceptible to an override by the regional Council of Governments (COG) for the area (the Southern California Association of Governments, SCAG, for example). This approach, too, failed to garner sufficient support.[69] Industry, local governments, and environmentalists were all troubled by this particular attempt at compromise over the complex siting issue. Resistance to any diminution of local siting authority was combined in this case with suspicion of the regional planning bodies. Few wanted to give the COGs power to site controversial waste management facilities over the explicit objections of local governments and a concerned citizenry.[70]

By late 1981, it was increasingly clear to political observers in Sacramento that any decisions to shift the basic authority over siting would have to be deferred. Consensus over some form of preemption or override simply did not yet exist. Assemblywoman Sally Tanner (D-El Monte), Chair of the Assembly's Committee on Consumer Protection and Toxic Materials, devised an effective compromise. Tanner's proposal was soon supported by the Brown administration. It called for formation of a new Hazardous Waste Management Council. A.B. 1543 gained extensive support, and easily passed both houses of the legislature. Governor Brown signed this bill into law on March 2, 1982.[71]

The Hazardous Waste Management Council

Essentially, A.B. 1543 gained time procedurally in order to allow a true substantive consensus to evolve over the changes needed in siting authority in California. The sixteen-member council is responsible for studying all aspects of the siting problem and devising a statewide plan to successfully locate needed hazardous waste facilities. A draft plan is due by April 1, 1983. Council members come from diverse backgrounds, thereby representing the several different constituencies whose interests are involved in hazardous waste management and facility siting: municipal and county governments, regional government associations, state legislators, environmentalists, state agency heads, hazardous waste-generating industries, and the waste management industry.[72] Sponsors of A.B. 1543 argued that a consensus achieved by this diverse group should stand a reasonable chance of subsequent adoption by the legislature. In essence, a representative council with a broad mandate to submit recommendations to the legislature was seen as the only politically and administratively feasible structure through which to address California's siting problem.

The council has to attempt to resolve the inherent conflict between the need for new hazardous waste management facilities and the public opposition to such facilities which emerges once they are proposed for specific locations. A.B. 1543's provisions draw on other states' approaches to hazardous waste management. The council's plan is mandated to incorporate:

1. Meaningful participation in the siting and permitting process by the public and local government units;
2. Interaction between the local community and the facility developer, including an opportunity for negotiation and arbitration for settling disputes;
3. A statewide and regional hazardous waste facilities needs assessment.[73]

The council is required to make recommendations for legislative, administrative, and economic mechanisms to facilitate siting new hazardous waste facilities. It is also to assess various techniques to compensate residents for injury and damages caused by the handling and disposal of hazardous wastes.[74]

Throughout its deliberations, the council is to ensure meaningful public participation. The underlying principle of A.B. 1543 was that acceptance by the public, especially by those living near an existing or proposed facility, is essential for any plan to gain political support and be successful.

By late 1982, the Hazardous Waste Management Council was engaged in monthly public meetings. Assisted by a small staff of its own and by its ability to draw on the considerable staff resources of the Toxic Substances Control Division within the Department of Health Services, the council's plan and its recommendations for modifications to existing siting authority were due to be prepared by April 1983. Consensus over the need for preemption remained illusory, however, as the various interests represented on this body continued to disagree (if gently in public) over such fundamental principles as the ability of today's joint state-local decision-making system to site new hazardous waste facilities. While A.B. 1543 had indeed shifted the arena for disagreement over preemption from the floor of the legislature to the periodic meetings of the new ad hoc council, this maneuver was proving no panacea. Knowledgeable observers believed that the council, in the end, would probably recommend at most some form of limited, tightly constrained new state authority to override local rejection of certain waste management facilities (landfills, perhaps). More likely was a fine tuning of the existing system. Few believed that the council would recommend full-fledged state preemptive authority. Council members in 1982 had been observing attempts to site new hazardous waste treatment facilities (described above), and to plan for more effective hazardous waste management in critical regions of the state. The success of these ongoing siting efforts will influence the degree to which the council recommends changes to the existing siting process.

State-Local Interaction in Siting

Regionalism is another unique aspect of California's approach to hazardous waste management. Reflecting this state's size and diversity, a special state-local intergovernmental effort was begun in the seven counties of Southern California in early 1981. The Southern California Hazardous Waste Management Project involves all the relevant interest groups—state and local, public and private—in this enormous region. The project's dynamics illustrate disagreement among these groups over basic principles and priorities in hazardous waste policy, especially over the relative roles and importance of land disposal versus treatment of wastes and over the need for long-range planning and site selection. As such, this effort is a microcosm of statewide hazardous waste politics in California.

In 1980 four of the five Class I hazardous waste land disposal facilities in Southern California closed unexpectedly within a few months of one another. As a consequence, a sense of urgency permeated discussions by local officials and waste management planners. The Palos Verdes landfill operated by the Los Angeles County Sanitation Districts in Los Angeles County (see figure 8.1) reached its engineered capacity. It was closed and capped, and monitoring wells were installed.

The 243-acre Calabasas landfill in the San Fernando Valley area of the same county, also operated by the sanitation districts, had opened in 1965. It received a wide range of waste types which were commingled with refuse or buried in the landfill. In 1980, based on detailed analysis of the site's geology, the districts concluded that the site's underlying geology did not meet state standards established in 1972, seven years after Calabasas had opened.[75] Nearby residents objected to continued use of this facility for hazardous waste disposal.[76] The

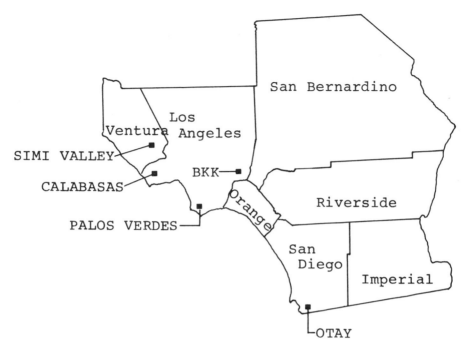

Figure 8.1. The Southern California region and Class I disposal sites, 1980

combination of geological assessment and local opposition led to a decision in 1980 to reclassify Calabasas to Class II (solid and municipal wastes only).

The Simi Valley landfill, operated by the Ventura County Regional Sanitation District on private land which they leased from the Union Oil Company, included 254 acres on which Class I hazardous waste disposal was allowed. Opened in 1971, by 1980 Simi Valley had received 6,500 tons of hazardous wastes, primarily pesticide containers, solvents, sewage sludge, cyanides, and refinery wastes. When the site operators reevaluated the area to see if it could meet new federal standards under RCRA, they concluded that geological conditions were inadequate to prevent possible groundwater contamination. As a result, the Class I portion of this site was closed in November 1980.[77]

Finally, the Otay landfill in San Diego County was closed by its operators in 1980 in the face of new EPA financial requirements. While the site was adequate technically, its small size—only about 5,000 tons/year—did not justify full financial coverage, and it was closed instead.[78]

This left only one landfill open for the off-site disposal of hazardous wastes in Southern California: the large facility owned and operated by the BKK Corporation in West Covina (also shown on figure 8.1). The BKK landfill itself seemed increasingly vulnerable to closure by a variety of political and environmental pressures. Nearby residents were vociferously demanding that it be closed. Odors and potentially dangerous air emissions (vinyl chloride, for example) could be detected in the vicinity.[79] The residents of West Covina organized politically. They attempted (unsuccessfully, but losing only narrowly) to recall the city council, and they placed on the November 1980 ballot a referendum item which, if passed, would have closed the landfill to new toxic materials. This measure also lost by a narrow margin, possibly because it was coupled on the West Covina ballot with a provision to raise the city's taxes by about $1 million annually to make up the deficit if the landfill were closed.[80]

Political officials throughout Southern California reacted sharply to a situation they found intolerable—their entire region having to rely on a single landfill for its off-site hazardous waste disposal, a landfill that itself was vulnerable to sudden closure. A series of meetings took place in late 1980 and early 1981, sparked by the personal efforts of Victorville City Council member Peggy Sartor, Los Angeles Mayor Tom Bradley, and a few other elected officials.[81] Several principles emerged from these sessions:

1. A cooperative state-local effort was necessary. Neither the counties and cities on the one hand, nor the state agencies on the other, could be successful on their own in coping with the siting and landfill crises.
2. A regional perspective encompassing all seven counties of Southern California was essential. Los Angeles County, in which approximately 75 percent of all the region's hazardous wastes are generated, could not manage these wastes entirely on its own; but neither could Los Angeles expect to ship all these wastes to somewhere else in the region.
3. While a growing proportion of the region's wastes would be amenable to treatment, significant capacity within the region for land disposal of toxic substances remained imperative. Reliance on a single, potentially vulnerable, disposal site was deemed unacceptable as a long-term public policy. As Mayor Bradley said in a November 1980 public hearing in Los Angeles: "It is certainly not in the best interest of Southern California to force industry to transport its wastes either 120 or 190 miles to the closest large Class I disposal sites if BKK landfill is closed."[82]
4. Clear criteria for selection of new sites had to be devised, and then applied throughout the region. For reasons of equity, the initial phase of the study was designed to identify some plausible locations in each of the seven counties, as a set from which the final selection could then be made.
5. A comprehensive approach to hazardous waste management was essential, encompassing air quality, surface and groundwater pollution, solid waste management, and public health.
6. While government would take the lead in this planning and siting effort, extensive participation by the concerned public, the waste generating industries, and the waste management industry was imperative to eventual success in siting. In the final analysis, the private sector would own and operate the new waste management facilities, and public consent in their siting would be required as part of the local permit review process.

By early 1981, these guiding principles provided the framework for organization of the Southern California Hazardous Waste Management Project, announced formally to the public in February 1981. The Department of Health Services assumed lead responsibility for this effort, consistent with its overall statewide role. The Water Resources Control Board also played an important role. Within the region, local governments were represented through a fourteen-member Policy Advisory Committee (PAC), established under the auspices of the Southern California Association of Governments (SCAG) and the San Diego Association of Governments (SANDAG). SCAG and SANDAG were responsible for involvement of local governments and for public participation in the project. The PAC included one elected county supervisor and one mayor or city council member from each of the region's seven counties. Overall project management was formalized in 1982 under a Program Review Board (PRB) chaired by the Director of Health Services and including the heads of the Air Resources Board, Water Resources Control Board, Waste Management Board, U.S. Environmental Protection Agency (Region IX), and the chair of the PAC (Chair of the Riverside County Board of Supervisors). These two groups met periodically to set policy and review progress in the effort. DHS appointed a project manager responsible to the PRB and the PAC.

Funds obtained from the U.S. Environmental Protection Agency, through grants to the state, and from the Los Angeles County Sanitation Districts were used to carry out specific project activities. Criteria for siting a new land disposal facility were developed.[83] Applied throughout the region through hydrogeologic and population analyses, these criteria were used to identify 448 potential locations, which were then further refined down to 28, 4 in each of the 7 counties. Public participation efforts were initiated, and a large Technical Advisory Committee composed of representatives from waste generating industries, waste management firms, and local governments' health and engineering departments was formed to review the project's technical conclusions.

By the end of 1981, two developments had profoundly altered this project's initial rationale. First, passage of S.B. 501, sponsored by Senator Boatwright, eliminated the BKK landfill's vulnerability to closure in the face of local political pressure.[84] This bill, designed explicitly to deal with the West Covina situation, removed from local governments their power unilaterally to close an operating hazardous waste landfill.

> . . . no local governing body . . . may enact, issue, enforce, suspend, revoke, or modify any ordinance, regulation, law, license, or permit relating to a facility that accepts both hazardous wastes and other solid wastes, such as municipal refuse, . . . and was operating as of January 1, 1981, . . . so as to prohibit or unreasonably regulate the operation of, or the disposal, treatment, or recovery of resources from solid wastes at any such facility.[85]

Whatever the extent of local political pressures, now the state had to concur before a licensed facility could be shut. In essence, the immediate hazardous waste crisis in Southern California was now ended. The search for a new landfill, while still important, could take place without an aura of impending catastrophe. Local officials, except in West Covina, began to pay perceptibly less attention to hazardous waste management, which no longer dominated their own agendas for action.

Second, the state in 1981 devised its new approach to hazardous waste treatment and the phaseout of land disposal of certain categories of wastes. The October 1981 executive order and the subsequent DHS regulations changed the focus of statewide policy. Landfills were no longer the sole answer, nor even the priority siting task. Energies were directed instead toward siting new treatment facilities needed to implement the intended landfill phaseout program.

These twin developments allowed some of the inherent inconsistencies between state and local priorities in hazardous waste management to begin to surface in the Southern California project. Increasingly it became apparent that local and state interests regarding treatment versus disposal were not identical, and might at times even be antithetical. These conflicting values surfaced in political discourse.

Local elected officials placed their priorities on availability of a network of adequate hazardous waste management facilities, in order to maintain economic vitality in the region and avoid illegal dumping. In Mayor Bradley's words:

> There is a need for sound planning to present the hard choices that must be made as we deal with the question of providing the facilities necessary to handle the enormous and growing volume of hazardous wastes. . . . Public fears of exposure to toxic materials have jeopardized the continuance of any hazardous waste disposal site. This is particularly true in Los Angeles County where there is a great public opposition to the continuance of BKK, the one landfill site now remaining in all of Southern California. I'm greatly concerned that the closure of the remaining hazardous waste disposal sites will endanger the economic viability of many local industries which provide thousands of jobs for our people. In addition, it is widely feared that certain waste generators will find it increasingly attractive to illegally dispose of waste materials in storm drains and sewers or on the land. . . . The most effective way to discourage illegal disposal is the provision of convenient hazardous waste disposal and processing sites.[86]

Sensitive to the pressures being placed on West Covina's elected officials by their constituents, and concerned about the apparent increase in illegal waste disposal in response to the sole landfill situation and the increasing costs of disposal at the BKK landfill, officials throughout Southern California felt compelled to continue the search within the region for a new supplement (or eventual replacement) for the BKK facility.

To the local officials, treatment of wastes was indeed desirable. They supported the emerging state policy. In Mayor Bradley's words again: "We must look for other ways of disposing these wastes. But I think for the moment, as a practical matter, we've got to preserve that particular option [landfills] until we find adequate alternatives to deal with these wastes in some other fashion."[87] Ending reliance on the West Covina landfill was imperative to the local officials, however. This was their dominant objective. The PAC voted unanimously in favor of an accelerated search to locate a supplemental landfill site. Similarly SCAG's Executive Committee pressed state officials to continue the process of developing new hazardous waste disposal capacity in Southern California. To them, this was at the center of prudent public policy formulation.

Influential figures in the state environmental agencies, however, increasingly adopted a different set of policy priorities in late 1981 and early 1982. While they recognized that treatment residuals would still have to be placed in the ground, their priorities rested with the siting of new treatment facilities, not new landfills. With treatment of 40–45 percent of the total waste stream foreseen in the new phaseout regulations, state officials believed that the remaining landfill capacity at BKK-West Covina would be adequate to meet the region's needs. In the meantime, the region's reliance on a sole landfill increased pressures to site new treatment facilities. To them, the more remote Kettleman and Casmalia landfills were adequate to supplement the facility in West Covina, and to keep its disposal fees within reasonable limits.[88] Should the West Covina landfill have to close in the future, these officials felt that the two outlying facilities could take up the slack until another site were located closer to the primary areas of waste generation.

These state-local policy differences were kept reasonably muted during 1982. With local acquiescence, state agencies refocused the Southern California Hazardous Waste Management Project on treatment facility requirements rather than on the search for a new regional landfill. Project participants prepared a comprehensive statement of regional goals and policies designed to guide the overall effort at hazardous waste management.[89] This innovative policy statement is shown as table 8.6. The statement was approved by SCAG and SANDAG and was formally adopted by several of the counties and cities in the region.

The project also stimulated preparation by the Department of Health Services of siting criteria for waste treatment technologies.[90] In contrast to the criteria earlier devised for new land disposal sites, these criteria pointed toward siting treatment facilities near the sources of waste generation where the new facilities would be compatible with existing industrial uses. The equity implications of requiring urban areas to accept waste treatment facilities, and remote areas to accept waste disposal facilities, enhanced the probability of successful siting for both.

Another important component of the project was a comprehensive regional analysis of waste generation, of existing and projected treatment and disposal capabilities, and of the resultant requirement for new hazardous waste treatment and land disposal facilities. An analysis of these complicated topics was completed in draft by late 1982.[91] This "Needs Assessment" became the focus

Table 8.6. Regional goals and policies for the safe management of hazardous wastes

Goals	Policies
Develop a comprehensive program for the safe management of all hazardous wastes.	Promote comprehensive planning. Provide adequate treatment and disposal capacity for all generators. Improve safety precautions and emergency response capabilities. Strengthen law enforcement.
Promote the use of proven and safe hazardous waste management technologies.	Encourage waste recycling, reduction, and treatment. Reduce dependence on land disposal.
Accelerate the siting and permitting of safe waste management facilities.	Develop appropriate siting criteria. Facilitate the local siting process. Consolidate the permitting process.
Involve the public in decisions on hazardous waste issues.	Educate the public on all issues concerning hazardous wastes. Consult the public on the siting of facilities. Involve the public in hazardous waste management planning.
Encourage cities and counties to ensure the safe management of hazardous wastes within their jurisdictions.	Promote local responsibility for locally generated hazardous wastes. Encourage local hazardous waste management planning and programs.
Encourage industry to construct and manage a system of high quality waste management facilities.	Promote cooperative planning by industry and local government. Encourage the equitable distribution of responsibilities and risks.

Source: Southern California Hazardous Waste Management Project, *Regional Goals and Policies for the Safe Management of Hazardous Wastes* (June 1982).

for the continuing debate over the need for new land disposal capacity, as the competing parties set forth their differing arguments. The project's revised scope also included an implementation strategy designed to identify the institutional, legal, financial, and political issues involved in siting new facilities in Southern California. This document concentrates on the state's regulatory responsibilities and on the counties' role in planning for hazardous waste management. Siting was viewed in the context of a continuing balance of local and state authority.

The project engaged in extensive public participation efforts designed to involve key groups in the siting process. Some fifty meetings were held with groups as diverse as the Los Angeles County Chamber of Commerce, the Sierra Club, the League of Women Voters, and the Imperial Valley Association of Governments. Five public workshops were held to reach others not associated with defined interest groups. Extensive contacts were made with the communications media to publicize the siting issue. All of this activity enhanced the public's understanding of the hazardous waste problem, although the intensity of public interest associated with an actual siting proposal in their own community could not be replicated in these informational sessions.[92]

By the close of 1982, the project was heading toward the long-deferred confrontation over the issue of seeking a site for a new landfill. Political pressures were growing in the complicated transition from the Democratic Brown to the Republican Deukmejian administration. The original project design called for field geologic reconnaissance of the twenty-eight potential locations (underway in 1982) to be followed by actual geological investigation—drilling bore holes, for example—at the one or two most likely sites. Local officials were pressing for this step to be taken, as part of their overall goal of supplementing existing reliance on a single off-site toxic waste facility. They believed that continued reliance on the BKK landfill alone, with its potential to charge high disposal fees, would lead to increased illegal disposal of hazardous wastes in Southern California: in the deserts, along the freeways, in municipal trash, and especially in the sewers.

State agencies in late 1982 were continuing to argue in favor of new treatment facilities; as noted in this chapter, they had achieved several siting successes. The economics of treatment, however, were intensely complex and the issues were highly political. For example, the BKK Corporation was both the principal corporate sponsor of new treatment facilities and the only operator of a Class I toxic waste landfill within Southern California. Any new competitor for the inexpensive land disposal of potentially treatable wastes might undercut the profitability calculations for BKK's Wilmington treatment center—the very facility needed to implement the state's landfill phaseout activity.

In summary, state environmental and public health agencies were increasingly locked into an unusual alliance with the corporate operator of the sole hazardous waste landfill in Southern California, an alliance which essentially required the absence of any new land disposal competition in order to construct the treatment capacity needed to implement the state's innovative program to curtail land disposal. This alliance of health regulators and corporate landfill operators was arrayed against a group of local elected officials pressing for additional land disposal capacity within the region.

The new Deukmejian administration will have to cope with these pressures and eventually sort out some kind of acceptable compromise. For example, it might be conceivable to identify a new landfill in the region, but link it explicitly to the state's treatment policies by restricting it to handling only the residuals from waste treatment. The new state regulations require that some 40 percent of the total waste stream be precluded from land disposal, undergoing treatment instead. Under this compromise, all wastes in Southern California would either undergo treatment or would be sent to an existing landfill (primarily to BKK-West Covina). This would avoid price competition between cheap land disposal of certain treatable wastes not covered by the state's regulations and the more expensive management of these wastes at the proposed new treatment centers. Such new residuals repositories might be somewhat easier to site (though by no means guaranteed) than would a traditional Class I facility, since they would be receiving only stabilized or dewatered wastes. And the amount of wastes headed for West Covina would be greatly reduced from present levels; most would be treated instead.

In sum, the political pressures in Southern California illustrate many of the dynamics typical of hazardous waste management and of siting hazardous waste facilities. A search for effective compromise was continuing in the face of apparently irreconcilable interests: state versus local, treatment versus land disposal, government versus the private sector, waste generators versus waste managers, competitive waste management firms versus one another, residents near hazardous waste facilities versus those living elsewhere. While the potential for politicization and conflict remains high, state and local agencies are searching for consensus through the Southern California Hazardous Waste Management Project.

Summary: Politics and Policy Development

Hazardous waste politics in California certainly illustrates the three themes identified in the introductory chapter to this volume: severity, complexity, and uncertainty. The tensions and agreements, and the successes and failures, mirror the national confusion over effective hazardous waste regulation.

California's ability to lead the country in new environmental policy directions—so evident previously in such diverse topics as coastal zone management, energy conservation, landfill standards, and the environmental review process—has been apparent in its new regulations to require hazardous waste treatment in place of cheaper but environmentally damaging land disposal. A national bill patterned on the California approach to hazardous waste management has been introduced in the United States Senate. This bill would require the Environmental Protection Agency to explore the advantages and costs of a nationwide ban on the land disposal of certain treatable wastes. In contrast, in the tortuous political arena of facility siting, one dominated by reliance on private sector initiatives, California has gingerly followed the initiatives of many other states. The themes of severity, complexity, and uncertainty help explain these apparent contradictions.

Severity

California has a severe hazardous waste problem. Massive volumes of hazardous waste are being generated in the state's enormous industrial complex (4,500 firms are "major generators of hazardous wastes");[93] huge amounts of pesticides are applied in its agricultural heartland; the state has its share of dangerous abandoned dumpsites—the legacy of past practices of unsound waste disposal; over 800 on-site and off-site facilities in the state handle over 1,000 kilograms of hazardous waste each;[94] and local residents express concern about environmental health and health dangers from existing licensed landfills.

In late 1980 and early 1981, Southern California suddenly became dependent on a single off-site disposal facility, one which was itself vulnerable to sudden closure. This led immediately to the perception that a hazardous waste crisis existed in the region. While the severity of the problem as measured in tons of waste or in incidents of illegal disposal may well have been somewhat less than in other industrial states, the waste disposal problem was perceived as acute by an aware public and by many articulate and powerful local elected officials. As a result, hazardous waste issues assumed a central position on Southern California's political agenda. Because of this region's size and importance, state-level institutions acted as well.

Officials responded aggressively to the threatening situation. They created (with state agencies) the Southern California Hazardous Waste Management Project. The state legislature passed a new bill precluding an affected local community—West Covina—from unilaterally closing an operating hazardous waste facility. The legislature considered a preemption bill to enforce the siting of new facilities in the face of public opposition, but then decided to defer decision on this matter until the Hazardous Waste Management Council had reviewed the situation. And the state adopted its new waste treatment policies. A perceived hazardous waste problem of unusual severity led to a governmental response unprecedented in its scope and rapidity.

By 1982, however, the crisis atmosphere had ended, especially in Southern California. Officials still expressed concern for effective hazardous waste management, but in far less frenetic terms. The stimulus for immediate response had lessened considerably, and gradualism had returned to the fore.

California's actions on the technical and scientific aspects of hazardous waste management cannot easily be explained in terms of severity of the problem, however. Other states seem to be deeper (especially in per capita terms) into the morass of abandoned dumps and dangerous disposal practices, yet they have not come anywhere near to California in adopting new technical approaches to the problem. Per capita budget allocations for hazardous waste management in California have been lower than in many other states. Instead, as suggested earlier, the principal explanation for this state's decision to focus on treatment technologies can be attributed to the leadership characteristics of the governor and his OAT staff, and to the impressive scientific professionalism within the Department of Health Services. (The quality of legislative professionalism in Sacramento contributed to support of the treatment policy with land disposal phaseout, though not to its initial stimulus.)

Complexity

Hazardous waste politics in California fall under this rubric as well. The state's environmental bureaucracy is highly fragmented. Statewide, regional, and local components each have degrees of autonomy. Institutional complexity was not of sufficient magnitude to forestall creation of the new policy of land disposal constraint, nor to block its initial implementation through promulgation of regulations and siting of new treatment facilities. However, serious problems of effective program implementation remain. Many of the needed regulations have not yet been promulgated. Permit issuance remains slow. Enforcement and inspection efforts are weak. In sum, much remains to be done—and the complexity of institutional arrangements renders implementation especially difficult.

Differences in viewpoint among the various state agencies generally were kept within acceptable limits as the state devised its new hazardous waste policies. An interagency task force, for example, helped respond to the October 1981 executive order. All the key state agencies cooperated in the Southern California project. Strong cooperation among the Department of Health Services, the Office of Appropriate Technology, the local air quality management districts, and the regional water quality control boards was crucial to the permit approvals granted to several new treatment facilities. While the Air Resources Board moved more rapidly than the DHS might have desired in proposing regulatory measures for toxic air emissions, the two state agencies agreed in the public hearings to work together to pursue the same general objectives.

State and local intergovernmental tensions are more evident, and have continued to hinder progress in devising a state siting policy. State-local disagreements were similarly evident in the Southern California project. Lobbying by the private corporations responsible for hazardous waste generation and waste management and for the siting of new waste management facilities adds an additional and dynamic dimension to the political scene, especially with respect to the battle over landfill siting in Southern California. Public interest groups do intermittent lobbying on toxic issues.

One other characteristic of institutional complexity deserves note in the California context: the state's relative autonomy from the federal system. While national laws such as RCRA obviously apply in California as they do elsewhere in the country, and while EPA funding has been important to the conduct of this state's toxics programs, California's size renders it independent from certain aspects of national policy development. As one official put it in an informal conversation with the author: "Washington, D.C., is three hours and 3,000 miles away."[95] How many other states, one wonders, could ban the land disposal of a large proportion of their hazardous wastes without worrying that waste generators would simply ship their loads out of state? And how many other states would even contemplate adopting a cancer policy on their own? The detailed and convoluted impacts of changing federal regulations so vital in so many states seem far less evident in California.

Uncertainty

Chapter 1 also points to the uncertainty of policy outcomes as a third basic characteristic of the hazardous waste issue. Its author questions the extent of society's commitment to effective hazardous waste management, and notes the significance of ambiguity regarding scientific and epidemiological information. Uncertainties about the adequacy of hazardous waste facilities to protect the public and the environment are also prominent in this analysis.

These topics are indeed important in California. At times, uncertainty has been addressed directly in the policy debate. Concern over the adequacy of "secure" landfills to protect the environment and public health lies at the heart of the phaseout regulations.[96] There is no guarantee either that the new treatment technologies will work as their proponents believe. And the manifold complications of scientific risk assessment have provided the central focus for discussion of the cancer policy documents, the revisions to the California Assessment Manual, the ARB's regulation of toxic air emissions, and the legislative debate over S.B. 1777.

Most significant, perhaps, is the uncertainty in late 1982 over the impact the new administration will have on hazardous waste management in general, and on the land disposal regulations in particular. Some observers, especially in industrial cities, have expressed a private conviction that these regulations will not be enforced fully once Jerry Brown and OAT have disappeared from the scene. Deadlines, they believe, will slip. The pressure will be off. As an article in *Hazardous Waste News* put it just after the election: "With the election of a Republican Administration in California, many political observers are predicting a deemphasis of the Golden State's role as a trend setter in the area of toxic and hazardous waste regulation."[97]

The land disposal regulations will have the force of regulatory law in January 1983, however, and strong support seems to exist in both houses of the Democratic-controlled legislature. Moreover, the governor-elect has been reported as favorably inclined toward effective regulation of hazardous wastes. The December 1982 issue of the *California Journal*, devoted entirely to interpreting the election results and predicting the Deukmejian governorship, summarized this issue as follows:

> The major environmental initiative pledged by Deukmejian [during the campaign] was to protect the public from hazardous waste disposal, a position no one seems likely to oppose. "I intend to make protection of the public and the environment from the improper storage, treatment, transportation and disposal of hazardous waste materials an important priority in my administration," he said. "I support development of state policies to reduce dependence on land disposal of hazardous wastes in favor of safer forms of disposal, such as recycling or treatment facilities."[98]

However, an editorial in the *Los Angeles Times* on December 1, 1982, while dealing explicitly with the new administration's approach to water policy, might have applied to toxic substances as well:

> There are enough uncertainties in the water picture to satisfy the most ardent mystery fan. One major unknown is the positions that Governor-elect George Deukmejian will bring to office. Philosophically, he is pro-develop-

ment, and his record on environmental matters as attorney general was undistinguished.[99]

Months will pass in 1983 before Californians and the nation will be able to see clearly where this state's innovative public policies toward hazardous waste management and environmental protection actually are headed.

The Politics of Hazardous Waste Policy

This analysis of California's approach to decisions on toxic substances and hazardous wastes has concentrated on the complex interaction of technological and political factors. Basically, politics has been dominant—in the personalities and institutions that have shaped the new policy on land disposal, and in the inability to reach consensus on appropriate intergovernmental roles in siting new hazardous waste facilities. The political pressures brought on state and local decision makers by corporate lobbyists have at times been overwhelming, and have never been absent. Yet scientific and technological expertise has played a vital role as well. Perhaps, on balance, this state's fundamental successes to date in approaching the difficult challenge of effective hazardous waste management can be attributed to this complex and dynamic blending of science and politics.

9. The Politics of Public Participation in Hazardous Waste Management

Walter A. Rosenbaum

Deep in the twilight of the Carter administration, even before Ronald Reagan unloosed his regulatory reformers upon federal waste programs, the glimmer of an imminent conflict over the future influence of citizens in the programs was already evident. One who saw the light early was James M. Scheuer, a New York Democratic member of the House Committee on Interstate and Foreign Commerce. "This is not a good environment for us to be asking for public participation funds," he explained in June 1980 to those hopeful that Congress might restore $647,000 excised from EPA's budget for citizen involvement in implementing the Toxic Substances Control Act (1976). "This is just an unhappy environment in which we are all functioning," he added.[1]

Things would get worse. Within a year, growing congressional coolness toward funding citizen activities in environmental programs—a public role often mandated by Congress itself—would be followed by the Reagan administration's pervasive efforts to rewrite and redirect citizen programs in ways that severely diminished the federal resources available for citizen involvement in hazardous waste and other environmental programs. Environmentalists called the Reagan plan a covert attack upon themselves and environmental regulations. In November 1981 the influential Conservation Foundation was warning its members that the new administration was using "sophisticated and numerous opportunities to constrict the interchange of information between citizens and their government" with "potentially serious effects on the political system as well as on environmental protection."[2] The Reagan administration justified its actions as a necessary reform to save money, reduce inefficiency, and streamline sluggish administrative procedures.

Stripped to essentials, the controversy was another skirmish in a decade-long running battle over the role of citizen groups in hazardous waste and other environmental programs. The new controversy illuminates enduring aspects of citizen involvement programs in hazardous waste management. First, these programs—like similar ones throughout the federal government—are constantly embattled and usually threatened by emasculation if not extinction. Despite periods of momentary vitality like the early Carter years, the programs—and, consequently, a major source of citizen influence in environmental policy—face a future as uncertain as the federal government's general response to the nation's hazardous waste ills. Second, the quality of citizen influence through such programs depends, in good measure, on administrative design. In the early

1980s, a major issue is whether the states possess the will and resources to fashion an administrative framework for vigorous citizen involvement when the federal government is rapidly retreating from its own commitments to a strong citizen role in hazardous waste management. Here is manifest one aspect of the complexity in the hazardous waste policy process. Citizen involvement programs are federalized and pluralized. Their character is shaped by the institutional and political interactions among federal, state, and local governments and by relations between these governments and the multiplicity of organized interests that are the constituency for public involvement programs. Finally, the severity of hazardous waste issues shapes the character of public involvement in several important respects: it has accelerated the formation of organized citizen groups seeking administrative access to hazardous waste policy making, intensified public opinion about hazardous waste, and complicated the search for effective management strategies for hazardous wastes.

These circumstances are presently affecting citizen participation in three major federal hazardous waste programs: the Toxic Substances Control Act of 1976 (TSCA), the Resource Conservation and Recovery Act of 1976 (RCRA), and the Comprehensive Environmental Response, Compensation, and Liability Act of 1980 ("Superfund"). It is useful to begin by describing the political factors exposing public participation to attack and revision, then to examine how the hazardous waste problem itself affects citizen activity. The impact of current administrative changes upon programs in Washington and the states will be explored. Finally, the significance of the presently volatile public mood about hazardous waste will be examined and some implications of these various conditions for the future will be suggested.

Participation's Precarious Prospects

During the 1970s, Congress and federal agencies wrote a host of potentially innovative public involvement programs into environmental law and then left them to a precarious future.[3] In most agencies, the programs have come to epitomize administrative marginality. Unpredictably or indifferently supported by Congress and their own administrative stewards, highly esteemed by a few bureaucratic managers, they are usually regarded as embellishments rather than essentials to substantive programs. The citizen programs have a constituency— primarily among environmental groups and others gaining advantage from the programs—but lack strong bureaucratic or congressional roots. "Public participation is always the first thing to go in a budget crunch," complained one state official to this author. "I've been working in citizen participation for years and you can almost predict that budgets and staff for c.p. will be bled early if something has to go." There are several reasons why such programs so often survive at the edge of extinction.

First, the programs arise from inconsistent and sometimes fleeting congressional motivations. It is often unclear whether Congress intended to invigorate or entomb citizen participation by voting it into law. Many congressmen, not excluding members of committees sponsoring statutes with citizen participation provisions, voted as a symbolic gesture—an apparently riskless affirmation of public rights to hold bureaucracy accountable—undertaken without much reflection about what they intended.[4] Often they cannot, or will not, say what they meant; the legislative history of public participation laws frequently betrays virtually nothing about congressional intent. For instance, in 1700 pages of legislative history concerning the Federal Water Pollution Control Act Amendments (1972), legislation embodying one of the broadest public involvement mandates in any federal law, one finds only two unedifying sentences concerning the objectives of that program.[5] Also, congressmen sympathetic to the environmental movement, or at least respectful of its growing political muscle in the early 1970s, often voted for citizen activities to assure environmental organizations a structural access to programs concerning them. (Environmental groups often deny that the programs were intended to create privileged access. Such groups and their ideological allies, nonetheless, dominate most of the programs. More candid environmentalists defend the program bias as a necessary counterweight to the influence enjoyed through other means by the regulated interests.) Congressional liberals, generally sympathetic to expanded public access in the administrative process, usually supported the programs on principle. And some congressmen voted for environmental laws oblivious to the citizen activities involved.

The consequences of such mixed motives were predictable. Early congressional supporters of citizen programs often wavered, or turned hostile, when citizen activists endangered or scuttled various beloved legislative projects—nuclear power plants, dams, flood control projects, and offshore oil exploration, for instance. Critics of environmentalists and environmental regulatory programs written in the early 1970s often vented their wrath on both the substantive programs and the procedural arrangements they perceived were providing a political buttress for the regulations. After the United States stumbled into the energy crisis of the mid-1970s, congressional apostles of expansive new energy production often viewed citizen involvement programs that delayed energy production or exploration—for instance, litigation against power plant siting initiated by public interest groups under expanded "standing to sue" in environmental laws—as endangering national security. The usually dependable support by the congressional liberal bloc for generous public participation in administration was substantially weakened between 1978 and 1980 by losses to Republicans riding the electoral backlash against Jimmy Carter into Congress. By the late 1970s, congressional tolerance for public involvement programs was languishing.[6] Seasoned bureaucrats soon learned the lesson. Rather than risk an uncertain fate for such programs during appropriations hearings, agencies determined to protect their citizen programs camouflaged them, if possible, in budget requests. EPA's Office of Hazardous Waste Management cloaked its citizen activities under the guise of "Information and Integration."[7]

A second factor leaving these programs continually vulnerable to change is the enormous discretionary authority given agencies in writing guidelines or regulations to implement vague legislative mandates for citizen activities, and the considerable budgetary powers wielded over programs by the Office of Management and Budget (OMB). In broad terms, citizen programs are well within the ambit of what Richard Nathan has called "the administrative presidency"—the domain of operationally effective presidential influence within the administrative branch created by the president's own administrative powers and that of administrative managers appointed by him.[8] The frequent obscurity of legislative intent and the absence of dependable legislative oversight in these citizen programs provide the opportunity for the intervention of the administrative presidency in the program management. Moreover, most citizen programs require state implementation; in effect, two orders of administrative discretion exist over program structures. Congress is usually silent about such federalized administration. How much responsibility should the federal government assume in financing the state participation programs? What staffing requirements, if any, should state agencies achieve for compliance with citizen activities? How much public involvement is enough? Funding levels also matter. Administrative guidelines are little more than paper promises without the budgetary resources to implement them. The president and Congress can powerfully shape the character of citizen programs using only the budgetary tools of requests and appropriations. This invests the OMB, the president's most important budgetary agency, with particularly great influence on citizen programs through its customary staff functions of recommending budget ceilings for agency programs. OMB preferences do not always govern presidential decisions, but OMB's own stance toward public participation can—and in recent years has—influenced strongly the quality of these programs.

While these circumstances usually prevail in almost any public involvement program—and particularly for those appearing in the 1970s—the character of public activities in hazardous waste management is also affected by qualities we have generally come to associate with the hazardous waste issue itself.

Hazardous Waste and Participation

The nation's hazardous waste problem is severe in several senses that matter for public involvement. Hazardous waste is a highly visible and strongly charged public issue. Unlike many other environmental problems, it can be made dramatically and emotionally relevant to a broad cross-section of communities where it appears. The discovery of an uncontrolled hazardous waste site, or the proposal to locate even a controlled site, is often enough to raise within a community the specter of public danger. This is the type of "safety crisis" which, William Lowrance notes, incites almost immediate and intense public involvement. "What is special about safety crises," he comments, "is that they arise with little warning; the implications for their well-being are taken personally and immediately by the affected public; the details of the technical assessments are often comprehensible only to a few specialists; and priorities are often difficult to establish, because perceptions of the problems differ so much."[9]

The Love Canal tragedy, and other chemical horror stories reported by the media, haunt public discussion of hazardous waste policy. Public reaction to such waste tends to be a strong initial fear of proximity to the materials and enormous community hostility to the prospect of a hazardous waste site remaining, or appearing, near the community. Regardless of the actual risks entailed in the transport or management of a hazardous substance, those convinced they will be affected are usually disinclined to believe reassurances of their safety, no matter the source. After studying twenty-one controversies over attempts to site hazardous wastes in a number of states, consultants to the EPA warned the agency in 1981 that it would have to adopt special new strategies in implementing RCRA to avoid intense citizen opposition:

> Should present approaches to siting facilities continue, the data of this project indicate that the prospects for successful sitings in most regions of the country are dubious at best, and grim at worst. National publicity concerning abandoned sites has made citizens and local officials increasingly aware of hazardous waste problems. They are also likely to be increasingly aware of actions taken by others to stop sitings. Opposition will, in all likelihood, become more widespread and sophisticated.[10]

Well-organized, intense, and widespread public involvement in hazardous waste management, particularly at the local level, can be expected to grow as the major federal programs, only now being implemented through the states, increasingly touch local communities.

This volatile combination of high issue visibility, strong citizen interest, and widespread fear of hazardous waste poses a difficult dilemma for the public agencies involved. Without public understanding and cooperation, successful efforts to site new hazardous waste dumps, to clean up existing ones, or to transport dangerous chemicals may be impossible. Thus, public involvement strategies seem almost essential to good program management. However, agencies seeking public involvement run the risk of encountering, or creating, an organized and strongly motivated public suspicious of agency recommendations about hazardous site locations and unsatisfied with agency assurances of the safety in recommended siting procedures. Public involvement programs, consequently, become a gamble against public prejudice—or, perhaps, an act of faith in the possibilities of public education—with considerable and costly risks to the agencies undertaking them.

Complexity, both technical and organizational, also affects public involvement in hazardous waste management. Public officials usually expect that citizen involvement activities will accomplish, if nothing else, a major part of the formidable task in educating the public about the implications of the technical issues inherent in hazardous waste management. Concerned citizens need enough information to make prudent judgments about policy options: data on the nature of toxics, the range of risks entailed in their use, methods of disposal or destruction, and criteria for selecting management practices. Given the inherent technical complexity of the hazardous substance problem, it seems almost essential that a public involvement program contain generous resources for production and dissemination of such information together with opportunities for organized citizen education. The organizational complexity of hazardous waste management is also evident in public involvement programs. As in the regulatory elements of TSCA, RCRA, and Superfund, the citizen components require collaboration among federal, regional, state, and local governmental agencies. EPA's national office, for instance, is expected to establish guidelines for public involvement programs mandated under the laws and then to assure that responsible state and local agencies comply with the guidance. EPA's regional offices act as mediators between Washington and the states in the process. Since guidelines are often late and tentative (EPA's guidance for the "community relations" segment of Superfund appeared more than two years after the law's enactment), the responsible agencies are often left to interpret the vagaries of the law for themselves. States disagree among themselves, and with EPA, about the appropriate criteria for program compliance. State and local governments often place citizen involvement in widely different priorities among the multitude of other statutory imperatives in TSCA, RCRA, and Superfund. And EPA itself, as will soon be evident, has drastically changed its participation priorities. Also, a vast array of organized groups, ranging across the entire federal domain from highly visible national organizations to a multiplicity of ad hoc local organizations, claim a right to be consulted and included in the design and implementation of citizen involvement programs. Moreover, regulated interests attempt to influence the administration of these programs in whatever manner is calculated to their advantage. In short, implementing public participation strategies is itself a policy process which embraces a great plurality of public and private interests, all actively seeking to shape administrative discretion in interpreting the laws to their different (and often conflicting) ends. Participation procedures are thus continually politicized and seldom durable.

Finally, uncertainty intensified by the Reagan reforms clouds the future of the public role in hazardous waste and other environmental programs. Unlike the Carter administration, its successor has substantively and symbolically relegated public participation to the margins of administrative attention. The Reagan agenda of "regulatory reform" contained an implicit agenda of participatory reform which soon materialized as a rapid, pervasive withdrawal of resources for the implementation of programs under TSCA, RCRA, Superfund, and most other environmental programs. Reagan spokesmen in the EPA and the Department of the Interior deny the existence of a participation "hit list" but many concerned observers think otherwise. Among the state administrators interviewed for this study—the state officials responsible for citizen activities under TSCA, RCRA, and Superfund in the thirteen states most seriously affected by hazardous waste—more than half suggested that the Reagan administration was reducing or eliminating public participation as an important component for future EPA and state programs. In any event, the burden of implementing the citizen activities required under these federal laws has now largely been displaced to the states together with the job of finding the required human and material resources for the programs. This responsibility occurs at a time of shrinking state revenues induced by prolonged recession, diminished federal aid, and other fiscal constraints that mitigate against program expansion. Further, many of the states most gravely afflicted by hazardous waste problems have only begun to build a citizen participation staff; reliance upon existing resources means the use of scarce resources. In short, the changes induced in Washington and the states—changes in citizen programs only modestly underwritten in the best of times—currently raise plausible doubts about the future vitality of governmental efforts to encourage a strong citizen role in waste management.

The impact of the present changes in the citizen programs and the implications for the future can best be appreciated by briefly describing the administrative development of these programs in the 1970s, with particular attention to the major impact, and often striking discontinuities, of program management during the Carter and Reagan years.

The Emergence of Public Participation

The 1970s was a decade of extraordinary innovation in arrangements for citizen involvement in the federal administrative process. During this decade, more than 60 percent of all current provisions for public participation in federal statutes were written into the law. These provisions frequently created statutory standards for public involvement, and statutory obligations upon agencies to encourage such involvement, far exceeding more traditional mandates.

An Innovative Decade

The new public involvement procedures were particularly innovative in the scope of the public to be reached by agencies, the variety of techniques which were permitted, and the range of administrative actions opened to public access —all substantially transcending provisions in the Administrative Procedures Act, the traditional standard for public participation in administrative process.[11] During this era were also written the "Government in the Sunshine Act," major amendments to the Freedom of Information Act, and other legislation intended to promote broad and sustained access to information about the conduct of administrative officials.[12]

In no domain of federal legislation were congressionally ordained public involvement programs more common than in environmental regulation. Beginning with the Clean Air Act (1970), Congress almost routinely added generous latitude for public participation in all subsequent major environmental enactments. This is apparent, in different ways, upon examination of the major hazardous waste programs created by Congress in the 1970s. The Toxic Substances Control Act requires, among other public provisions, that EPA issue public notice whenever it receives preliminary notice of a chemical manufacturer's intent to produce a potentially toxic substance; when chemical manufacturers submit to EPA test data on new production chemicals; when EPA makes a preliminary determination of safety for new chemicals, and much more.[13] The nation's major hazardous waste management legislation, the Resource Conservation and Recovery Act, contains a broad mandate for the EPA's promotion of citizen activism throughout its regulatory programs. In language almost identical to that of earlier water pollution legislation, the law placed responsibility for initiating and sustaining public involvement clearly on both the federal and state regulatory agencies involved:

> Public participation in the development, revision, implementation and enforcement of any regulation, guideline, information or program under this Act shall be provided for, encouraged and assisted by the Administrator and the States. The Administrator, in cooperation with the States, shall develop and publish minimum guidelines for public participation in such process.[14]

Finally, the first major federal legislation intended to clean up the nation's most dangerous abandoned waste sites, the Comprehensive Environmental Response, Compensation, and Liability Act (1980) encouraged federal and state implementing agencies to initiate "community relations programs" to assure that citizens affected by hazardous waste removal actions were informed and involved in the remedial work.[15]

Programs mandated are still not programs implemented. It remained for EPA regulations and guidelines—the flesh and blood on the statutory skeleton —to define the actual programs' character. Congressional funding, together with citizen responses, would also contribute to program definition. Nonetheless, the new legislation at least provided opportunities for innovation in citizen programs. The programs were judicially enforceable. And a vocal, politically influential constituency of organized citizen groups, particularly environmental organizations in the case of solid waste management, were determined to apply pressure on the EPA and the states to assure program implementation. The programs had begun with reasonably benign congressional regard, an aggressive constituency, and at least some agency tolerance.

A Brief and Fragile Vitality

The EPA and the states undertook their public involvement responsibilities dutifully, if not enthusiastically, under the Nixon and Ford administrations. Several experimental programs, most notably to create "workshops" across the nation to educate citizen leaders about the objectives of the Clean Air Act and the Clean Water Act, were undertaken from 1970 to 1974. The White House remained largely indifferent to the programs while the Office of Management and Budget, taking an adversarial stance that was to persist to the present, customarily treated almost all forms of new citizen involvement as marginally productive and unjustifiably expensive. In the Watergate years, however, citizen programs had strong congressional allies; environmentalists were still enjoying their newly acquired political muscle. EPA and the states were largely prepared to accord public involvement a place, though sometimes reluctantly, in environmental management.

The federal hazardous waste programs arrived almost simultaneously with Jimmy Carter's accession to the White House. Carter, the first president with opportunity to influence citizen activism in these programs, took a number of actions well into the final year of his term to demonstrate his administration's commitment to public involvement throughout the federal administration and particularly in the multitude of regulatory programs within the jurisdiction of EPA.[16]

Among Carter's general measures was an executive order in March 1978 directing administrative agencies to give the public "an early and meaningful opportunity to participate in the development of agency regulations"; more than thirty-four agencies eventually reviewed and, in some instances, revised their procedures for regulatory enactments to comply with the executive order. The president continually enlarged the influence of the White House Office of Consumer Affairs until it became the administration's most militant advocate of citizen activism within the executive branch. Equally significant were administrative arrangements created by EPA Administrator Douglas Costle. A Special Advisor to the Administrator for Public Participation appeared, creating for the first time in the agency a high-echelon overview of participation activities through an individual directly accessible to the administrator. The special assistant's high administrative visibility was important for the actual work accomplished and for the symbolic emphasis upon citizen activism as a salient issue within EPA. In 1979, the agency proposed and later enacted regulations for the first time establishing agencywide guidelines for citizen involvement in all program offices. Among the important provisions of the new regulations was a requirement that all EPA program offices allocate specific fiscal and personnel resources for public involvement activities whose objectives were to be consistent with the agency's general program goals.[17]

An important manifestation of this concern for citizen activism in the hazardous waste programs were budget allocations for citizen programs growing from $500,000 in FY 1978 to $700,000 in FY 1980 in the agency's Office of Pesticides and Toxic Substances.[18] Other evidence was the designation of a small staff within the office with primary responsibility for public involvement (the Office of Toxics Integration). The most ambitious public activity initiated by the office was a national "Waste Alert" program, conducted in 1978–79 by a broad coalition of environmental and public interest goups including the American Public Health Association, the League of Women Voters, the National Wildlife Federation, and the Izaak Walton League. The program, intended to involve citizens in the reform and implementation of RCRA, included regional workshops to which were invited leaders of national, state, and regional groups concerned with the law; the "workshops" were a prelude to a second and third wave of state and community workshops whose ultimate intent was to educate citizens about the nature, risks, and distribution of hazardous substances in their states. The "Waste Alert" continued a tradition, begun in the early 1970s, of "contracting out" citizen involvement programs to major national environmental and public interest groups—an indication of the dominance such groups enjoyed in the programs and the wide acceptance of public involvement within the agency.

By the middle of the Carter administration's final year these actions, however well intentioned, had still not opened most crucial regulatory procedures under TSCA, RCRA, or Superfund to public involvement. In fact, the Carter administration was retreating from its earlier zeal for participation and, in so doing, had begun the erosion of institutional support vastly accelerating under Carter's successor.

The Submergence of Participation

EPA's waning attention to public participation, particularly in waste management regulation, was less a conscious decision than a product of uncongenial, converging circumstances in the later Carter administration. Nonetheless, the effect was to weaken the administrative structures and to deplete the resources available for citizen activities in most environmental areas.

The End of the Carter Era

Citizen activity was constricted in waste management programs by the early 1980s largely because EPA had barely begun to implement these waste programs. In February and May 1980, for instance, the EPA finally issued regulations to implement Title C of RCRA, containing that legislation's important regulatory provisions for hazardous waste transportation and disposal. These new regulations established the standards by which state hazardous waste transportation and disposal programs would be authorized. Only after these regulations were issued, and state programs were approved, could the complicated job of permitting hazardous waste transportation and disposal be started in the states and citizen activism be initiated in the process. EPA's approach to TSCA was even more glacial.[19] By July 1980 EPA had started regulatory action to determine the safety of only five classes of chemicals among thousands within the scope of TSCA's requirements. Only in February 1980 did EPA make its first request to chemical manufacturers, as required under Section

8(a), for information about the health and environmental effects of 2,300 chemicals whose safety EPA eventually had to determine. While public notice, comment, and hearings were required during the EPA's determination of chemical safety, the procedures themselves and all the related activities to which public participation might be relevant were, in effect, still in limbo four years after TSCA's passage. And Superfund had yet to be enacted. The results of these program delays under the Carter administration was the confinement of citizen activism largely to the drafting of regulations for the programs' implementation and to a variety of workshops and other public information actions associated with the "Waste Alert."

Although citizen involvement was active, and sometimes influential, in this early EPA rulemaking, it gave few indications of the agency's future commitments to assisting those many other important aspects of citizen involvement that depended more upon the agency's initiative. (Agency rulemaking, largely governed by the Administrative Procedures Act, required public notice, comment, and hearing; the initiative in such instances must usually come from citizen groups.) Moreover, group activism at this early stage in program development was largely confined to representatives of the major national environmental and public interest groups, such as the National Resources Defense Council and the League of Women Voters; the breadth and depth of citizen activism at the state and local level remained unknown. Also, citizen activity during the Carter administration provided scant evidence for the strength of citizen involvement during the later, crucial implementation of programs. Even in these relatively routine forms of public involvement, however, the agency betrayed a conservatism toward innovation. While the agency had determined that it possessed the authority to fund public intervenors in rulemaking under TSCA—a rather experimental procedure also attempted in a few other federal agencies—it had spent only $1,500 to assist one public interest group in such procedures between 1976 and 1980.[20]

The most significant evidence of the Carter administration's weakening support for citizen activities in the hazardous waste programs was the OMB's deletion of almost $700,000 in FY 1982 for public involvement activities associated with the new Office of Pesticides and Toxic Substances—the largest identifiable public involvement funding in the agency's whole hazardous waste activities, and one intended largely to fund public information materials for the programs.[21] Behind this sudden evaporation of the office's participation budget was the lingering antipathy toward the newer citizen programs simmering in OMB throughout the Carter administration. Earlier, White House preferences had largely prevailed; OMB grudgingly permitted moderate funding for citizen activities. But in his final year, Carter and his advisors were absorbed by his reelection campaign, the Iranian crisis, the worsening national economy, and the rising tide of congressional support for some extensive budget cutting for FYs 1980 and 1981. The mercurial congressional mood about public involvement was increasingly critical. One portent of this shifting temper was evident as early as mid-1979 when Congress rejected modest appropriations to fund public intervenors in proceedings of the EPA, the Federal Energy Regulatory Commission, and the Nuclear Regulatory Commission.[22] Internal budgetary juggling enabled EPA to restore a substantial portion of the lost congressional funds but the warning was clear: officials expected, quite correctly, that future agency resources for the citizen programs in hazardous waste, and perhaps for all other environmental programs, would be problematical.

Thus, the Reagan administration with its ambitious agenda for reform in citizen programs came to Washington at a decisive and opportune moment. "Community relations" under Superfund were yet to be defined. Citizen involvement in state programs to be implemented under RCRA, including participation in the whole process of actually issuing permits for waste transportation and disposal, were yet to be defined and implemented. The long, complicated process required of EPA to identify, test, and control the manufacture of hazardous substances under TSCA had barely started and the public's future role in those procedures was undetermined. In short, the institutional and political structure defining the real opportunities for public participation in federal hazardous waste programs in the first half of the 1980s was— no matter what paper commitments existed in regulations or guidelines—still largely unsettled during the critical phase of program implementation.

The Reagan Onslaught

"We had a staff of 10 and a projected budget of $800,000," remarked one individual responsible for EPA's public activities under TSCA. "Then one day in January 1981 I got word that we were being discontinued. Everything. The citizen participation programs right now do not exist." Similar stories would be common throughout the EPA's numerous program offices during the first eighteen months of the new administration. With the precision and rapidity of a carefully premeditated strategy, the EPA's new administrator, Anne Gorsuch, initiated through her staff pervasive changes in the agency's basic structures for citizen involvement in its many environmental regulations. In hazardous waste, as other programs, the new strategy had two objectives: (1) to reduce severely the financial and personnel support available for public involvement activities at EPA's Washington and regional offices; and (2) to displace concurrently to the states an increasing responsibility for organizing and supporting citizen involvement activities in hazardous waste management. Environmental and public interest groups, traditionally the clientele of the endangered programs, protested that the actions were intended to terminate the programs. EPA's leadership and White House spokesmen defended the program as the application of the "new federalism" to citizen activities, an example of the Reagan administration's determination to restore greater authority and influence over regulatory programs to the states.

These changes raised fundamental questions about the future opportunity structure for citizen influence in hazardous waste policy through the mid-1980s:

(a) how much opportunity for citizen involvement remains in EPA's own offices? (b) how energetically would EPA encourage the states to assume a major role in encouraging citizen involvement in their own hazardous waste programs? and (c) would the states assume the considerable burden necessary to promote vigorous citizen involvement? The early impact of the Reagan reforms on EPA can be described from information obtained directly through EPA's program offices and from citizen groups familiar with the programs. An early insight into the consequences of these changes for the states can be gleaned from interviews with state officials responsible for public involvement in hazardous waste programs among states with the most serious waste problems.[23] Attention will focus principally upon state activities associated with RCRA, the federal waste program with the broadest and most discretionary participation mandate, and upon Superfund, which is intended to deal quickly and incisively with the nation's most dangerous abandoned waste sites.

Citizen Noninvolvement at EPA

In the early 1980s, the EPA appeared to be rapidly, if not precipitously, eliminating the human and material resources that in the past largely sustained its citizen activities. If the agency's leadership were not, in fact, hostile to citizen involvement in hazardous waste and other environmental programs, it was nonetheless dismantling the administrative apparatus upon which its commitment to such involvement was traditionally measured. It was an agency moving its program largely in reverse from the late 1970s. In the process, the modest public participation programs only recently created in hazardous waste management seem destined to virtual extinction.

Evidence of these changes is abundant. Among the most important at EPA's headquarters were:

1. The total elimination of the public involvement budget, and all the staff, committed to citizen activities in the program offices responsible for TSCA and RCRA.
2. The elimination of the Special Advisor to the Administrator for Public Participation, a position created under EPA's prior administrator, Douglas Costle.
3. The cancellation of authority for all program offices to award contracts for citizen "workshops" similar to TSCA's earlier "Waste Alert" program sponsored by environmental and public interest groups.
4. A severe reduction in the funding and personnel for public information materials produced in the agency's Office of Public Affairs—a major source of hazardous waste information for citizens in the past.
5. A provision in the proposed 1983 Reagan budget that would substantially reduce the fees paid by the federal government to attorneys for public interest groups suing the EPA for alleged failure to enforce environmental laws. These awards, presently permitted under TSCA and a multitude of

other environmental regulations, are a means of encouraging participation in the administrative process by groups otherwise unable to finance such litigation but representing potentially important viewpoints. Environmental groups have used these opportunities to considerable advantage in influencing EPA's prior hazardous waste policies.[24]

Officials responsible for directing the "Community Relations" program associated with Superfund have been extremely wary about encouraging almost any formal public involvement, such as public hearings, during determinations of remedial action for dangerous waste sites. The Director of EPA's Office of Solid Waste and Emergency Response, William N. Hedeman, Jr., reminded EPA's regional administrators in a mid-1982 memorandum that the community relations program was intended "to deal constructively with public response to Superfund actions—not to provoke undue concern through unnecessary hearings or inflammatory publications."[25] This aversion to public hearings is rooted in EPA's current conviction that hearings are likely to "politicize purely technical issues," incite inappropriate public fears, and obstruct proper solutions to the most acute waste site problems.[26] Thus, another official in the same office explained, EPA will continue to resist pressure from environmental groups to create public hearings as a routine part of Superfund actions:

> We have had pressure from the EDF [Environmental Defense Fund]. They want public hearings but EPA does not. Public hearings largely become media events. We prefer more informal events without so much publicity. EDF also wants more active citizen participation funded through EPA. Funding of any environmental groups for such activities, like workshops, is out.

These EPA actions leave any further responsibility for encouraging such citizen involvement largely to the states—a proper situation in the perspective of Reagan's "new federalism." The state response is particularly important for RCRA because, as a recent EPA consultant observed, the states "now play the pivotal governmental role and intend to continue" while EPA "intends for states to implement RCRA and has shown no desire to become directly involved in siting."[27] Stated differently, the state response to citizen involvement mandates may determine, in large part, the quality of citizen access to the strategic governmental processes for the future administration of RCRA's important Section C.

The State Response

Active citizen programs require an administrative investment of money and manpower. By late 1982, most states were in the early phases of administering the permit programs of waste sites under RCRA's Section C. A significant early measure of state concern for citizen participation, and a suggestive indicator of citizen access to these programs, is the trend in state funding and

Table 9.1. Reported changes in budget and personnel for public involvement in RCRA programs among states with the most serious hazardous waste problems, 1979–82

	Budgets	Personnel
Increasing	6 (46%)	9 (69%)
Decreasing	2 (15%)	1 (8%)
Unchanged	4 (31%)	3 (23%)
Don't know	1 (8%)	—
TOTAL	13 (100%)	13 (100%)

Source: Compiled by the author.

personnel for citizen activities. These early permits are especially important in gauging citizen influence because during the first five years most hazardous waste sites will be permitted and, consequently, most of the crucial policy decisions about site locations, operation conditions, and enforcement levels will be made. For this reason, the General Accounting Office has recommended that during this early phase the largest program resources for public participation should be committed.[28]

Measured by money and manpower, the thirteen states with the most serious hazardous waste problems show a very uneven, and often sluggish, response to their growing responsibilities for site permits and public involvement in the permitting process. This is evident in table 9.1, which indicates reported changes over the last three years—in effect, during the first years of most state programs—for public participation resources as reported by appropriate agency officials in the thirteen states. The data are, at best, ambiguous: while more than half the states had increased their administrative personnel involved in citizen activities, only six (46 percent) had also increased their public involvement budgets. More revealing are reported levels of activity (public hearings, workshops, public notice, and information) currently undertaken and anticipated. Table 9.2 summarizes the reported changes in state participation activities in 1981–82 and anticipated changes for the following year, 1983. The data in table 9.2 suggest what conversations with the reporting officials confirm: after an initial two- to three-year period in which citizen activities were provided a place in agency plans, the programs have stabilized and seem likely to expand in a relatively small proportion of the surveyed states (about 25 percent).

This early stabilizing in state participation programs occurs at a time when most state officials report a significant decrease in EPA support for the state activities. Among the state officials who expressed knowledge about EPA's current public involvement activities, almost 70 percent (nine of thirteen) suggested that in various ways EPA was withdrawing its support for citizen involvement under RCRA. While it is premature to predict how well current state programs will encourage citizen involvement in RCRA's permit procedures, it is apparent that the state programs have largely stabilized while citizen organizations con-

Table 9.2. Reported changes in public involvement activities for RCRA programs among states with the most serious hazardous waste problems, 1982

Activities compared to those of 1981:		
Less active	1	(8%)
Same as last year	5	(39%)
More active	6	(46%)
Same, very active	1	(8%)
Total:	13	(101%)
Changes in activities anticipated for 1983:		
Not much change; no change	7	(44%)
New staff	3	(19%)
More activity	4	(25%)
More state support	1	(6%)
Less activity likely	1	(6%)
Total :	16	(100%)

Note: Multiple responses included.
Source: Compiled by the author.

cerned with waste permits and related issues are explosively growing at the state and local level. Many of the thirteen states have apparently not anticipated the actual burden that will fall upon them when responding to these proliferating citizen groups. And many of the state programs still emphasize a relatively passive citizen role: programs often focus upon informing the public and allowing it some voice in technical planning but restraining it from influence in the full range of technical and policy issues to be resolved in hazardous waste site permits.

The Explosion of Citizen Activism

An almost universal aspect of reports from all state agencies surveyed was mention of the rapid growth within the last year of citizen groups concerned with hazardous waste management at local and state levels. The growth rates among such groups vary from state to state but remain high in most states. Pennsylvania is among the leaders. An illustration of the growth and impact of these groups on RCRA is found in data assembled by Pennsylvania's Department of Environmental Resources. During the period 1981 to 1982, the department's Bureau of Solid Waste Management received applications for hazardous waste site permits under RCRA; about 75 percent of these applications involved groups protesting some element of the permit application or the administrative determination made on the application. As table 9.3 illustrates, these groups represented not only citizens but often a coalition of citizens and local officials. Does this citizen activism have an impact upon the permit process? Indirect evidence suggests the impact may be very substantial. Among the 94 permit applications associated with group protest in Pennsylvania, about

Table 9.3. Composition of groups protesting hazardous waste permits considered under RCRA provisions in Pennsylvania, 1981–82

Public agencies	3	(3%)
Citizens	56	(60%)
Citizens and municipal officials	28	(30%)
Municipal officials	4	(4%)
Not ascertained	3	(3%)
Total:	94	(100%)

Source: Compiled by the author.

46 percent (forty-two) were rejected, withdrawn, appealed, or otherwise delayed. Although citizen activism was not solely responsible for all these permit delays and rejections—in many cases the responsible state officials were convinced before the public activity that applications had serious flaws—the public activism was clearly influential, if not decisive, in a great number of permit determinations.

Concurrently with this multiplication of ad hoc local groups has been a growth of statewide organizations concerned with hazardous waste issues or, in some instances, the involvement in hazardous waste problems by existing state organizations, such as the Audubon Society or Common Cause, active in other public policy matters. Eleven of the thirteen states surveyed reported an increase in state group activism. This growing activism among state groups is encouraged by a concurrent increase in the number of environmental groups now attempting to increase public awareness of hazardous wastes. Emerging gradually is a network of organizations, reaching deeply and broadly across the federal structure, with a mounting capability to disseminate information and to mobilize large numbers of citizens in the waste policy process.

One influential national organization committed entirely to citizen activism in hazardous waste policy is the Citizens Clearing House for Hazardous Wastes, started in 1981 and, significantly, a direct result of the Love Canal controversy. The Clearing House, formed by Lois Gibbs as a result of her experiences as a Love Canal housewife during the New York incident, has several purposes: to provide general information about hazardous waste to interested citizens, to inform existing citizen groups about opportunities to influence hazardous waste policy, and to offer expert assistance for groups concerned with highly technical waste issues. Although the primary purpose of the Clearing House is to obtain and distribute information about hazardous waste rather than to organize group action, the Clearing House's activity undoubtedly becomes a catalyst to the formation of citizen groups as well as a cue for their program and tactics. Another organization, formed for a somewhat different purpose but having much the same effect on hazardous waste politics, is the Citizen Participation

Fund recently started by the Lincoln Filene Center at Tufts University. The fund—which introduced its first public solicitation letter with a reference to the boiling emotions generated among local citizens by the Love Canal incident—is attempting to underwrite a national campaign to encourage greater citizen activism among all major policy areas at all governmental levels. A very high priority for the fund is a program to offset the Reagan administration's drastically diminished assistance to citizen activities in the national waste management process.

Other groups, long active in environmental affairs, are developing specialized staff, publications, and activities to emphasize hazardous waste issues. The Environmental Action Foundation, for example, has recently published *Exposure*, a biweekly national newspaper focused exclusively upon hazardous waste topics. Most of the nationally active environmental groups, such as the Conservation Foundation, Sierra Club, and National Wildlife Federation, now have staff with specific responsibilities in hazardous waste policy. The League of Women Voters has recently initiated a national campaign of hazardous waste education for its members and, through them, for the local communities where they live. This gradual diffusion of hazardous waste activism within the group structure of the environmental and public interest movement is likely to continue. This trend, together with growth of other organizations concerned with hazardous waste at all governmental levels, represents an ongoing institutionalization of the hazardous waste issue likely to provide a firm base for future citizen activism in hazardous waste policy procedures.

Conclusion: The Dangerous Prospects of Confrontational Politics

Evident even at this early phase in the development of national hazardous waste policies are several trends in public participation which have important implications.

First, state governments are becoming an increasingly significant arena—perhaps *the* most important arena—for public activism in hazardous waste management. The growing salience of the state setting arises partially from the determined efforts of the Reagan administration to invest state governments with increasing authority and discretion in implementing federal, as well as state, regulatory policies for hazardous waste. Also, there is now a "new federalism" in public participation policy: the states are largely left to be the trustees for public participation commitments made in federal hazardous waste laws by Washington's de facto abdication of this responsibility. Thus, converging upon the states simultaneously are new authority and new opportunities for public involvement in hazardous waste policy. In the past, national environmental groups, like most other interest groups concerned with federal environmental policy, confined most of their attention and resources within the federal government. Now, as regulatory authority flows downward and states become strategic settings for crucial policy decisions concerning hazardous waste, these national organizations are becoming aware of the need to bring sufficient influence on the states to produce effective public access—which means effective environmental group access—to the new policy processes. In light of increased public activism already evident at state and local levels, one can expect growing pressure upon the states for citizen involvement in these policy areas.

All this would not seem particularly ominous if the states appeared alert to the increasing importance of the citizen activism and prepared to create the resources necessary to accommodate this rising citizen concern for hazardous waste issues. However, a second important trend is what one does *not* generally find among the thirteen states surveyed in this study. While three or four states are increasing budgets, personnel, and activities committed to public activism in the near future, most of the surveyed states are not anticipating program expansion, are not investing heavily in public participation activities generally, are apparently not giving public activism a high priority with the hazardous waste programs under RCRA or Superfund. Further, there are many indirect suggestions that the states, with a few exceptions, have a strong inclination to keep public activism quarantined from the more substantive, substantial elements of hazardous waste policy decisions. Some states appear to believe that "public participation" consists largely of informing the public about hazardous waste issues rather than soliciting public opinions and responding to them through policy formulation. It was this that EPA's consultants had in mind when they noted that among the states they studied—including many in this survey—the tendency was to keep the public from providing "substantive input to technical and non-technical aspects of governmental decision-making" on hazardous waste.[29] Thus, many states seem headed for an impending, and probably acrimonious confrontation with citizen groups not only about the substance of hazardous waste policy but also about the role of citizens in formulating that policy.

Such confrontation sustained across many states can present state officials with a difficult and possibly dangerous situation they may not deliberately will and may not even anticipate. State agencies now assuming responsibility under federal and state hazardous waste laws often have scant experience with hazardous waste management or public involvement programs. There is, as yet, little institutional learning about the impact of citizen concern with hazardous substances and the procedures for coping with citizen involvement. Further, state agencies recognize that they cannot permit public opinion or citizen group activism to paralyze essential action to protect public health or safety from hazardous substances. Agencies will have an almost inevitable and legitimate concern that public participation not degenerate into agency immobility in the face of an aroused, and possibly ignorant, public which demands governmental action inconsistent with sound technical handling of hazardous waste issues. Thus, agencies may instinctively seek to constrain public influence in the policy process, not from callousness toward citizen interests but from a professional sense of responsibility for protecting the public interest. In this manner, the hazardous waste issue forces upon program managers an extraordinarily difficult and perhaps unsolvable task in balancing the need for responsiveness to citizen concern with the capacity to act effectively in environmental management. Especially if public officials believe that "public involvement" means arousing public fears and irrational public responses to waste issues, they will have a natural inclination to shun citizen activism whenever possible.

Clearly, no solution to these problems of citizen participation is easy, or predictable, or universally valid in all states. However, the most prudent course in dealing with the new citizen activism in waste policy would seem to lie in avoiding the highly disruptive confrontations that might occur if agencies fail to make a genuine effort to include citizens in a broad and meaningful range of policy decisions. The failure of citizen involvement in agency programs is likely to mean, at the very least, more litigation as groups seek to use the courts to obtain the kind of recognition and influence which eludes them through participation procedures. Regardless of the final outcome, litigation commonly means protracted, expensive, and often cumbersome resolution of policy issues that might have been more satisfactorily settled in all respects through citizen involvement procedures. Further, confrontations over citizen participation breed the antagonisms between citizens and agencies that inhibit agency abilities to educate citizens adequately so that they can aid in implementing regulatory policies. The alternative—creating many different opportunities for citizen involvement at crucial phases in hazardous waste policy formulation—has no attached guarantees. The risks remain real that public involvement may, despite the best agency intentions and soundly conceived programs, still work against the agency or its programs. But opportunities for better outcomes are, at least, real and attractive.

Perhaps the one certainty is that the need to accomplish meaningful citizen involvement in the policy process while simultaneously dealing adequately with environmental degradation is more imperative in hazardous waste than in any other major environmental area. The ability to do one task almost presupposes the ability to accomplish the other. And the need to accomplish both swiftly is compelling.

10. State Roles In Siting Hazardous Waste Disposal Facilities: From State Preemption to Local Veto

Susan G. Hadden, Joan Veillette,
and Thomas Brandt

The Siting Problem

Selection of a site for a hazardous waste disposal facility (hereafter called a HWDF) is the first and perhaps the most intractable problem in the always difficult process of hazardous waste disposal.[1] Siting engenders public opposition because it concentrates costs in the area where the facility is located, although benefits are dispersed among all users of the products generating the wastes; because it often pits state against local governments; and because it allows for such a wide variation in the nature of the relationship between public and private entities. Siting of other kinds of facilities has always been a political issue, since the very act of choosing a specific site imposes differential costs and benefits on citizens. Cities compete for major manufacturing plants as sources of employment and revenue which are, of course, denied to cities not selected by industries. Before RCRA, when HWDFs did not have to meet stringent requirements, they were usually sited near waste generators and were largely ignored by local and state authorities. However, with the passage of RCRA, together with the extensive media attention on Love Canal, a less favorable image has been created in the public mind; the result has been a reverse Chamber of Commerce response, often referred to as "the NIMBY syndrome" (or, Not In My Back Yard). Indeed, a 1980 public opinion survey conducted for the Council on Environmental Quality by Resources for the Future reported that "a majority of the respondents endorsed new, secure, regularly inspected waste disposal facilities, but only if located over 100 miles from their homes."[2]

The major reason for concern about the proximity of HWDFs is possible risks to health and safety. Yet similar risks are often accepted in other circumstances. HWDFs are seen to provide benefits to all users of products which generate hazardous wastes while imposing their costs on only a limited group in their immediate vicinities. Furthermore, HWDFs are unlike many other noxious neighbors in that they provide few if any jobs for the local community[3] and may not provide as much property tax revenue as alternate industrial uses. Thus siting of HWDFs redistributes benefits away from individuals in host communities and especially from those nearest the facility site.

Siting is thus an important subclass of those issues that are properly public because the market cannot effectively treat them. HWDFs have significant externalities: in addition to the obvious risks to health, they may reduce property values, generate malodorous fumes, and attract industries that generate wastes. In addition, the market provides little incentive for proper operation of a waste disposal facility; ill-effects on health are seldom attributable directly to the facility and in most cases do not occur for twenty or more years. Finally, the imbalance of costs and benefits suggests the need for an intermediary.

Siting also raises the issue of the proper and most effective relationship between public and private sectors. Selection of sites and the operation of facilities by the private sector have resulted in abuses in the past. The response to these abuses was the passage of RCRA and TSCA and the promulgation of regulations. Ensuring compliance with the regulations is costly for both sides and requires continuing intervention. At the same time, public opposition has delayed location of new facilities, creating an equal or worse risk to health and increasing costs to the private sector. States can assume roles ranging from passive observers of private actions to competitors of the private sector that own and operate public HWDFs.

In addition to creating conflicts about these issues that are essentially ideological and therefore irresolvable, decisions to site HWDFs are politically difficult because of the continuing technical uncertainties about the long-run effects and effectiveness of disposal methods and sites. Decisions that have both highly technical and difficult political dimensions have often been delegated to regulatory authorities by elected officials anxious to be relieved of them.[4] With the passage of RCRA and the promulgation of regulations governing the technical qualifications of sites, therefore, it was natural that siting decisions be placed in the hands of regulatory agencies. As public awareness of the implications of HWDF siting grew, however, affected citizens demanded more active roles in siting decisions for themselves and for local elected officials who could be held more directly accountable.

These demands did not lessen the intensity or basic irresolvability of the conflicts; if anything, they expanded their scope by pitting local governments against state entities responsible for ensuring the availability of disposal sites. Most states whose citizens demanded special HWDF siting laws created complex procedures incorporating a multiplicity of checks and balances. Their responses offer a classic example of Lowi's interest group liberalism:

> Traditional, progressivistic expansions of representation are predicated on the assumption that law is authoritative and that therefore one must seek to expand participation in the making of laws. The "new representation" extends the principle of representation over into administration, since it is predicated on the assumption that lawmaking bodies and conventional procedures cannot and ought not make law.[5] This may be the most debilitative of all features of interest-group liberalism. . . . It impairs the self-correctiveness of positive law by the very flexibility of its broad policies and by the bargaining,

co-optation, and incrementalism of its implementing processes. . . . Interest-group liberalism seeks pluralistic government, in which there is no formal specification of means or of ends. In pluralistic government there is therefore no substance. . . . There is only process.[6]

The emphasis on process helps to explain both the diversity and the complexity of the siting laws of the different states. In this chapter we attempt to order this complexity by considering state laws as results of choices on three policy dimensions: state role, local power, and public participation. The final portion of the chapter considers the state laws as responses to the conflicts engendered by siting.[7]

Policy Options in Siting Laws: Complexity and Uncertainty

Under the Resource Conservation and Recovery Act of 1976, responsibility for siting HWDFs belongs to the states. EPA has promulgated rules concerning permitting of sites to ensure the protection of public health and the environment; similar requirements are of course embodied in those state programs authorized by EPA. Siting itself, however, has been left entirely to the states. One important reason is that the states can more easily tailor programs to local needs. Another, equally important, is that RCRA did not empower EPA either to build and operate disposal facilities or to acquire land to lease to private firms. States have the power of eminent domain and other land-use powers that are crucial to the operation of a siting program.[8]

Since compliance with RCRA did not force states to develop consistent siting programs, there is wide variation in the current policies. As we have noted, most states implicitly or explicitly delegated this authority to the executive agencies responsible for overseeing other aspects of RCRA. Some states, however, came under pressure to pass special legislation that would define the siting process more explicitly. As of July 1982, twenty-two states had passed siting laws; they are Arizona, Connecticut, Florida, Georgia, Indiana, Kansas, Kentucky, Maryland, Massachusetts, Michigan, Minnesota, Nebraska, New Jersey, New York, North Carolina, Ohio, Oregon, Pennsylvania, Tennessee, Utah, Washington, and Wisconsin.[9]

Although the common purpose of the laws is to increase the likelihood that HWDFs can be sited, each state has adopted a slightly different way of doing this. The tendency of states to write laws that differ in form is exacerbated by the complexity of the siting process, which entails developing a statewide hazardous waste plan, identification of potential sites, selection and permitting of sites, operation of the facility, and closure. At each of these stages, states have many options. Table 10.1 presents a "Chinese menu" of policies out of which states may select one or more from each group, representing a stage of the policy process. If there are only five policy options for each of five stages of the siting process, state laws could take literally millions of forms.

Table 10.1. Policy options in siting laws

Planning	Selecting sites

–By siting board
    advisory or formal
    members represent specific groups
 	
–By departments
 	
Permitting sites
 	
–Permits issued by board
–Permits issued by department
    (different departments for different
    kinds of sites)
–Permits reviewed by board
–Permits reviewed by local officials
–Permits may be vetoed by local officials
–State legislature issues permits
 	
Assurances for localities
 	
–Contingency plans required
–Post-closure site restoration
–Financial assurances
–Strong state inspection
–Incentives
    proportion of facility use
    direct payment from state
–Arbitration and negotiation prior to
siting
–Continuing local involvement
–Insurance of property values
	–Board selects
	–Developers select
	–State narrows field but developers
	  select
	–Technical criteria are selected
	–Social, economic, land-use criteria
	  established
	
	Operating sites
	
	–State condemns land, buys land
	–State operates site
	–State contracts for site operator
	–Private operation of sites
	–Inspection (frequency, stringency)
	–Operators licensed or certified
	–Operators not licensed or certified

Source: Compiled by the authors.

The complexity of individual state laws resulting from this programmatic abundance is exemplified by figure 10.1, which presents a flow chart of only part of Wisconsin's siting law, passed in 1982. Some analytical parsimony may be obtained despite this complexity by considering that all options are exercised along three dimensions of policy: state initiative, local power, and public participation. We consider each in turn.

State Initiative

Perhaps the most important series of choices a state can make determine how strong and direct a role it will play in siting of HWDFs. One indication that a state is determined to seize control of a process that has often been very fragmented is the creation of a new siting board. Administrative fragmentation is the rule rather than the exception in state solid waste management programs. As noted in chapter 7, for example, Texas has four agencies which are responsible for parts of the hazardous waste programs alone, not including the Transportation Department which oversees wastes when they are being moved from place to place. The fragmentation is due, in part, to the fact that state programs

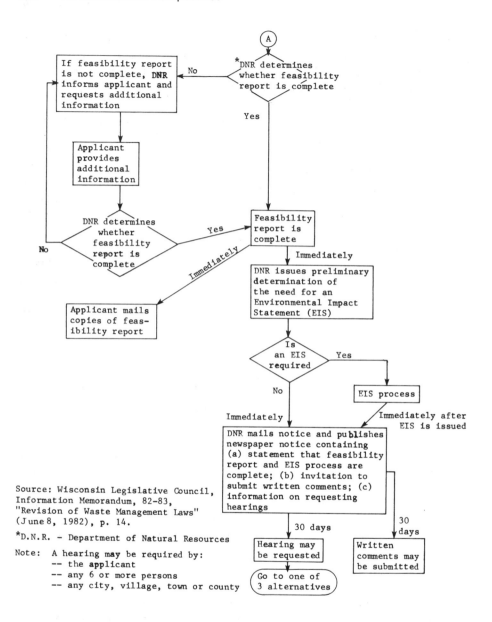

Source: Wisconsin Legislative Council,
Information Memorandum, 82-83,
"Revision of Waste Management Laws"
(June 8, 1982), p. 14.

*D.N.R. - Department of Natural Resources

Note: A hearing may be required by:
        -- the applicant
        -- any 6 or more persons
        -- any city, village, town or county

Figure 10.1. Flow diagram of Wisconsin's siting process

are the products of diverse laws that have evolved at different levels of government over a period of some twenty years. It may also reflect reluctance for any one agency to have to resolve the deep conflicts inherent in siting; fragmentation is one kind of "process" response of interest group liberalism.

Since administrative consolidation is so often politically difficult, many states have obtained coordination among agencies through memoranda of understanding. Thus, implementation has tended to remain as it was before. The new siting boards, on the other hand, have been given policy-making powers or advisory powers that will allow states to develop more coherent overall policies and to ensure that siting decisions are equitable. Whether they contain public members or consist entirely of agency representatives, siting boards and authorities are at least a signal that a state intends to "do something" about siting.

Another way in which states can take a strong role is by assuming the power to initiate a siting decision. In many states, permitting agencies must take a passive role, waiting for private operators to investigate sites and apply for a permit for them. EPA has distinguished, within these passive roles, between states that use data provided by private industry and states that use site data generated by themselves or by independent third parties.[10]

If a state must depend on potential operators both to find sites and to determine whether they meet criteria, the state role is limited to approving or disapproving essentially random sites. Even in this case, public hearings mandated by RCRA do provide an opportunity to challenge the validity of the operator's data. However, the state can help point potential operators in the desired directions if it can compile a list of appropriate sites or can generate data about the sites.

Twelve states have passed laws that place the initiative for siting in the hands of state agencies. The power to initiate the siting process, which involves direct selection of desirable site locations, can provide several advantages to the state. First, public control over the location of new facilities can allow the incorporation of social, political, and economic factors as well as technical ones in the site selection process. A second advantage of a state-initiated siting process is the possibility of allowing the public to influence the criteria for selection of sites. Presumably, public participation at this early and less salient stage of the process might help to reduce the likelihood of local opposition once a site is actually selected.

One difficulty that could arise in a process in which the state actively selects site locations is that private enterprise may not find it economical to develop and operate in chosen areas. When other locations are more profitable (although perhaps not as desirable when other social costs are considered), the state could conceivably be left with an inoperable siting plan. It is difficult to assess the magnitude of this problem as most siting initiatives are in an embryonic stage. It appears, however, that states whose siting initiatives have responded to the

Table 10.2. State siting programs

	State creates new siting board	State agency may purchase site	State agency may operate site	State agency initiates siting process
Arizona		X	X	X
Connecticut	X			X
Florida	X[a]			
Georgia	X	X	X	X
Indiana	X			
Kansas	X[b]	X	X	X
Kentucky				
Maryland	X	X	X	X
Massachusetts	X			
Michigan	X			
Minnesota	X	X		X
Nebraska				
New Jersey	X	X[c]		X
New York	X			X
North Carolina	X	X		X
Ohio	X			
Oregon		X		
Pennsylvania	X			X
Utah	X	X	X	X
Washington	X	X	X[e]	X
Wisconsin	X	X[d]	X[d]	

pressing need for a new facility have not experienced difficulty in attracting private interest.

Another important consideration in adopting an active state role in locating sites is the added cost involved. When the siting process is initiated by developers, the costs associated with locating the site (data generation, manpower, etc.) are absorbed by the private sector. These costs would be transferred to the public if the state chose to take an active role in siting.

A final way in which states can take a strong role is through the purchase of a site or operation of a facility. Seven states permit the relevant agency or board to purchase a site; five permit its operation. The possibility of direct state ownership, acquisition, construction, and/or operation of a facility complements the power to initiate siting. If approval of a needed facility is blocked by political or economic factors, direct state action may avert an impasse and ensure that the hazardous waste plan is followed. Another advantage of direct state ownership of disposal facilities is the permanence and accountability obtained by assigning responsibility for facility care and maintenance to the public sector. State ownership can be particularly important after closure to ensure long-term care and maintenance.

The disadvantages of empowering the state to construct, own, and/or operate facilities are the direct costs involved in the initial investment and the indirect

Table 10.2. State siting programs (continued)

	Local veto	Local participation on state siting board	Local involvement in construction permit issuance	Local regulations exist	Local review board exists
Arizona					
Connecticut				X^g	
Florida					X
Georgia					
Indiana		X			
Kansas		X			
Kentucky	X^f		X		
Maryland		X^a			
Massachusetts		X	X	X	X
Michigan		X		X^g	
Minnesota		X			X
Nebraska	X		X		X
New Jersey					
New York	X	X			
North Carolina					
Ohio					
Oregon					
Pennsylvania		X	X		
Utah		X			
Washington				X	
Wisconsin		X		X	X

costs associated with a politically undesirable activity that private enterprise might have both the expertise and desire to undertake. It is for this reason that several states, upon actively locating and acquiring a site, have leased the property for private development and facility operation.

Table 10.2 lists fifteen program options states have exercised in writing their siting laws; the first four of these are the factors just described that contribute to a strong state role. From the table, the reader can determine what states have adopted particular programs.

The two major factors we have discussed are siting initiative and state ownership or operation of a HWDF. A third important determinant of the state's role is the existence of a local veto. EPA believes that even with relatively weak controls, local governments have been able to delay siting attempts or frustrate operations. When local controls over zoning and waste planning include veto power over site development, EPA contends that the primary role of local jurisdictions will be to block siting attempts.[11] In the scoring scheme presented here, therefore, the local veto is included as a *negative* factor in a strong state role. Only Kentucky, Nebraska, and New York include a local veto in their siting laws, although as indicated above, other states do provide localities with other kinds of oversight.

Table 10.2. State siting programs (continued)

	No local hearings	Property owners notified about public hearing	Public participation included in			Construction permit review
			Planning	Site selection	Need for environmental certif.	
Arizona						
Connecticut		X		X	X	
Florida			X^a			
Georgia	X					
Indiana		X	X			
Kansas						
Kentucky						
Maryland		X		X	X	
Massachusetts		X			X	X
Michigan			X		X	
Minnesota		X	X	X	X	
Nebraska						
New Jersey						
New York		X				
North Carolina						
Ohio					X	
Oregon						
Pennsylvania						
Utah	X		X			
Washington						
Wisconsin	X		X^a	X^a		

a. Advisory.
b. Off-site disposal facilities only.
c. Temporarily. State cannot hold beyond a short period.
d. Only the recycling authority has this power.
e. Very hazardous waste.
f. Landfills only.
g. Subject to override.
Source: Compiled by the authors.

Table 10.3 puts the three aspects of state role—siting initiative, state owner-ship or operation, and local veto—together on a continuum from active to passive, most active on the left. Legislation which provides the strongest public role in siting empowers the state to initiate or participate actively in locating sites, to construct, own, lease and/or operate disposal facilities, and to preempt local ordinances and requirements which might impede the siting process. Of the twenty-two states that we are considering, six provide this comprehensive state initiative.

Local Powers

The second important dimension of choice in state siting laws is the extent of power granted to the affected localities. At each stage in the siting process,

Table 10.3. A continuum of state roles from state initiative to local initiative

State initiative ◄——► Local initiative

state initiative	state initiative	developer initiative	developer initiative	state initiative	developer initiative	state initiative	developer initiative
state pre-emption of local laws	state pre-emption of local laws	state pre-emption of local laws	state pre-emption of local laws	local veto	local veto	local veto	local veto
state / own construct facilities	private own / construct operate	state own / construct operate	private own / construct operate	state own / construct operate	state own / construct operate	private own / construct operate	private own / construct operate

Active ◄——► Reactive

Arizona	Minnesota	Michigan	Connecticut		Nebraska	New York	Kentucky
Georgia	Pennsylvania	Ohio	Florida				
Kansas	New Jersey	Oregon	Indiana				
Maryland			Massachusetts				
N.C.			Tennessee				
Utah			Wisconsin				
Washington							

Source: Compiled by the authors.

representatives of the locality can be included or excluded from the formal decision-making process.

Table 10.2 also includes information on local powers. Six states provide for temporary members of their siting boards; these temporary members represent affected communities and participate only in those parts of the deliberations of the siting boards that relate to the sites in their areas. Three others have different mechanisms for ensuring that the views of proposed host communities are represented on siting boards; all nine are shown in column 6 of the table. In addition, five states provide that local regulations, including zoning and other restrictions, be satisfied in siting decisions, while five states allow local bodies, sometimes specially constituted local siting boards, to participate in or issue construction permits for HWDFs. This in effect means that local bodies are involved in permitting facilities. Finally, New York and Nebraska allow for localities to override state decisions; Kentucky's local veto applies only to landfill sites.

Massachusetts' siting law provides for a careful division of powers that leaves the state with considerable control over the siting process without sacrificing local interests. Once a site is proposed by a developer, a local assessment committee is established to represent the host community. Negotiations between the developer and the committee are used to determine appropriate compensation for site acceptance. A State Site Safety Council (SSC), comprised of nine state commissioners, six representatives of state professional organizations, and seven members of the public, oversees the negotiation process. The SSC assesses the socioeconomic impact of the project, awards technical assistance grants to local assessment committees, determines appropriate compensation from the developer to the host and abutting communities, and determines services to be provided to the host community by the state. Impasses in the negotiation process are to be resolved through binding arbitration.

By summer 1982, the SSC had received three proposals for disposal facilities —a solvent recovery plant, a multipurpose facility, and an aqueous treatment plant. A site had been selected for the solvent recovery plant and negotiations had begun between the developer and the community.[12] Because of the potential for undermining a plan and for delaying indefinitely the selection of any HWDF sites, EPA believes that local power is inconsistent with a strong state role. However, unless there has been local involvement early in the siting process, localities may use their powers to impede implementation of the plan by hindering operation of facilities.[13] States (of which Massachusetts, Michigan, and Minnesota seem to be good examples) can grant localities power without sacrificing their abilities to coordinate siting with an overall hazardous waste strategy.

Public Participation

The third dimension of policy choice for states is the degree of public participation involved. As we have noted, RCRA requires public participation in the siting of HWDFs; the minimum requirement is that a public hearing be held. Most states comply only with this minimum requirement, but the same forces that call forth special siting laws often demand incorporation of more elaborate means of public participation.

The relationship between the public participation dimension of policy choices and the other two dimensions—state role and local powers—is complex. On the one hand, incorporation of public participation into the site selection process can attenuate state control. One example of this is the provision in four states for individuals to sue to stop construction of HWDFs approved according to the provisions of the state's siting law. On the other hand, inclusion of the public can legitimize a site selection process and thereby both decrease opposition and increase state control. Many of the siting laws attempt to resolve this contradiction by a complex process that both includes participation and circumscribes it. For example, Maryland's process gives a citizen siting board power to approve a permit, but places most other powers in an executive agency.

Since public hearings are the most common means of public participation, many states have sought to enhance the quality of participation by improving the hearing process. Florida and Kansas require that notification be given through local newspapers to all governments within a three-mile radius of the proposed site; Connecticut requires that at least one public hearing be held in the evening in the host community. Six states require direct notification of affected property owners to ensure that they will know about the public hearing. Only three states do not require that a public hearing be held in the host community; we count this as a negative factor in public participation programs. In Kentucky, an extra hearing is conducted before an operator can legally be discharged from his responsibilities for closing and post-closure monitoring of a facility.

Another common form of public participation, which we have already discussed in another context, is the siting board. Although some states, such as Georgia, have established coordinating bodies that do not include citizen representatives, most use boards both for coordinating and for participation. Among the powers that states have given to boards that include citizen members are: participation in development of the overall hazardous waste plan, participation in site selection, and oversight of the permitting process by providing certificates of need and/or environmental soundness.

Finally, appropriate incentives to localities can be developed with citizen involvement. Most of the states have a formula of benefits based on facility use, but Massachusetts involves citizens both at the state and local levels to determine what is appropriate compensation.

Table 10.2 includes information about the most common state provisions for public participation. Powers of siting boards are noted in the table only when boards have some citizen members.

Table 10.4. Summary of state siting laws

	State role	Local power	Public participation
Arizona	3		
Georgia	4		−1
Utah	4	1	1
Maryland	4	1	3
Kansas	3	1	
Minnesota	3	2	4
Connecticut	2	1	3
Pennsylvania	2	2	
Washington	2	1	
Massachusetts	1	4	2
Wisconsin	1	3	4
Michigan	1	2	3
Florida	1	2	1
New York	1	2	1
Indiana	1	1	2
Ohio	1		
Oregon	1	1	
Kentucky	−1	2	
Nebraska	−1	3	

Source: Computed by the authors.

Summary

Although table 10.2 lists fifteen different policy options, it does not begin to capture all the nuances of the various state siting laws. The table does illustrate our contention that, although policy choices have ramifications in all areas, it is possible to array most options along three coordinates. Table 10.4 presents a summary of the information in table 10.2 by assigning an arbitary score of 1 to each policy option incorporated in a state's siting law. Local veto is given a score of −1 in the state initiative column, since it detracts from state initiative, but a +1 in the local power column. The two states whose laws make no provision for local public hearings receive −1 score for that element because this makes public participation more difficult. A state whose law includes every positively scored program listed would thus have maximum scores of 4 for state role, 5 for local power, and 5 for public participation.

Table 1Q.4 shows that in many cases, states chose several options within the same general area, adopting most or all of the means of public participation, for example. These concentrations of programs along a policy coordinate are indicated by scores of 3 and 4 in the columns of table 10.4. The table thus summarizes this section, which has focused on options within the dimensions. The overall pattern of state policy, however, depends on the relations among the dimensions. It is to that subject we now turn.

Table 10.5. State siting policy and amount of waste generated

	High waste states	Low waste states	Total
Siting legislation	N.J., Ohio, Penn., N.Y., Mich., Tenn., Ind., N.C., Mass., Fla., Wis., Ga., Conn., Ky., Md., Minn.	Wash., Kans., Oreg., Neb., Ariz., Utah	
	(16)	(6)	(22)
No siting legislation	Ill., Calif., Texas, Va,. Mo., La., S.C., W.Va., Ala.	Iowa, Del., Miss., Color., Okla., R.I., Idaho, Maine, N.H., N.M., Mont., Vt., Nev., Alaska, Hawaii, N.Dak., S.Dak., Wyo., Ark.	
	(9)	(19)	(28)
Total	(25)	(25)	$N = (50)$

Source: Compiled by the authors. Data on waste generation is taken from the categorization used by Lester et al., in chapter 11, table 11.4.

Patterns of State Policy: Some Conclusions

Our opening section noted three kinds of conflicts that characterize siting decisions: conflicts among individuals about the distribution of costs and benefits of policies, conflicts between states and localities, and conflicts between public and private sectors. Passage of a siting law is itself an indication that the intensity of these conflicts had reached the critical point at which some political action was necessary. However, the difficulty of resolving the conflicts has led many states to pass complex, process-dominated siting laws.

We might hypothesize that the severity of a state's waste problem would be related to its level of policy activity in this arena. Table 10.5 adopts the high waste–low waste distinction made by Lester et al., in this volume to test this hypothesis. About two-thirds of the high waste states and only one-fourth of low waste states have enacted siting legislation. Thus the crudest measure of policy response—passage of the siting law—is correlated with amount of waste.

We have seen, however, that the laws vary widely, incorporating more than fifteen policy options along three major dimensions of choice. An examination of table 10.2 also suggests that there are some patterns in the policy choices although the artificiality and small range of the indices precludes any serious hypothesis testing. In table 10.4 we have regrouped the states to suggest some broader categories of state siting policies, starting at the top with the strongest state roles.

Not suprisingly, states that have the lowest score on state role, -1, have high scores on local powers. Conversely, Georgia, with the highest possible score for state role, makes virtually no provision for local power or for public participation.

We distinguish four groups of states. At the top are Georgia and Arizona, which have decided to solve their siting problems by action entirely at the state level. Arizona's score of 3 reflects the absence of a new siting board or agency, which was not needed under the state's unique system of having siting decisions made by the legislature. These states have responded to siting problems in the most straightforward, and some might say heavy handed, way. They seem to have emphasized resolution of the public-private conflict—that is, finding a site—at the expense of other local and individual concerns.

The second group of states has also opted for a strong state role, but their laws appear to reflect an attempt to respond to more of the complex conflicts surrounding siting. Most of these laws incorporate policy options from all three dimensions of choice. Utah, with relatively low scores on local power and public participation, seems to represent a transition from the first group, but Maryland, Minnesota, and, to a lesser extent, Connecticut incorporate a large number of program options in their very complex and innovative laws. These laws address all three of the kinds of conflicts surrounding siting and provide flexibility to respond to changes in the intensity of any of them.

These complex laws have the potential for creating a siting process that is so cumbersome that it fails to achieve its purpose. Lowi suggests that the process is itself the important feature of the laws, legitimizing by its very complexity any decisions that may finally be made. The states in the second group may have state initiative features strong enough to achieve closure; Maryland, for example, has successfully sited a HWDF in Baltimore County using its new process.

Massachusetts, whose siting law also incorporates a variety of features, is a transition to the next group of states, with laws characterized by a significantly lower amount of state initiative than the previous two groups. Massachusetts chose this reduced state role in a deliberate effort to give the state credibility as the mediator between host communities and HWDF operators. This pattern may also prove to be successful; as noted above, one (nonlandfill) site has tentatively been designated in Massachusetts. The score on public participation does not represent Massachusetts' commitment to the public; our scoring procedure simply did not incorporate many of the innovative features of the state's law since they are unique.

The laws of New York and Michigan also reflect understanding of the complexities of siting, and include options along all three dimensions. They have resolved the state-local conflict more in favor of localities than the states in the second group. However, Michigan does retain some state control because local members of the siting board are only temporary members. This is a feature that Michigan seems to have pioneered, and which has been adopted by several other states as a compromise between state and local needs. New York's law incorporates three state initiative programs but also includes a local veto, which effectively transfers power to localities. This ambivalence may well be the result of policy makers' attempts to balance their preference for a strong state role against the strong local pressures generated by events at Love Canal.

The fourth group of states is made up of the two states that provide for a local veto, and otherwise give strong evidence of emphasizing local power. The absence of scores on public participation probably reflects the fact that localities include the public in making their decisions, reducing the need for the states to make specific provisions for public participation. These laws are designed primarily to resolve state-local conflict; surely their emphasis on local power undermines the states' abilities to plan for hazardous waste and therefore may exacerbate rather than resolve public-private and distributive conflicts.

Since our scoring system incorporated only some of the policy options available, individual states' positions may not be completely accurate in table 10.4. However, we believe that regardless of the status of any particular state, it is accurate to distinguish the four general patterns of state response. These divergent kinds of responses are the result of special circumstances in each state, and of differences in state constitutions and state political styles. They also reflect the different intensities or relevancies of the three different kinds of conflicts that surround siting.

Despite their differences, virtually all the siting laws can be seen to fit the pattern in U.S. policy in which the response to multiple, intense conflicts and to scientific and technological uncertainty is the creation of complex decision-making procedures. Table 10.4 appears to reflect a trade-off between legitimacy of decisions, achieved here by selecting a multiplicity of options on all dimensions and thereby creating a complex siting process, and efficiency in implementing a long-range hazardous waste disposal plan. (The resolution of this problem among states without special siting laws has not been explored here, although we might guess that their procedures may be more dominated by the states, and especially by technology-oriented agencies.) The complex processes raise the cost of waste disposal; since the most intense public opposition comes in response to proposed landfill sites, the cost of that form of disposal may well rise disproportionately. This may provide waste generators with incentives to use other, often safer forms of disposal, and to reduce the amount of waste generated. As a result, these siting procedures may indirectly achieve their purpose of obtaining socially acceptable forms and places for the disposal of hazardous wastes. On the other hand, increased complexity in the siting process, together with increased local opposition to hazardous waste sites, may ultimately encourage more illicit disposal of hazardous wastes. Clearly, the siting procedures discussed above (and the role of public participation in that process) reflect the complexity in the decision-making process and extreme uncertainty in outcomes so characteristic of this severe public policy problem.

11. A Comparative Perspective on State Hazardous Waste Regulation

James P. Lester, James L. Franke,
Ann O'M. Bowman,
and Kenneth W. Kramer

The purpose of this chapter is to examine the fifty American states' policy response to the hazardous waste problem and to evaluate the utility of several indicators hypothesized to be sources of their responses. Three related research questions are pursued. First, are measures of technological, resource, and political factors within a state significantly associated with variation in hazardous waste regulation? Second, what is the relative influence of these factors upon state policy in this area? Finally, does their relative influence depend upon such conditioning influences as regional differences (South or non-South) and severity of the policy problem (high or low waste states)?

We first review the relevant theoretical literature and operationalize our variables before we test several hypotheses suggested by the literature on environmental politics and by the preceding case studies.

Explaining Environmental Policy Formulation: Some Theoretical Considerations

Students of comparative state politics have attempted to develop various models which explain the different processes by which public policies are formulated and particular results are produced.[1] These models may be labeled as follows: Model I,–the "technological pressures" model; Model II,–the "resources" model; Model III,–the "political demands" model; and Model IV, the "administrative-organizational factors" model. Table 11.1 details each of these models in terms of conventionally employed indicators of various phenomena.

The "technological pressures" model suggests that rapid and concentrated population growth, extensive industrialization (especially a greater reliance upon the petrochemical and metallurgical industries), and steadily increasing rates of public consumption of goods and services create severe pollution problems which, in turn, bring about strong pressures for environmental protection policies. Thus, an obvious source of regulatory policy differences among the states is the severity of the pollution problem itself.[2]

Recent studies suggest, however, that state and local environmental control policies are either not strongly correlated with the actual level of pollution or

Table 11.1. Determinants of state hazardous waste policy: Concepts, model, and indicators

Antecedent variables	Independent variables	Dependent variable
Prior environmental context	Political influences	State policy
A. Technological pressures	A. Political demands	A. State hazardous waste policy, 1976–79
1. Industrialization, 1975	1. Democratic party strength, 1978	
2. Pollution potential	2. Democratic party strength, 1976–79	
3. Chemical waste generation	3. Interparty competition, 1978	
4. Hazardous waste sites	4. Interparty competition, 1976–79	
B. System resources	B. Administrative-organizational factors	
1. Personal income, 1975	1. Legislative professionalism, 1976	
2. Spending, pc[a] 1975	2. Bureaucratic consolidation, 1974	
3. Median income, 1975	3. Bureaucratic consolidation (weighted)	
4. Percentage poor, 1975		

a. pc = per capita
Source: Compiled by the authors.

that more refined indicators of pollution severity are needed before a final assessment of their effect on state policy is known.[3] We feel that the group of indicators used in the subsequent analysis is more sensitive to the parameters of the problem than those used previously. We therefore expect indicators of technological pressure, particularly chemical waste generation and the total number of hazardous waste sites in the state, to affect a state's propensity toward hazardous waste regulation. Thus:

> *Hypothesis 1:* The greater the technological pressure (or severity of the pollution problem), the greater the regulation of hazardous waste.

Four indicators of this model will be evaluated: (1) percent of the labor force in manufacturing (i.e., industrialization); (2) pollution potential; (3) a state's ranking in terms of chemical waste generation; and (4) the total number of hazardous waste sites in each state.[4]

The "resources" model posits a direct relationship between the socioeconomic resource base of a political system (national, state, or local) and levels of policy outputs. For example, it is often suggested that the level of a state's economic wealth sets limits on, or provides opportunities for, the provision of public goods and services by a government for its constituents. This consideration is often overlooked by those who assume that the failure of government to act in the environmental sphere is caused by states' backwardness or nonresponsiveness to the public policy problem of environmental pollution.[5] Thomas R. Dye presents the strongest argument for the "resources" perspective. He finds that states' degree of economic development accounts for a significant portion of the variance in policy outputs across a variety of state functions (i.e., state aid to education, welfare, transportation, natural resources, etc.).[6]

Tests of this perspective in the environmental policy area have generally concluded that political variables are largely insignificant predictors of pollution control while industrialization, urbanization, and (to a lesser extent) median income account for the greater proportion of the variance.[7] Hence, it is suggested that environmental (hazardous waste) policy is more aggressive in states that can "afford" stringent business regulation. Thus:

> *Hypothesis 2:* The greater the wealth of a state, the greater the regulation of hazardous waste.

In order to evaluate the resource model of hazardous waste regulation, four conventionally used measures of state economic wealth will be utilized: (1) personal income per capita; (2) state spending per capita; (3) median family income; and (4) percent of the state population below the poverty line.[8]

A third model is based on possibly the most common generalization in the environmental politics and policy literature; environmental policy making is, to a large extent, explained by political party differences.[9] For example, Dunlap and Gale argue that there are important reasons for expecting significant partisan differences to emerge on environmental issues:

On the one hand, pro-environmental measures generally are opposed by business and industry, entail an extension of governmental regulation and intervention, and imply the need for "radical" rather than "incremental" policies. On the other hand, traditionally the Republican party, relative to the Democratic, has maintained a more pro-business orientation, a greater opposition to the extension of the governmental power, and a less innovative posture toward the use of governmental action to solve societal problems.[10]

This hypothesis, linking Democratic partisanship with support for environmental policy, has been tested and generally affirmed through the examination of state legislative and congressional voting behavior[11] and public opinion data.[12]

To our knowledge, only one study of environmental policy constitutes a comparative state project in the sense that all fifty states are included in the analysis.[13] That study, along with fifty-state comparisons in other policy domains, suggests a much different aggregate relationship between Democratic partisanship and environmental policy activity. We would still expect partisanship and policy outputs to be significantly related. However, we would not expect Democratic partisanship in state legislatures to be associated with high levels of policy activity. Comparative state studies of other policy domains have consistently found Democratic party strength to be negatively related to policy outputs. This finding has, of course, been due to the anomalous character of the southern states. Legislatures in these states have traditionally been dominated by Democrats who ideologically have little in common with their counterparts elsewhere. At the same time, as a group southern states have normally been outliers at the low end of a variety of policy output measures. Their outlying nature in comparative analyses in terms of both the independent and dependent variables has substantially affected, or perhaps distorted, the results obtained. We have no basis upon which to expect the relationship to be any different in the environmental policy area. That is, with the inclusion of all fifty states, we expect the opposite relationship between Democratic partisanship and environmental policy activity to that found for single states, a subset of the fifty states, and the U.S. Congress. Thus:

Hypothesis 3a: The greater the Democratic party strength in the state, the less the regulation of hazardous waste.

Although it has not received much attention in the environmental policy literature, comparative state research more generally emphasizes the policy relevant role of interparty competition. Competition between the political parties, rather than legislative domination by one or the other, is what results in policy activity. As legislative control is more closely contested, policy overtures are made to various interest groups in hope of obtaining electoral support and thereby greater legislative hegemony.[14] In terms of the regulation of hazardous waste, this explanation suggests that policy activity levels would be low where a single party dominates the state legislature and high where the parties are narrowly divided. Thus:

Hypothesis 3b: The greater the interparty competition, the greater the regulation of hazardous waste.

Measures of the political demands model are Ranney's index of Democratic party strength and a folded Ranney index of interparty competition. Both point (1978) and period (1976—79) indices were utilized.[15] These two measures are based on four components: (1) the average percentage of the popular vote won by the Democratic gubernatorial candidates; (2) the average percentage of the seats in the state senate held by the Democrats; (3) the average percentage of the seats in the state house of representatives held by the Democrats; and (4) the percentage of all terms for governor, senate, and house in which the Democrats had control. Each state's percentages were averaged to develop an index of Democratic party strength.[16]

A fourth explanation focuses upon administrative and legislative reforms as potential explanations of public policy outputs. For example, the movement for administrative reform advocated consolidation of administrative agencies into a small number of departments, organized by functions whose heads are appointed and controlled by the governor. This reorganization would, it was argued, help to eliminate jurisdictional overlap, jealousies, and conflicts between multiple agencies in a specific functional area. Moreover, consolidation of the bureaucracy would serve to increase the governor's span of control and as a consequence, the chief executive would be more able to mobilize the bureaucracy in order to carry out immediate policy objectives.[17]

Very few studies have empirically examined the linkage between state bureaucratic structure and policy.[18] Since bureaucratic decision makers play an increasingly important role in the policy process, the organizational context within which they operate deserves careful attention in any policy analysis.[19] It appears that a concentration of authority, fiscal resources, and technical expertise—as well as the ability of an organization to provide scientific information for bureaucratic decision making in the environmental (hazardous waste) area—may strongly affect the formulation of policy.[20] On the other hand, institutional fragmentation of the environmental bureaucracy is reported to have a negative impact on pollution control regulations.[21] In this regard, it is argued that bureaucratic fragmentation of authority and decentralization of fiscal resources works to the advantage of those seeking to prevent the enactment of environmental legislation or that such bureaucratic fragmentation greatly limits both technical expertise and fiscal resources for dealing with the hazardous waste problem.[22]

It is therefore suggested that creating a new, single-purpose pollution control agency may well encourage the adoption of a tougher regulatory approach. Consolidating pollution control authority into a single agency allows, for example, the public to hold one state organization accountable for pollution problems of any type and makes it easier for the public to know where to go with complaints.[23] At the same time, consolidation directly promotes environmental policy formulation by concentrating both fiscal resources and bureaucratic expertise into a single agency which may then be applied to the particular environmental problem at hand. Thus:

Hypothesis 4a: The more consolidated the state environmental bureaucracy, the greater the regulation of hazardous waste.

Our two measures of bureaucratic capability are based upon weighted and unweighted indices measuring reorganization (or consolidation) of the state environmental bureaucracy. The unweighted measure of bureaucratic consolidation is based on a four-category variable, with higher scores reflecting greater consolidation. The second index of bureaucratic consolidation weights each state's structure by the number of years it has been in place up to 1976.[24] Weighting has the effect of increasing the scores of states whose reorganization took place very early and minimizing the scores of those states whose reorganization was only recently accomplished. States with a longer history of environmental reorganization are more likely to have established jurisdiction in this area, to have accumulated the necessary technical expertise, and to have more concentrated fiscal resources available for dealing with hazardous waste (or other environmental problems).

A second body of reform literature focuses upon the state legislature. The ability of a "professional" (versus an "unprofessional") state legislature to be more responsive to the needs and concerns of its citizens has been the focus of several empirical studies.[25] Reformers argue that "professionalism" will result in legislatures which are generally innovative in many different areas of public policy, generous in spending and services, and "interventionist" in the sense of having powers and responsibilities of broad scope.[26]

Legislative reformers have hoped that reforms would lead to policies better designed to cope with pressing state problems. However, comparative state policy research generally has shown that these reforms have not had much of a policy impact.[27] Nevertheless, policy analyses which do not consider the potential influence of state legislative capabilities assume that political party differences affect state policies independently of organizational influences. Presumably, this view of state politics omits a very important linkage in determining state policy outputs.[28]

Moreover, case study evidence suggests that legislative advisory capabilities in the science and technology area are instrumental in facilitating state policy formulation involving complex issues such as hazardous waste management.[29] Thus:

Hypothesis 4b: The more professional the state legislature, the greater the regulation of hazardous waste.

An updated version of Grumm's legislative professionalism index will be used to assess the impact of legislative capabilities. The measure of legislative professionalism is based upon a number of variables reflecting the degree to which a state's legislature is "professional" or "nonprofessional."[30]

Finally, it is important to recognize that certain interactive effects may be operating in the policy process.[31] Not only are these explanatory variables affecting policy outputs, but they may also be influencing one another. Unlike most comparative state analyses which assess policy activity only in terms of economic resources and political and/or administrative phenomena, we will also be examining the role played by the severity of the problem itself. As was mentioned above, the direct impact of problem severity upon regulation will be evaluated. Also, there is reason to believe that the nature of the problem may interact with other elements in the normal policy-making process so as to influence the focus on policy activity. Ingram et al. found evidence of this type of phenomenon in their examination of water resources policy in the Four Corners states. They suggest that when the water resource problem is less pronounced (or when solutions are not politically feasible), the focus of policy activity inevitably moves outside the legislature to special commissions, executive agencies, the courts, and so forth. In these instances, "legislatures will act structurally, reassigning decision-making responsibility elsewhere."[32] However, they further reasoned that once the water resource problem assumes crisis proportions (or becomes politically "ripe"), the incentives for policy action become increasingly legislative. What this suggests is that policy problems must achieve a certain level of substantive concern before they become part of the busy legislative agenda.

Our data, specifically the administrative-organizational indicators, will permit a test of this hypothesis. Thus:

Hypothesis 5: Where the hazardous waste problem is more severe, regulation will increase as a function of the professional capacity of the state legislature.

Hypothesis 6: Where the hazardous waste problem is less severe, regulation will increase as a function of bureaucratic capabilities.

The Dependent Variable: State Regulation of Hazardous Waste

Appropriate measures of state policy outputs are a continuing source of controversy in the comparative state literature. For the most part, fiscal indicators—expenditures—have been utilized in these analyses. While fiscal output measures have certain practical (ease of access) and statistical (interval data) advantages, they have been roundly criticized on theoretical grounds as providing an overly narrow operationalization of policy activity.[33] More to the point, it has convincingly been argued that economic output measures tend to bias comparative analyses in favor of economic input measures.[34] The dominant role played by socioeconomic factors in multivariate analyses—at the expense of political indicators—has been described as more of a measurement artifact than a reflection of the actual dynamics of the policy process. Likewise, in the environmental literature it has been suggested that socioeconomic factors may be more important in determining the level of state spending for environmental programs than in determining the type of state environmental policies enacted.[35] Hence, it is frequently suggested that as nonfiscal indicators of policy activity are developed, a more accurate picture of the policy role played by political and administrative factors will be possible.[36] Their predictions notwithstanding, researchers using nonfiscal output measures have generally not produced findings which differ radically from those obtained using fiscal indicators.[37]

The measure of policy outputs used in this analysis is nonfiscal in nature. It assesses state policy for the regulation of hazardous waste and is based on the results of a National Wildlife Federation (NWF) nationwide survey of state toxic substances programs.[38] Forty-three specific questions were addressed to the commissioners of state health and environmental agencies in early 1979. These questions ascertain the extent of state legislation for the regulation of hazardous waste in five key areas: (1) the definition of hazardous waste; (2) the extent of state regulatory authority for hazardous waste management; (3) specific hazardous waste management provisions; (4) state siting provisions; and (5) provisions for state monitoring and enforcement. The fifty states were scored and ranked across all these items.[39]

The relatively few studies which have employed nonfiscal indicators of policy outputs fall into two categories: those which assess the redistributive impact of public policy,[40] and those which focus upon a special policy subset labeled innovations.[41] Our dependent variable clearly falls closer to the latter category.[42] As was suggested above, however, comparative state examinations of the determinants of adopting innovative policies substantially agree with the economic explanation offered for policy outputs measured in expenditure terms. Jack Walker's earlier conclusion that innovative states tend to be larger, wealthier, and more industrialized still provides an adequate characterization of these research findings.[43]

Method

The basic objective of our analysis is to determine why some states more actively regulate environmental problems than do others. The four models introduced above provide alternative explanations of the determinants of state regulatory activities. Initially, these explanations will be separately considered by examining the predictive capacity of multiple indicators of each phenomenon introduced above. This preliminary exercise will be used to gain insights which will facilitate the eventual selection of single indicators of each model for an assessment of the policy process in a multivariate framework.

Throughout the analysis, careful attention will be paid to possible distortion introduced by the eleven southern states. As was previously noted, collectively the southern states have consistently displayed outlier characteristics in terms of both the independent and dependent variables in fifty-state policy analyses. Comparative researchers have become increasingly sensitive to the analytical implications of the South's unusual characteristics. Since this analysis breaks unfamiliar ground in the examination of hazardous waste policy due, among other things, to its comparative focus, we do not wish to present a description of the policy process which is unduly influenced by a small number of unusual observations. Thus, at each stage of the analysis, the results obtained for all fifty states will be compared with those obtained for the thirty-nine nonsouthern states.

We will also be interested in the conditioning effect, if any, which the severity of the hazardous waste problem has upon the policy process. In the multivariate portion of the analysis, the parameters of the process will therefore be evaluated where the problem is and is not acute.

Analysis and Results

Table 11.2 presents the results of ranking the fifty states according to their summated scores on the dependent variable. The most interesting aspect of these rankings is the relatively high activity levels of most of the southern states. Nine of the eleven southern states are above the median. As was mentioned above, one of the consistent findings in the comparative state literature has been that the South collectively has lagged behind the other states in terms of policy outputs. The more nearly normal distribution of the southern states on the dependent variable suggests that the explanatory model in this policy area will differ from that found elsewhere.

Table 11.3 examines the four alternative explanations of environmental policy activity in terms of the multiple indicators of each model for the fifty states. The best bivariate predictors of environmental regulation are the indicators of technological pressures and the administrative-organizational model. Those states which have severe hazardous waste problems are reacting by way of more extensive regulatory activities in this area (hypothesis 1). Also, those states with professional legislatures and/or consolidated bureaucratic responsibility for pollution-related problems are more likely to be active in this policy area (hypotheses 4a and 4b). The political demand explanation is less compelling. Only the longitudinal measure of Democratic party strength is significantly associated with policy activity (hypothesis 3a).[44] Interparty competition is not a policy relevant factor (hypothesis 3b). None of the economic resource indicators are significant predictors of hazardous waste policy (hypothesis 2).

As we suggested above, the more normal distribution of the southern states in terms of the dependent variable has had profound effects. The performance of the economic and political models demonstrates the impact of this empirical

Table 11.2. State ranking for hazardous waste regulation, 1979

State	Rank	State	Rank
→ California	1	Rhode Island	26
Maryland	2	Michigan	27
Tennessee	3	Montana	28
South Carolina	4	Wisconsin	29
Washington	5	Massachusetts	30
Delaware	6	Minnesota	31
Oregon	7	→ New Mexico	32
Connecticut	8	North Dakota	33
Florida	9	Iowa	34
Ohio	10	Utah	35
Vermont	11	Alaska	36
New Jersey	12	Hawaii	37
→ Texas	13	Missouri	38
North Carolina	14	Nebraska	39
Pennsylvania	15	Colorado	40
New York	16	Kentucky	41
Illinois	17	West Virginia	42
Louisiana	18	Maine	43
Alabama	19	Mississippi	44
Kansas	20	Idaho	45
Oklahoma	21	Georgia	46
Arkansas	22	New Hampshire	47
Virginia	23	Nevada	48
Wyoming	24	South Dakota	49
Indiana	25	→ Arizona	50

Source: Kenneth S. Kamlet, *Toxic Substances Programs in U.S. States and Territories: How Well Do They Work?* (Washington, D.C.: National Wildlife Federation, 1979), reprinted with permission.

phenomenon. Most comparative state policy research has consistently demonstrated two things. On the one hand, economic resources have been shown to be strongly related to policy outputs. Such is not the case in these data. On the other hand, Democratic party strength has usually been negatively related to policy outputs while interparty competition has been a positive predictor. The relationship here is positive. These earlier findings have, again, been due to the unusual—that is to say poor—performance of the southern states. While the political and economic complexion of the South relative to the rest of the nation has not changed markedly during the past decade, our survey data suggest that their collective reaction to this policy problem has.[45] As a result, the bivariate role of resources is negated and the impact of Democratic strength is now in a positive direction.

Table 11.3 also presents a comparison of the coefficients for all the states with those of the nonsouthern states in order to evaluate the influence of the South. Some differences are apparent.

In terms of the political demand characteristics, we find ourselves in the

Table 11.3. Pearson product-moment correlation coefficients (r) showing the relationship between selected independent variables and states' scores on hazardous waste regulation

	All states ($N = 50$)	NonSouth ($N = 39$)
Model I (Technological pressures)		
Industrialization, 1975	0.42 ***	0.42 **
Pollution potential	0.31 **	0.34 ***
Chemical waste generation[a]	0.49 ***	0.46 ***
Total waste sites in states	0.35 **	0.37 **
Model II (Resources)		
Personal income, 1975	0.12	0.30 *
Per capita spending, 1975	0.02	0.10
Median family income, 1975	0.14	0.29 *
Percentage poor persons, 1975[a]	−0.17	−0.35 **
Model III (Political demands)		
Democratic Party strength, 1978	0.19 *	0.19 *
Democratic Party strength, 1976–79[a]	0.25 *	0.27
Interparty competition, 1978	−0.11	−0.08
Interparty competition, 1976–79	−0.17	−0.19
Model IV (Administrative-organizational factors)		
Legislative professionalism, 1976[a]	0.43 ***	0.48 ***
Bureaucratic consolidation, 1974	0.25 *	0.31 *
Bureaucratic consolidation (weighted)[a]	0.30 **	0.41 **

a. Variable selected for inclusion in the subsequent analysis.
* $p \leq 0.05$
** $p \leq 0.01$
*** $p \leq 0.001$

unusual position of wondering whether the overwhelmingly Democratic majorities in southern legislatures are artificially inflating the relationship between politics and policy in a positive direction. The fifty-state associations of policy activity and the four measures of political demands are virtually identical to those found in the nonsouthern states. Again, only the longitudinal measure of partisanship is marginally significant.

There are differences in the magnitude and significance of the coefficients reflecting the relationships between the resource indicators and hazardous waste policy. Whereas none of these variables is significant when all states are considered, three of the four become significant with the exclusion of the South. This reflects the fact that in the fifty-state situation, unlike other relatively poor states, the South was active in policy terms. With the exclusion of these states, the explanatory power of economic resources necessarily increases.

There is some marginal change in the relationship between the administrative-organizational capabilities and policy. For example, the magnitude of the relationship between bureaucratic consolidation and policy increases in the nonsouthern states. The indicators of technological pressures, on the other hand, remain essentially the same as that found for all fifty states.

The next step in the analysis is to incorporate these four models into a more comprehensive explanation of the processes leading to state regulatory activity in this policy area. For multivariate purposes, we choose the best bivariate predictor of policy activity from each of the models; percent poor reflects the resource model; political demands are represented by Democratic party strength for the period 1976–79; technological pressures are indicated by chemical waste generation; and to reflect both the legislative and bureaucratic aspects of the administrative-organizational model, legislative professionalism and the weighted index of bureaucratic consolidation will be utilized.

Table 11.4 considers the processes at work in the fifty states as well as those operating in the twenty-five high and low waste-generating states. In each case the four independent variables are regressed upon hazardous waste regulation. For the fifty-state regression, legislative professionalism and Democratic party strength are significant predictors of the dependent variable. However, when we categorically control for the severity of the problem, the policy processes diverge. Where the problem is more severe—in high waste-generating states— the policy response is differentiated substantially in terms of legislative professionalism (hypothesis 5). In fact, the nonsignificance of the overall regression reflects the fact that the other three independent variables contribute virtually nothing to an explanation of regulatory activity. In these states, hazardous waste regulation is not a partisan issue. Neither Democratic nor Republican controlled legislatures are any more likely to be taking policy action. Likewise, the existence of a centralized bureaucratic mechanism for dealing with the environmental problem makes no difference. Rather, the influence of these latter factors, especially bureaucratic consolidation (hypothesis 6) is important only where the problem is less severe—in low waste-generating states.

When we consider all the states, it therefore appears that where the chemical waste problem is severe, it is a policy problem to be dealt with by state legislatures. Where it is not particularly acute, hazardous waste regulation is more likely to be promoted by bureaucratic initiatives and partisan political activity.

Table 11.5 examines the character of the policy process in the thirty-nine nonsouthern states. The results confirm our suspicion that the fifty-state description is noticeably distorted by the unusual nature of the southern states. Legislative professionalism still plays an important role. As anticipated, in the absence of overwhelmingly Democratic southern legislatures, the explanatory power of the partisan variable becomes nonsignificant. Likewise while not statistically significant, we see an increased impact of the resource and bureaucratic variables.

One final issue remains: the influence of problem severity upon the manner in which hazardous waste policy is enacted in the thirty-nine nonsouthern states. Once again, the states were divided into high and low waste categories and separate regression analyses were carried out.[46] These results are also presented in table 11.5. As was the case with the fifty-state analysis, the policy processes differ in the high and low waste states.

In those nonsouthern states where the hazardous waste problem is more severe, policy action is primarily a function of resources and legislative profes-

Table 11.4. Multiple regression analysis of hazardous waste regulation: All states, by waste generation

Independent variables	All states (N = 50)			High waste states[a] (N = 25)			Low waste states[b] (N = 25)		
	b	Beta	F	b	Beta	F	b	Beta	F
Percentage poor	-0.29	-0.20	1.86	-0.41	-0.29	0.93	-0.41	-0.31	3.69
Democratic party strength	12.23	0.34	6.06*	10.82	0.27	1.05	14.18	0.41	6.65*
Legislative professionalism	2.24	0.34	7.27**	2.43	0.47	4.75*	2.58	0.18	1.36
Bureaucratic consolidation	0.14	0.19	1.96	-0.16	-0.26	1.06	0.48	0.61	15.88**
Multiple R	0.57			0.48			0.74		
R^2	0.32			0.23			0.55		
$\bar{R}^2$	0.26			0.07			0.46		
F ratio	5.28**			1.47			6.08**		

a. These states are (in order of the severity of chemical waste generation): N.J., Ohio, Ill., Calif., Pa., Tex., N.Y., Mich., Tenn., Ind., N.C., Va., Mo., La., S.C., Mass., Fla., Wis., W.Va., Ga., Conn., Ky., Ala., Md., and Minn.

b. These states are Wash., Iowa, Kans., Del., Miss., Ark., Colo., Okla., Oreg., R.I., Idaho, Maine, Nebr., Ariz., N.H., Utah, N.Mex., Mont., Vt., Nev., Alaska, Hawaii, N.Dak., S.Dak., and Wyo.

* $p \leq 0.05$
** $p \leq 0.01$

Table 11.5. Multiple regression analysis of hazardous waste regulation: Nonsouthern states, by waste generation

Independent variables	All nonsouthern states[a] (N = 39)			High waste states[b] (N = 19)			Low waste states[c] (N = 20)		
	b	Beta	F	b	Beta	F	b	Beta	F
Percentage poor	−0.51	−0.26	3.59	−1.19	−0.60	7.62*	−0.36	−0.21	1.47
Democratic party strength	9.61	0.23	3.18	16.14	0.39	3.50	16.32	0.44	6.03*
Legislative professionalism	2.07	0.35	6.48*	2.45	0.53	7.99*	2.54	0.19	1.23
Bureaucratic consolidation	0.19	0.25	3.31	−0.21	−0.32	2.27	8.38	0.70	15.70**
Multiple R		0.65			0.72			0.75	
R^2		0.42			0.52			0.57	
$\bar{R}^2$		0.35			0.39			0.45	
F ratio		6.18***			3.86*			4.93**	

a. These states are (in order of the severity of chemical waste generation): N.J., Ohio, Ill., Calif., Pa., N.Y., Mich., Ind., Mo., Mass., Wis., W. Va., Conn., Ky., Md., Minn., Wash., Iowa, Kans., Del., Colo., Okla., Oreg., R.I., Idaho, Maine, Nebr., Ariz., N.H., Utah, N.Mex., Mont., Vt., Nev., Alaska, Hawaii, N.Dak., S.Dak., Wyo.

b. These states are N.J., Ohio, Ill., Calif., Pa., N.Y., Mich., Ind., Mo., Mass., Wis., W.Va., Conn., Ky., Md., Minn., Wash., Iowa, Kans.

c. These states are Del., Colo., Okla., Oreg., R.I., Idaho, Maine, Nebr., Ariz., N.H., Utah, N.Mex., Mont., Vt., Nev., Alaska, Hawaii, N.Dak., S. Dak., Wyo.

* $p \le 0.05$
** $p \le 0.01$

sionalism. Nine of the eleven southern states previously fell into the high waste category. Although their policy activity levels were shown to be comparable to those of the other states in that group, these states were significantly less affluent. Hence, with their removal from the analysis, the explanatory role of resources becomes significant.[47]

The situation in the nonsouthern, low waste states is somewhat more clear cut. In the low waste states, bureaucratic consolidation is the dominant factor in the policy process. To a much lesser extent, Democratic party strength is also important. The highest levels of policy activity occur in those states that are both dominated by Democratic lawmakers and have consolidated bureaucracies.[48]

The different policy processes at work in the high and low waste nonsouthern states provide interesting insights concerning the role played by the two administrative-organizational variables. Levels of bureaucratic consolidation and policy activity are significantly lower among low waste-generating states than they are for the high waste group. Hence, bureaucratic consolidation is a significant bivariate predictor of policy across all thirty-nine states. Increasing levels of bureaucratic consolidation in low waste states accounts for most of the explained variance in policy activity. No such relationship holds in high waste states. This suggests that bureaucratic consolidation only makes a difference up to a certain point. Apparently, in the face of a severe hazardous waste problem, increasing bureaucratic consolidation past a certain point has no policy implications. Rather, given a severe problem and relatively consolidated bureaucracies, higher levels of policy activity occur with increased resource availability and more professional legislatures. These findings support hypotheses 5 and 6 above.

Conclusions

This chapter has examined the determinants of hazardous waste policy in the American states. The data have been examined in a variety of ways. An analysis of all fifty states has been carried out as well as one of the thirty-nine nonsouthern states. In turn, each of these analyses controlled for the severity of the hazardous waste problem by dividing the states into high and low waste-generating groups. This methodology, while perhaps cumbersome at times, is nevertheless justifiable for at least two reasons: one reason has to do with analytic rigor; the other with the applied consequences of hazardous waste research.

As a general rule, when an already small number of cases contains a series of observations—the southern states—whose characteristics uniformly differ from those of the majority of the cases, analytic prudence requires that their impact be assessed. Central tendencies in the data may, after all, reflect processes applicable to most of the observations or to actually very few. While perhaps obvious from a statistical point of view, comparative researchers have too frequently ignored the influence of the South upon theories of public policy. We found, for example, that the inclusion of the southern states in the analysis specified a role for partisanship which was not applicable to the remaining thirty-nine states. As a result, we have given the thirty-nine state analysis considerably more weight than that involving all the states.

We also feel that the methodology contributes to the applied utility of this, or any other, policy analysis. From a practitioner's perspective, the important issue involves those aspects of the policy-making process which, if manipulated, result in policy activity. Hence, special care has been taken to point out that policy relevant factors differ somewhat depending upon the character of the states under consideration. In some states—high waste—the impetus for policy activity is apparently legislative, while in others—low waste—it is bureaucratic. While it would have been more straightforward to phrase our analysis in terms of all the states, the implications of our results would not be applicable to many of those states.

The conclusions from this analysis are outlined below.

Technological Pressures

The American states are clearly responding to the existence of the hazardous waste problem (hypothesis 1). While our data do not permit an evaluation of the qualitative nature of their responses, we are able to say that the quantitative character of state regulatory activities is indeed a function of the severity of the problem itself. This is encouraging and even perhaps surprising to policy analysts who suggest that environmental policy outputs are not strongly related to objective conditions or the need for policy.

Administrative-Organizational Capabilities

State organizational capabilities, or a consolidated environmental bureaucracy and (especially) a professional legislature, promote the development of hazardous waste regulation. The importance of legislative professionalism held in both the fifty- and the thirty-nine-state analyses, as well as for high waste-generating states. Bureaucratic consolidation played an overwhelming role in low waste states.

The interplay between legislative professionalism and bureaucratic consolidation on the one hand, and the severity of the hazardous waste problem on the other suggests a type of threshold phenomena may be at work. On the one hand, the severity of the hazardous waste problem may influence the focus of policy activity (hypotheses 5 and 6). That is, once the problem reaches a certain threshold of severity it is taken seriously by the state legislatures; it becomes part of the legislative agenda. Up to that point, other matters consume legislative attention and policy activity is focused bureaucratically. On the other hand, a similar threshold situation involving legislative professionalism rather than problem severity may be operating. As a group, those states where the problem is more severe also have more professional legislatures. Thus, the fact that the problem is handled legislatively where it is more severe may be due to the capacity of the legislature itself. By the same token bureaucratic initiatives could be due to the absence of a professional legislature rather than the absence of a severe problem. Regardless of whether legislative professionalism or problem severity is the crucial factor, we are able to say that where these two phenomena are present, increasing centralization of the bureaucratic mechanism for dealing with the hazardous waste problem has no measurable policy consequences.

Political Demands

In some respects, the most troublesome findings in the data involve the impact of partisanship upon hazardous waste policy activity. One of the most prevalent generalizations in the environmental politics literature is that Democratic lawmakers more actively support the protection of the environment than do Republicans (hypothesis 3a). Thus, from the perspective of that literature, our rejection of this hypothesis is surprising in and of itself. We are furthermore in something of an awkward position because we are rejecting the argument; yet empirically, in the fifty-state analysis, it is supported. Our rejection is based upon the thirty-nine-state analysis which excludes the southern states. Were we to argue that Democratic strength promotes environmental regulation, such a conclusion would be based largely on the impact of the eleven Democratic southern states. In effect, we would be equating liberal policy making with southern legislators. Southern Democrats are, on the whole, decidedly more conservative in policy terms than their counterparts elsewhere. This has been demonstrated repeatedly in the comparative state literature. Considerably more work needs to be done before we can conclude that this policy area constitutes an exception to this long-standing rule.

Resources

The impact of economic resources upon hazardous waste policy was only significant in nonsouthern high waste-generating states (hypothesis 2). The resource model, of course, says nothing about the policy process per se. What it may indicate is that these wealthier states are not hesitant to place regulations, by whatever means, upon their industry which poorer states would avoid.

Implications

Our goal throughout has been to integrate our environmental policy work into the larger comparative state paradigm. That is, the contending comparative state explanations of policy activity were evaluated vis-à-vis the regulation of hazardous waste. In so doing, we illustrated that some prevailing generalizations in the environmental area are subject to modification within a fifty, or even thirty-nine, state context. The most notable of these is the relationship between political partisanship and policy activity. We would strongly emphasize, however, that this does not necessarily deny that there are policy relevant ideological differences between liberals and conservatives concerning the regulation of hazardous waste. Our intuition suggests that pronounced differences do exist. Rather, what our findings imply is that in terms of the conventional comparative state methodology, such differences are not empirically evident. This may be due to certain quirks in the methodology (i.e., the southern states) or the way in which these ideological differences are assessed (i.e., partisanship). Regardless, as viewed from a comparative perspective, the generalization arrived at above is that the partisan complexion of state legislators plays an inferior role in the regulation of hazardous waste to that of legislative and administrative phenomena and, in certain situations, to fiscal resources.

In the course of the analysis, several empirical findings emerged which also conflict with the mainstream comparative state literature: the restricted role of economic factors, the importance of legislative and bureaucratic considerations, and the relatively high activity levels of the southern states. Given the recent emergence of this policy area, it would obviously be premature to conclude that, for whatever reasons, environmental policy making in the fifty states fundamentally differs from that of the myriad other policy realms investigated by comparative researchers. Rather, we feel obliged to point out possible rival explanations for our conflicting findings. Three possible explanations may account for the differences between these findings and other comparative work: (1) the policy arena itself; (2) the manner in which the dependent variable is measured; and (3) the qualitative nature of the states' responses.

Aside from the contextual factors discussed above, the threat of federal intervention may account for the unusual policy activity levels of the southern states. As we mentioned at the outset, the states are prompted by the national government (through RCRA) to implement a hazardous waste control program. If the states refuse to act, EPA is authorized to formulate and implement the program itself. Therefore, the states have little choice concerning whether or not hazardous waste will be regulated. They may only determine who will do the regulating and the ultimate extent of the regulatory activities. Given the usual aversion of more conservative state governments to federal intervention, it is not surprising that the southern states have taken policy action. The RCRA mandate requires that all states (in order to administer the program themselves) adopt standards at least comparable to those of EPA.

The federal mandate (RCRA) thus ensures that a minimum level of regulatory activity will be undertaken regardless of state-level parameters. The development of regulatory policy by normally inactive states—principally southern states—is assured. This does not then lead to the conclusion that the absence of broad scale economic constraints upon policy is artifactual. The federal mandate only accounts for minimum activity levels. It does not account for the wide range of activity above that level. If the economic explanation is controlling in this policy area as it is in many others, it should still account for the remaining variance across the states. It does not. Relatively poor states, especially nine of the eleven southern states, have gone considerably beyond the minimum commitment to hazardous waste regulation required by the federal government. A shortage of economic resources does not constrain this activity. Rather, these states are clearly reacting to the severity of the problem and solutions are being formulated by state legislative and bureaucratic organizations.

A second alternative explanation of our contradictory findings involves the way in which we measured the dependent variable. As opposed to the conventional reliance upon programmatic expenditures, we examined regulatory policy enactments. Unlike other researchers utilizing nonfiscal policy outputs measures, we did find them to be more closely related to political and organizational factors than they were to socioeconomic characteristics of the states.

It would be tempting to conclude that through the use of regulatory legislation we have developed a more sensitive measure of policy outputs which overcomes the bias of expenditure outputs in the direction of economic inputs. However, what is more likely to have occurred is that the nonfiscal output measure, rather than being somewhat more sensitive to the policy process generally, is itself biased in the direction of noneconomic inputs. Bias, however, should be understood to suggest incomplete rather than inaccurate measurement. Both fiscal and nonfiscal output measures reflect facets of policy action. Until more is known concerning how they relate to each other and how well they reflect actual program implementation and impact, it seems unwise to advocate an exclusive reliance upon one or the other.

Finally, a third possible explanation is that some of the fifty states have sought to avoid federal intervention by adopting what are essentially "paper programs" for the regulation of hazardous wastes. Thus, the observed "overperformance" by the Deep South states may, in fact, reflect symbolic responses to the hazardous waste problem. Such a policy response allows the southern states to achieve "substantial equivalence" to RCRA and avoids federal intervention for failure to comply. This strategy would help to explain their scores in this area versus their underperformance in many other policy areas.

Based on our case studies of Texas and Florida, there is recent evidence to support this rival explanation. For example, as of October 1982 Texas is one of only five states to achieve Phase II-a and Phase II-b interim authorization. Yet, as chapter 7 points out, the Texas response is described as a "structural solution" or a vague, "let the bureaucracy decide" approach which, as a result, leaves the hazardous waste issue largely unresolved in that state. If the Texas response to the hazardous waste problem is typical of the Deep South (or the other forty-nine states), then what appears to be overperformance may, in fact, later reveal itself to be less than innovative behavior.

However, these rival explanations must remain tentative and await further evidence before any firm conclusions are drawn. More contemporary analyses may provide much insight into the responsible management of hazardous waste by detailing the *qualitative* nature of states' responses, and understanding the contextual (i.e., internal) and intergovernmental (i.e., external) factors associated with that action (or inaction).

IV. Conclusions

12. The Politics of Hazardous Waste Regulation: Theoretical and Practical Implications

Ann O'M. Bowman

Institutional Challenges

The thematic unity of this research collection can be encapsulated in this statement: hazardous waste management presents a tremendous challenge to governmental institutions. The following list of hazardous waste incidents demonstrates this reality.

Chemical tank cars from a derailed freight train in Louisiana force 3300 people from their homes and cause extensive ground contamination.[1]

The illegal dumping of PCB-laden transformer oil along two hundred miles of North Carolina roadside leads to a state and federal cleanup effort that involves a massive relocation and burial of the toxic soils in a landfill, triggering widespread, highly visible protest activity.[2]

The discovery of the build-up (nine times the allowable amount) of toxic sludge at a hazardous waste treatment plant in downtown Waterbury, Connecticut, touches off an exercise in responsibility evasion by state and federal environmental officials.[3]

In Indiana, a voluntary cleanup of a 60,000-plus barrel waste dump is funded by twenty-four waste generators in exchange for an Environmental Protection Agency (EPA) pledge of no further litigation.[4]

EPA's refusal to intervene in an Arkansas neighborhood experiencing soil contamination, even after an assessment by the Public Health Service that serious health risks were evident, leads to charges from the Environmental Defense Fund that EPA has diminished its standards for protecting the public from dangerous wastes.[5]

The municipally funded cleanup of a sixty-acre Philadelphia dumpsite (city-owned land, reportedly utilized by waste disposal companies that bribed city workers to ignore the illegal dumping) is completed.[6] The city is attempting to recoup the costs, estimated to be in the $9 million range, by identifying and subsequently negotiating payment from or suing the producers of these toxic wastes, now liable for their ultimate disposition.

A jury awards $58 million to forty-seven railroad workers in Missouri who were exposed to dioxin—a carcinogen of terrifying proportions—when a tank car loaded with 30,000 gallons of chemical solvent ruptured.[7]

New York continues its ban on commercial fishing in the Hudson River when tests on striped bass show concentrations of polychlorinated biphenyls and dioxins exceeding the Food and Drug Administration's allowable levels.[8]

Each of these incidents represents a different facet of the hazardous waste issue: transportation accidents, illegal disposal, careless enforcement, public-private sector cooperation, agency and interest group relationships, innovative local government efforts, liability and judicial branch involvement, and health risks and economic loss due to contamination. The incidents occurred within a two-month period. While each one is discrete, the localized magnitude and their devastating cumulative impact galvanizes public concern. With interest piqued, the public turns to government for remedial action.

The governmental policy response to the toxic waste challenge thus far has been mixed, at best. If one applies a reasonable evaluative criterion—anticipatory action to protect public health and safety—the performance of relevant governmental institutions is less sanguine. Decades of industrial chicanery, governmental myopia, consumer carelessness, and public indifference have culminated in the present situation: we face unrelenting problems armed with institutions and strategies of uncertain value. It is clear that hazardous waste policy making for the future demands both institutional strength and innovative solutions.

In the ensuing section, the preceding chapters will be summarized. A synthesis of these chapters, keyed to their political implications, will follow. In the concluding section, issues that are likely to dominate the future in hazardous waste policy will be identified and explored.

Some Theoretical Considerations: An Interpretative Examination of the Chapters

The themes of this research collection are set in chapter 1 where Lester argues that hazardous waste policy making has three dominant characteristics: the severity of the problem, the complexity of the political process, and the uncertainty of an effective policy response. Subsequent chapters develop and amplify these themes. Plainly put, hazardous waste generation, treatment, transportation, storage, and disposal constitute a severe environmental test for society.[9] Because hazardous waste regulation tends to entangle the EPA, state governments, local communities, industry, and the public, political complexity is assured. A workable intergovernmental activity mix remains elusive in this developing policy area and uncertainty inevitably results. This thematic orientation—severity, complexity, and uncertainty—forms the framework for the subsequent chapters.

Chapters 2 through 4 revolve around federal and intergovernmental perspectives, more specifically, the role of EPA in hazardous waste management. The inadequacy of EPA's performance in the implementation of TSCA and RCRA dominates the discussion in chapter 2. Implementation was the responsibility of the agency, but it was not a lockstep, smoothly unfolding process. Policy reversals, ambivalence, and delay impeded the installation of the programs. Riley links these consequences to four confounding factors. First, the broad goals of the legislation established a set of claimants whose very nature invited conflict. Multisource communication breakdowns occurred with EPA serving in a brokerage role attempting to mollify the contending parties. Extra-agency leadership (especially at the presidential level) set into motion a reordering of priorities and consequent resource allocation adjustments. These conditions diminished EPA's capacity to function. Implementation deficiencies resulted and program success became even more problematical. The primary interests of the opposition (in this case, those who gained some sort of competitive economic advantage from unregulated toxic wastes) were served.

Chapter 3 goes to the heart of bureaucratic organization and behavior in its "insider's" analysis of EPA's Superfund preimplementation planning project. Examination of this segment of EPA's operation produces an evaluation somewhat different from that of chapter 2. Not immune to internal bureaucratic sniping and rivalry, EPA nonetheless developed a credible record in preimplementation planning. The key structural units in the project were the agency-wide workgroups assigned to tackle myriad policy and operational issues. A reservoir of slack in terms of resources, commitment, and time meant that project managers were able to plan rationally for the implementation of the ensuing legislation. The design of the workgroups played a turf legitimizing role and a new bureaucratic unit—The Office of Hazardous Emergency Response —was created. Hence, when the Superfund legislation was signed into law, the EPA had a structure ready to accommodate it. Cohen and Tipermas report, however, that operational issues afforded a greater likelihood of successful adoption than did policy issues. A note of caution is necessary: preimplementation planning is not without drawbacks. The intrusion of unanticipated events (legislative compromises, a change in presidential leadership) can upset the balance and, if too disruptive, render portions of the previous effort ineffectual. Therefore, the interest of the preimplementation planners in controlling the ultimate policy product is significant.

The inherent intergovernmental tensions encompassed in the implementation of RCRA, especially those between federal regulators and state program administrators, are examined in chapter 4. RCRA is somewhat unique in that it strikes a midpoint between the national preemption and state discretion poles prevalent in other environmental legislation. Lieber's thesis is that EPA delays —a function of the complexity of the issue, public participation requirements, and poor management—have seriously hindered states' efforts to comply with RCRA. This has occurred despite tacit understandings that RCRA was to establish federal minimum standards and not to disrupt existing state programs. States merely were expected to bring their programs into compliance (first "substantial equivalence," then "full equivalence") with RCRA provisions.

In addition, federal incursion into hazardous waste control may have strained interstate relations. The enactment of strict regulations in one state may encourage the transportation of toxic wastes to states whose regulations lag, thus bolstering their reputations as dumping grounds. As a retaliatory measure to limit the inmigration of toxic wastes, states could impose exorbitant fees on the importation of these wastes for disposal. A significant intrastate problem involves hazardous waste disposal facility siting. These types of issues are likely to crowd court dockets in the 1980s. And this will occur against a backdrop of diminished federal funding.

The intergovernmental issues raised by Lieber provide a transition for the single state analyses in chapters 5 through 8. The underlying theme fueling these four chapters is that internal contextual factors have worked to produce a wide range of policy responses to the hazardous waste problem.

Florida, with its fragile environment and threatening hazardous waste problem, is the setting for chapter 5's pursuit of this question: how much governmental interference in the market will accomplish the goal of protecting the public while avoiding deleterious upsets of private sector economic activity? Williams and Matheny suggest that in Florida, competitive market pressures force small scale and medium-sized waste generators (who produce about 75 percent of Florida's toxic wastes) to resort to unsafe dumping as a means of minimizing disposal costs. Large scale producers of hazardous wastes are able to extort advantageous compromises at the policy formulation stage, hence their economic pressures are lessened. Therefore, as the critical model of social regulation suggests, it is the functioning of the private market that causes market failures, not external, manipulable factors.

In looking at the MITRE model to assess relative risks at hazardous waste sites, the tenets of the critical approach to regulation win out again. There is a basic incompatibility in the utilization of "objective" risk assessment mechanisms for what ultimately are politically and economically driven decisions. Institutionally, Florida's lack of commitment to toxic waste regulation is evident in that, despite its penchant for authorization legislation, Florida fails to fund adequately its implementation. Hazardous waste regulation remains a low priority item in terms of fiscal commitment.

Williams and Matheny reach the conclusion that hazardous waste management in Florida is inadequate and likely to worsen. Conventional regulatory prescriptions espoused by neoconservatives are likely to mold policy making in the near future. The specter of crisis management looms large in what is, in reality, a reactive decision-making posture.

In contrast to Florida's performance, New York has exercised strong state leadership in hazardous waste management by using conventional regulatory mechanisms. As mentioned in chapter 6, New York represents a special category due to the immediacy of the Love Canal episode. Worthley and Torkelson credit an aroused public with issue formulation and agenda setting, but note the quixotic nature of public attention to the hazardous waste issue. They speculate that parochial community interests are destined to clash with the larger public interests, especially in siting decisions.

The activities of industry are quixotic as well. Alongside instances of voluntary cleanups by toxic waste generators is evidence that illegal dumping continues unabated. Technological advances—incineration, recycling, waste exchanges— offer some amelioration of the problem, but tensions remain. Noteworthy are EPA and state agency conflicts that eventually envelop local government as well. Bisecting these relationships is the private sector. At virtually every stage of hazardous waste management, these overlapping interests collide. Worthley and Torkelson suggest that the public is the ultimate loser in these battles. They argue that the future may not be quite so grim if only because events will force the contentious parties to cooperate. Public-private partnerships are potential vehicles for problem solution. In addition, citizen mobilization will be a potent source of demand as the hazardous waste issue matures.

A splintered bureaucratic attack on the hazardous waste problem in Texas has produced a policy response that is minimal at best. As pointed out in chapter 7, four agencies share primary responsibility for implementation of hazardous waste policy. Texas generates tremendous amounts of toxic waste, and dangerous abandoned dumpsites dot the coastal region. The fragmented administrative structures pursue technological solutions to the problem. A certain acquiescence toward ongoing methods, a lack of fiscal and enforcement capability, and a reluctance to take legal action against violators, characterize toxic waste regulation in Texas.

Clearly, in chapter 7 Kramer is challenging the notion that Texas has been an innovator in hazardous waste policy development. He argues instead that, if anything, Texas has been a reluctant leader. Kramer labels the Texas response "reactive" to federal prodding. Critical actors in the issue area fall into two groups: state-level bureaucrats and industry interest groups. Their relationships condition policy making. The bureaucrats are in a favored position because other institutional actors (legislators, the governor) are plagued by structural weaknesses. Consequently, the prevailing bureaucratic perspective (among relevant administrators it is presumed to be of a technical/engineering nature) pervades hazardous waste policy making. Industrial interest groups are not a countervailing force, rather they appear to have developed symbiotic relationships with the program administrators. Their influence is readily demonstrated in their ability to promote or quash legislation, depending on its threatened impact on their interests.

Yet there are some limits on their power. For example, within the industrial coalition, on occasion, interests will diverge. In addition, indications are that their political influence is more frequently challenged by emerging adversary groups. Texas, as yet, has experienced no catastrophic toxic waste incident to stimulate widespread citizen attention. As a result, the role of public opinion remains somewhat dormant. The future in Texas appears to be incremental drift as the policy area evolves.

Hazardous waste management activity in Florida, New York, and Texas is dwarfed however by California's inventive efforts. Particularly prominent are California's initiation of a land disposal phaseout and the concomitant facilitation of alternative treatment mechanisms. The impetus for these endeavors springs from the interaction of gubernatorial leadership and innovative organizational activity.

Yet as Morell indicates in chapter 8, "innovative" is not an apt description of all of California's toxic waste regulatory activity. For example, the siting of hazardous waste facilities generates conventional state-local political battles. Deliberations concerning the locus of authority for siting decisions—state preemption at one extreme, local veto at the other—reflect the divisive reality of the toxic waste issue. In California, community rejection of proposed hazardous waste facilities led to industry sponsored state preemption legislation. This effort, and a subsequent compromise bill to authorize regional planning councils to override local vetos, were defeated in 1981. The legislature did approve the creation of a politically diverse task force to develop, among other items, a statewide siting plan. As Morell suggests, this deferral of decision making and shift of the arena will not make consensus on the proper state-local relationship any easier to achieve. Thus even in California, with its demonstrated penchant for innovation, important elements of the hazardous waste policy package remain unresolved.

The Florida, New York, Texas, and California cases demonstrate that states experience differing pressures and demands. That the states therefore exhibit differing reactions is not at all surprising and, from the viewpoint of the defenders of the American federal system, may be an advantage. States, if inclined, can design hazardous waste management programs tailored to specific conditions and circumstances.

In chapters 9 through 11, aggregate, comparative analysis replaces the case study method. Chapter 9 explores the role of the public in hazardous waste management. There is a haunting irony to the subject: at a time when citizen interest in toxic waste is escalating, the mechanisms for public involvement in policy formulation are diminishing. Unless this disparity is addressed by the states where the locus of decision making has shifted, confrontations will shape the future.

Rosenbaum traces the checkered history of citizen participation in environmental programs. In the 1970s, it was common to find provisions for public involvement appended to environmental legislation. During the decade, institutional support for citizen participation fluctuated, primarily downward. Even with President Carter's relatively sincere commitment, only modest outreach sorts of public involvement were developed. Information dissemination, primarily through workshops, marked the federal effort. The change in presidential leadership in 1980 set into motion two trends: a lessening of administration interest in citizen participation and a devolution of authority to the states for the operation of hazardous waste programs. This is the arena in which the toxic waste battles of the future will be fought.

Yet inclusion of the public in hazardous waste policy formulation is not without risk. Incorporation of citizen input into decision making may make implementation of a technically derived course of action more difficult. In other words, the state environmental agency may find that responsiveness to citizen concerns conflicts with efficient program management. Resolution of these conflicts could be quite volatile.

Chapter 10 isolates a central, highly political issue in hazardous waste regulation—the siting of disposal facilities. The problem is especially controversial in that the perceived interests of the local community clash with the broader public interests of the state. There is an inverse relationship between costs and benefits with the costs of siting decisions disproportionately borne by the residents, the benefits by the nonproximate public. The explosiveness of the issue (there are technical uncertainties, too) has caused politicians to delegate siting authority to state regulatory agencies. In chapter 10, three dimensions of policy responses are explored: state initiative, local power, and public participation.

Active and passive extremes mark the wide array of postures available within each dimension. For example, states may initiate the siting process or they may merely react to the applications of developers. Local governments may block the construction of a hazardous waste disposal facility or they may find their laws preempted by the state. The public may participate as members of siting boards or find their involvement limited to testifying at public hearings. A multitude of policy options exists within each of these dimensions, as well.

The Hadden, Veillette, and Brandt examination of states which have adopted special siting laws reveals four general patterns: (1) state dominance—the state has assumed virtually total responsibility in the siting process; (2) state/local mix—the state has retained a significant role in siting decisions but has allowed a blend of local government and public access to the process; (3) local/state mix—the state has reduced its role while enhancing that of the localities and the public; and (4) local dominance—the state has abdicated a leadership role, the critical decisions in the siting process are handled by local governments. In each pattern, early but limited public involvement may facilitate the siting process by legitimizing government action and perhaps coopting incipient opposition. In any event, complex decision-making procedures are the result despite the pattern emulated. And, as the researchers note, complexity pushes the siting price tag ever upward, spurring, perhaps, the development of alternative disposal techniques or greater illegal dumping.

In the single state case studies, the differing policy-making contexts were analyzed. Chapter 11, by Lester, Franke, Bowman, and Kramer, attempts to determine whether there is any aggregate relationship between economic, technical, and political variables and levels of hazardous waste policy adoption as of 1979.

Four models of the likely determinants of hazardous waste policy adoption are drawn from the literature on environmental policy and comparative state politics. The research question is simply this: how do technological pressures in the environment, the availability of economic resources, political demands, and administrative-organizational factors affect hazardous waste policy in the fifty states? The dependent variable is a policy adoption score—a nonfiscal measure of the extent to which a state regulates hazardous waste. For each model, several indicators capturing different aspects of the dimension are tested.

The best predictors of hazardous waste policy adoption appear to be techno-logical pressures and state organizational capabilities. States in which greater amounts of toxic wastes are generated are engaged in more extensive regulatory activity. The presence of a consolidated environmental bureaucracy and a professional legislature also promote the development of hazardous waste regulation. The researchers uncover an interesting interaction pattern between organizational capability and the severity of the hazardous waste problem in the fifty states. It appears that as the toxic waste problem attains a certain severity threshold, responsibility for regulatory activity shifts from the bureau-cracy to the legislature. The wealth of the state and political partisanship play decidedly less important roles in understanding hazardous waste policy adop-tion across the country. When southern states are removed from the analysis, however, economic resource variables become significantly linked to policy adoption. Dividing the nonsouthern states by level of waste generated (high/ low) reveals that the policy processes again differ in the two categories. In the high waste-generating states, adoption behavior is linked to resources and legislative professionalism. For the nonsouthern states with less severe toxic waste situations, bureaucratic consolidation and, to a lesser extent, Democratic party strength are factors associated with policy activity.

It is also instructive to reexamine the analytical framework presented in chapter 1. The model identified a series of variables assumed to be associated with the development and implementation of hazardous waste policy. While chapter 11 systematically examined the relationships between several of the independent variables and hazardous waste policy adoption at the state level, other chapters utilized components of the model in exploring particular aspects of the policy-making process. Of special interest are the pieces of the model which, based on the findings of the case studies, appear promising for further pursuit.

One hypothesized relationship which resounds throughout the chapters is the importance of resources upon policy implementation. Funding—agency re-sources, intergovernmental aid, state finances—is consistently related to policy activity. However, perhaps a more important resource is commitment. Commit-ment to responsible hazardous waste management by an array of actors criti-cally influences hazardous waste policy making and implementation. Thus, the role of policy commitment—the concept includes executive leadership, decision makers' dispositions, and public involvement—studied in chapters 2 through 8 is illustrative. At the national level (chapters 2 through 4), EPA's lackluster performance in fulfilling its charge under TSCA, RCRA, and Superfund has been attributed to the agency's cautious disposition as well as to inadequate funding. At the state level (chapters 5 through 8) similar constraints prevailed. For example, in California, gubernatorial leadership has been a spark, while in New York mobilized community-based citizen groups have prodded state action. The absence of this commitment by well-placed actors in Texas (or perhaps more accurately, the influence of industrial interests committed to minimal hazardous waste regulation) has negatively affected the policy response. And in Florida, despite general environmental concern in the legislature, legislative commitment to hazardous waste management has been limited. Therefore, resources, including policy commitment, clearly affect hazardous waste policy activity.

Future research might profit from the development of measures that tap the commitment dimension of the resources factor. A reservoir of commitment has special importance in that it may be able to offset somewhat a diminution in fiscal resources. From these preliminary indications, one might speculate that a concept such as "leadership intensity" has potential explanatory power. The locus of leadership is variable, however. Examination of the findings of this research collection will reveal the multisource nature of leadership—governmental, private sector, citizenry.

Other factors affect the likelihood that these resources will be amassed and channeled toward hazardous waste control. For example, media attention to the existence of dangerous waste sites can stimulate interest among potentially significant actors. And, once interest develops, reformed governmental institutions such as professional legislatures and consolidated bureaucracies may facilitate the formulation of hazardous waste policy. Structural reform has apparently improved the functional capacity of these institutions. One may discover that the absence of reformed structures, and more especially, the presence of a highly fragmented bureaucracy, is likely to frustrate the role of interorganizational communication. A multiplicity of agencies involved in this area, with differing perspectives and policy solutions, may adversely affect or delay positive government action.

Further dissecting the generalized dependent variable in the model (hazardous waste policy formulation and implementation) into specific components, e.g., disposal facility siting or the creation of cleanup contingency funds, elevates other variables to prominence. Citizen involvement conditions the siting process while public-private sector relations affect the latter. Or, if one were to isolate another element in the dependent variable, achievement of substantial equivalence with RCRA guidelines, interorganizational communication may become a significant variable. The point is that the various subsets of the dependent variable offer a fertile testing ground for hypotheses generated in this research.

Thus, the comprehensive model offers a basic understanding of the clusters of variables influencing hazardous waste policy formulation and implementation. The pieces may be combined and recombined to explain hazardous waste management across the country.

Central to each of these chapters has been the pervasive influence of politics in hazardous waste management. Whether the tensions are among levels of government, across governmental branches, between the public and private sectors, between the governmental apparatus and the citizenry, or across functional bureaucratic units, there can be little doubt that political considerations shape hazardous waste control. And while governmental influence in hazardous waste policy cannot be denied, evaluations of governmental performance challenge the adequacy of it. Despite pockets of innovation, the bulk of governmental activity has fallen short of optimal. In the next section, the role of politics in hazardous waste management will be examined.

Policy Implications: Hazardous Waste Regulation as a Political Activity

What the hazardous waste issue points up is the transference of a technological problem, with economic overtones, into the political arena. Hazardous materials exist in the environment naturally; however, of greater concern are those that are the end result or by-product of the production process. End results (e.g., pesticides) are utilized by consumers; by-products accumulate until disposed of. It is in the discard stage that these materials earn the designation "hazardous waste."

For years, a seemingly acceptable nongovernmental solution to the disposal of these toxic wastes was effected. Generators of hazardous wastes would dispose of them on-site (allowing them to accumulate in surface impoundments such as pits, ponds, and lagoons) or contract for their transport to an off-site disposal location. In either case, the hazardous by-product was conveniently handled by the private sector. If any governmental intervention occurred, it was limited to enforcement of general nuisance laws.

Intrinsic to this process, of course, is the temptation to secure economic advantage by instituting a low cost disposal mechanism. The allocation of resources to the discard stage would, rather obviously, cut into profits. Consequently, minimizing expense at this stage would return greater profitability to the enterprise. Any hint of future complications could be ignored in the pursuit of short-term economic advantage.

The growing awareness of the larger impact of this process—that low cost generally translates into unsafe disposal—forced governmental intrusion. Highly visible spills and accidents triggered public clamor for corrective action by government and turned what had been popularly thought of as the musings of doomsayers into a real political issue. The interaction of economically motivated disposal behavior, citizen mobilization, and government regulatory activity make the North Carolina PCB case worthy of exploration.

North Carolina, PCBs, and Politics

In 1978, over two hundred miles of roadsides in fourteen North Carolina counties were contaminated by the illegal spraying of polychlorinated biphenyl-laced oil.[10] Responsible for the dumping of the 31,000 gallons of waste oil were three truck drivers who had been hired by a large transformer repair company.[11] Illegal dumping was selected as a low cost means of evading compliance with proper disposal regulations. Thus a technological problem was solved by circumventing the politically determined remedies and substituting instead economically justified behavior.

But there was a potentially dangerous externality associated with the preferred disposal option: the threat to public health. Odors and seepage from the waste oil eventually attracted public attention and produced the predictable outcry for governmental intervention. The governmental response was two-pronged: to file charges against the truck drivers and the transformer repair company owner and to initiate procedures for the disposal of the highly toxic soil. Convictions of the perpetrators led to jail sentences, fines, and the filing of civil suits for damages. But the soil disposal issue presented no easy solution. The events associated with its pursuit illustrate some of the themes of the previous chapters.

Federal-state relations. North Carolina treated the affected roadsides with a mixture of activated charcoal and emulsified asphalt to immobilize the toxins. However, permanent, in-place treatment is not allowable under EPA regulations and the state's request for a waiver was denied. Incineration of the soil was rejected by EPA as too time consuming and expensive. Consequently, removal of the soil from the roadsides and transportation of it to a landfill was the disposal option selected.

The use of rational criteria in siting decisions. Examination of approximately one hundred potential landfill sites led to the selection of a 142-acre field in Warren County. What made the site attractive for disposal, according to public statements by North Carolina officials, was its proximity to the contaminated roadsides, its geological characteristics, and its scattered population settlements.

Public opposition to disposal facilities. Announcement of the selection triggered immediate reaction. Warren County, poor with a high proportion (64 percent) of black residents, does not possess the attributes typically associated with political mobilization.[12] Yet, driven by visions of the heavy hand of racism masquerading as scientific decision making, Warren County residents mounted extensive political protest. As their efforts to prevent the construction of the dump were rebuffed, they organized demonstrations at the landfill. Their ranks were reinforced by the involvement of established civil rights organizations. Demonstrations at the landfill led to the arrest of more than 500 protesters.

State-local relations. The Warren County Commission (which had adopted an ordinance prohibiting the PCB disposal) sought unsuccessfully in federal court to prevent the state's burial of the contaminated soil in the county.

Eventually, the county, in exchange for dropping its appeal of the lower court's negative decision, struck a deal with the state by which a 123-acre buffer area surrounding the landfill would be turned over to Warren County. The state continued to hold title to the nineteen-acre landfill area.

Technological treatment. Thirty-two thousand cubic yards of contaminated soil, collected by scraping the roadside in swaths two and a half feet wide and three inches deep, were loaded in tarpaulined dump trucks, hauled to the site, and dumped. The process took six weeks and cost the federal government $2.54 million with North Carolina providing the remaining 10 percent of the funds. Once dumping was completed, the landfill was capped and the scarred roadsides were to be regraded and reseeded. Yet, the risk associated with the disposal technique remains uncertain. Within EPA, there was some dispute as to the probability of the eventual leaking of toxic leachate.[13]

The lessons from the North Carolina PCB case can be summarized in the following manner:

Waste generators will pursue cost beneficial means of handling toxic by-products. At some point, illegal dumping becomes more attractive than safe disposal.

There is a critical need for adequate treatment, storage, and disposal of hazardous wastes. Treatment technology has lagged behind waste-generating capacity.

Treatment and disposal facility siting possesses tremendous salience to the proximate public. Siting proposals can split apart the established coalitional pattern of interest group activity and forge recombinations.

Hazardous waste policy, despite efforts at depoliticization—at making the regulation of hazardous waste the responsibility of scientists and administrators—remains a political issue.

Despite recent governmental catch-up efforts, inadequate regulatory mechanisms currently exist.

Hazardous waste control is not unijurisdictional; rather, it is a multigovernmental issue. A significant amount of cross-governmental tension is created as local, state, and national officials interact to address the problem, ironically precipitated, in part, by governmental failures.

Satisfactory public-private sector relationships remain underdeveloped.

It is the interconnections of these elements that structure the future in hazardous waste policy.

Future Issues and Policy Choices

Examination of the preceding list reveals, besides the interconnections, the prominence of politics in the process. Condensed and simplified, the central argument of this book is that responsible hazardous waste management is primarily a function of political activity. However, it would be incorrect to underestimate the power of economics and technology in conditioning the political process. Because hazardous waste is an emerging policy area, the opportunity exists to develop an effective management strategy. But these opportunities will be lost if remedies are narrowly focused. Significant political solutions must take into account their economic and technological implications. In the discussion to follow, the key political issues—inclusive of their economic and technological dimensions—will be presented.

Levels of Government and Hazardous Waste Policy

Because of the political nature of hazardous waste management, the involvement of governmental institutions is inevitable. The future portends a contracted role for the national government, an expanded one for state and local governments. The national government has provided the framework for attacks on the hazardous waste problem. Federal efforts such as TSCA, RCRA, and Superfund are aimed at ensuring that minimum standards of hazardous waste control exist throughout the nation. Without the federal impetus, some states would have continued to lag in policy development. Implementation of the manifest tracking system and myriad other regulations on facility permitting will allow states to develop equivalent programs and thus receive operational authority. As the number of state authorizations under RCRA increases, the federal role will be reduced to oversight and inspection.[14] EPA's leadership capabilities remain somewhat problematical given the reports of internal management dissension, especially between political appointees and mid-level career bureaucrats.[15] EPA's clashes with Congress regarding the agency's refusal to provide Superfund cleanup documents further erode organizational performance.[16] And the growing suspicions that *political* considerations played a part in the release of Superfund monies has created a flurry of congressional inquiries and investigations.[17] One outcome of the conflict has been a revival of congressional interest in restructuring EPA as an independent unit, governed by a commission.[18]

One facet of hazardous waste policy awaiting federal attention is that of compensation for victims of toxic pollution. Victim compensation provisions were stripped from the 1980 Superfund legislation in an effort to secure passage of the bill. As a compromise, Congress directed a panel of legal experts to study the compensation issue. Their report, released in 1982, was critical of existing legal remedies and recommended adjustments. Called for was a bilevel approach: one for smaller claims to be financed through a tax-generated federal fund, the other for larger claims to flow through the state court system.[19] The reforms enhance the position of the alleged victim and extend the liability for the alleged perpetrators. Observers predict intransigence from affected industries as the recommendations receive legislative airing.[20] Congressional unwillingness or inability to reach a consensus on the compensation questions could create still another void into which states might venture, if desirous.

Beyond oversight and inspection, federal involvement in the future will most likely aim at the encouragement of new technology through the provision of funds to finance research and development. In addition, the national government might attempt to spark management innovations by other jurisdictions through some type of competitive grant program. Too, the federal government will continue to be involved in regulating ocean dumping and in the interstate aspects of the hazardous waste problem, particularly transportation.

States bear the responsibility for implementing hazardous waste programs, and the future appears to be one of increased programmatic activity. Even if the Reagan administration's new federalism is short-lived, states will have the option of exhibiting greater policy leadership in the future. Reorganization of state environmental agencies, for example, tends to elevate the function of hazardous waste regulation by giving it a discrete bureaucratic niche. External pressure to construct treatment and disposal facilities will force even the most reluctant states to address the siting issue. Transfer centers—facilities where small generators can consolidate and store wastes temporarily—need to be established. State legislative attention is increasingly drawn to questions of hazardous materials transportation (carrier requirements and good samaritan legislation), workers' right-to-know about toxic substances in the workplace, financial assurances for long-term care of disposal sites, and the like.[21]

The option of enacting legislation to expand basic RCRA provisions or to address issues not covered in federal law may tax the capacities of some states. Those that do exercise leadership will be able to tailor the policy response to specific conditions and needs. Assuming the availability of adequate resources, state policy action may lead to more effective program operation.

A significant problem for state-level hazardous waste regulation revolves around interstate competition. State initiated endeavors to secure economic vitality by retaining present industry and enticing new industry to relocate carry environmental costs. Yet there is reluctance to adopt "too stringent" a set of hazardous waste controls because it could dispel the favorable business climate the state is attempting to create. This dilemma could be moderated somewhat by a regional approach to hazardous waste management. For example, an interstate compact produced the Northeast Hazardous Waste Coordinating Committee which functions as an information clearinghouse. Regionalism may be a casualty of the new federalism, however.

Local governments possess the very real problem of playing host to toxic waste facilities. Even though on-site disposal is likely to increase, the key issue that will dominate state-local relations will be the siting of hazardous waste treatment and disposal facilities.[22] Local government, due to its "creature of the state" nature, has little independent authority. The state will determine the extent to which local governments can influence siting decisions through zoning regulations or land-use plans.

The Public-Private Sector Relationship

In the preceding section, a "technology encouraging" role for the federal government was advocated. Thus one of the public-private sector interfaces could be government financial support of research and development in alternative treatment and disposal technologies. Some of the technological avenues seeming to be most attractive include: (1) waste reduction, (2) pretreatment, and (3) resource recovery. The first option involves the diminution of the amount of waste generated in the production process. The payoff for the installation of these modifications is that there will be less waste to dispose of, hence a cost savings will result.

But a redesign of the production process is not the only waste reduction alternative available. A range of pretreatment techniques—physical, chemical, biological—exist. Their goal is not confined to a reduction in the volume of waste but includes the isolation of reusable substances and the detoxification of waste streams as well. Interest in the development of pretreatment depends on incentives: directly, by means of federal research subsidies; indirectly, by a state's economic discouragement of landfilling.

While the potential of resource recovery is evident, its extensive use will first require a reorientation toward conceiving of toxic waste streams as valuable. Growing awareness of recycling opportunities (especially as costs of disposal and raw materials escalate) has led to the formation of waste exchanges and the emergence of a new occupation—waste brokerage. Thus far the impact is limited in that only about 10 percent of the available waste materials are exchanged.[23] Federal legislation to encourage these options would certainly speed their development and diffusion.

For states, their relationship with the private sector will be one of determining appropriate regulatory parameters. As noted previously, this is a sensitive exercise due to the political resources possessed by the regulated industries. Industry may be more likely to perceive of hazardous waste as an economic problem thus concluding that government regulation is counterproductive. The preferred policy solution emanating from that conclusion would be one in which the state would enact cost effectiveness measures such as pollution taxes and effluent charges.[24] Thus states will expend tremendous energies locating the proper balance between excessive regulation and no regulation.

New York, committed to regulation, is attempting to handle the problem by beefing up its enforcement capability.[25] A specially trained unit, the Bureau of Environmental Investigations, has been formed to uncover the illegal storage, transport, and disposal of toxic wastes. The intent of this legislatively funded effort is to convince violators that the cost of their behavior in fines and imprisonment makes compliance economically preferable. By increasing the certainty of getting caught and the severity of punishment if convicted, the rational violator should respond accordingly.

A continuing challenge for states will be funding. In a state experiencing a decline in revenue growth or wide competition for available resources, regulation can be an expensive bureaucratic activity. Therefore, pressures will mount to reduce the cost borne by the state. There are a number of ways of accomplishing this. For example, the existing regulations can be retrenched or selectively enforced. The outcome may be short-term savings, but in the absence of curative action by hazardous waste industries, it increases the probability of severe long-term consequences. Another strategy would be to shift some of the implementation and enforcement activities to local governments, but that virtually ensures that only the wealthy communities would have adequate programs. Any expectation of comparable statewide standards would be dashed. A different type of option would maintain current regulations and levels of enforcement but transfer the cost of such to the general public via a tax increase or to the relevant hazardous waste generators and handlers through fees. The public is unlikely to accept a tax increase without protest. And in a state trying to bolster its economic base, additional fees could provoke industrial departure. Yet there may be expanding sentiment that those who place the regulatory burden on the state ought to bear the cost of regulation.[26] Each of these alternatives contains negative consequences. Policy makers will be forced to sift through and weigh the relative advantages and disadvantages of each course of action. The task will not be easy but diminishing federal funds make it unavoidable.

In the public-private sector relationship, a central linkage exists between a state's environmental control agency and hazardous waste-related industries. Not necessarily an adversarial relationship, describing it as cooperative, however, might be misleading. Scant data exist on environmental agency capture by hazardous waste industry, but one might speculate that the linkage has not been regularized to the extent that capture has resulted.[27] Hazardous waste is a developing policy area, hence the struggle for influence in the policy process in the states continues.[28] As the states become the important arenas for policy formulation and implementation, the public-private sector relationship will be shaped accordingly.

And it may be that the salience of the hazardous waste issue is so great that a third party—the public—cannot be ignored. Preliminary research comparing the attitudes of state environmental agency personnel and chemical industry officials hints that the antagonism between regulator and the regulated may recede as they attempt to manipulate citizen opposition, at least in disposal facility siting.[29] Both sets of administrators endorse passive public involvement (e.g., testifying at administrative hearings, membership on siting committees) but reject more extreme measures such as voter ratification of site selection. The officials diverge, however, on attitudes toward more intermediate forms of citizen participation (e.g., the initiation of environmental lawsuits and the support of environmentally committed candidates). The percentage of state administrators favoring such activities is 42 percent and 33 percent respectively. For industry officials, those percentages are 14 percent and 5 percent.[30] Thus the expected conflict between the regulatory agency and the regulated industry may be submerged in pursuit of public mollification.

Citizen Involvement in Hazardous Waste Policy

One regulatory cost-cutter is coproduction—the involvement of the citizenry (the consumer) in the provision of a service.[31] An example of active coproduction would be the use of citizen volunteers (rather than employees) to monitor the contaminant level at a disposal facility. A more passive approach to coproduction involves corrective behavior on the part of residents, such as engaging in proper disposal of pesticide residues or waste oil.[32] The concept of coproduction meshes with the oft-heard demand for less government. The relative newness of hazardous waste regulatory activity may facilitate the installation of coproduction.

A more formal means of citizen participation is through the use of task forces, study committees, and the like. Creation of these groups buys time for the state, shifts responsibility, provides expertise at little cost, and ensures a source of support for whatever recommendations are made. These ad hoc groups generally are constructed to maximize the array of viewpoints and perspectives in the policy dialogue.

Citizen involvement in hazardous waste management is being encouraged by nongovernmental entities, more specifically, a variety of national environmental organizations. The Environmental Defense Fund, the Sierra Club, the Citizens Clearinghouse for Hazardous Waste, and the Environmental Action Foundation are spearheading grassroots efforts to influence hazardous waste policy. They provide technical information about toxic waste, inspect local dumpsites, and organize citizens for political action.[33] Coalitions of community-based environmental groups are undertaking citizen education projects as well. Increased activism heightens the visibility of the issue and, consequently, its unmanageability.

Concluding Remarks

From this examination of three policy dimensions, it appears that eight interrelated concerns will dominate state-level hazardous waste policy horizons:

- allocating resources for cleanup activity,
- commencing the negotiation of settlements with generators, handlers, and disposers of toxic wastes for cleanup cost sharing,
- enacting the appropriate balance of regulatory controls so as to encourage compliance while spurring the development of alternative technologies,
- apportioning the regulatory activity between state and local spheres,
- bolstering the enforcement effort,

— determining what role the public will play in hazardous waste policy formulation and implementation,

— entering into interstate agreements to address the potential regional effects of toxic wastes,

— designing more effective relationships with the private sector.

Whether or not the new federalism experiment prevails, it is clear that state governments cannot shrink from the hazardous waste challenge. Assuming that the devolution to the states of greater programmatic authority is not merely an aberration in an otherwise centralizing spiral, then interstate and intrastate issues will define the future. Desultory governmental behavior will no longer be tolerated by the public. The definitive need in state-level hazardous waste management in the 1980s will be, quite simply, effective political leadership.

Epilogue

Ann O'M. Bowman

The cascading events of early spring 1983 that swept the Environmental Protection Agency into the center of controversy underscore the political implications of hazardous waste management. While students of public administration have long since discarded the Wilsonian assumption that politics and administration are separable in the policy-making process, the revelations of the gross intrusion of politically motivated concerns into the implementation of hazardous waste policy has stunned observers. The resulting furor has elevated the hazardous waste issue to a special prominence.

In examining the events, four related issues seem to dominate the period: EPA's internal management operations, the agency's intra-branch links with the White House and its extra-branch links with Congress, EPA's connection to the hazardous waste industry, and the agency's relationship with state governments. The discussion to follow will highlight the significant factors in these four areas.

What had been a mere hint of intra-agency dissension—a whistle-blowing employee criticizing EPA's handling of hazardous waste programs—blossomed into a full-fledged dispute when the agency apparently launched a campaign of retaliation. Hugh B. Kaufman, a constant critic of EPA's management of toxic waste programs, ultimately negotiated a settlement with the agency after a Labor Department investigation sided with his allegations of harassment.[1] But Kaufman's charges of agency mismanagement sparked several congressional investigations. The resulting clamor, triggered by agency refusal to provide a congressional subcommittee with documents relating to the Superfund program, reached its apex when the House of Representatives cited EPA Administrator Anne Gorsuch Burford for contempt of Congress.[2] From that point the battle seemed to oscillate along a continuum whose end points were containment of the crisis and exploitation of it. Negotiations over access to the documents continued amid White House claims of executive privilege and congressional counterclaims of politically inspired manipulation of toxic waste programs and subsequent White House coverup efforts.[3]

Within the agency, shake-ups among top personnel occurred. Rita Lavelle, an assistant administrator of the toxic wastes program, was dismissed by President Reagan after charges lodged against her regarding conflict of interest and industry favoritism would not fade.[4] Two weeks later EPA's inspector general and the official in charge of administration for the agency resigned, both of them reportedly at the request of the White House.[5] The administration dispatched a team of replacement officials to EPA ostensibly to improve the management capacity of the agency. But these actions failed to quell the escalating disputes, especially as revelations surfaced suggesting a political rationale for the release of Superfund cleanup monies. Specifically, agency critics asserted that funding announcements and settlement agreements were propitiously timed, regardless of site severity, so as to aid senatorial and gubernatorial candidates who were administration supporters and hinder administration opponents.[6] Further, the disclosure of the compilation by agency leaders of a "hit list" of EPA officials considered extreme environmentalists or Democratic partisans and marked for demotion, transfer, and dismissal further damaged the agency's credibility.[7] Efforts to reverse the agency's tarnished image culminated in the highly publicized Times Beach, Missouri, buyout plan, where after years of delay EPA announced that the federal government would purchase property in the dioxin-contaminated community.[8]

The turmoil grew to such an extent that Burford's resignation became inevitable, coinciding with the arrangement of a compromise on the disputed documents by the Reagan administration and congressional subcommittees. Whether Burford was merely a pawn in a larger executive-legislative confrontation is not clear. After her resignation Burford claimed she had been willing to comply initially with subcommittee requests regarding document provision but was overruled by a White House–Justice Department plan.[9]

Before the controversy subsided, 13 top-level officials resigned from the agency. Many of these high-ranking officials had entered EPA with a decidedly anti-environmental perspective, intent upon diminishing governmental involvement in industry activity. Appointment of EPA's deputy administrator, John Hernandez, as acting chief of the agency was a stopgap measure at best. Almost immediately, reports circulated concerning the decline in research and development spending during his tenure as deputy and his propensity to minimize threats to public health from hazardous wastes.[10] What led to Hernandez's demise were allegations by EPA Region 5 personnel that he had interceded to allow Dow Chemical Company to modify an EPA report linking the company to dioxin contamination in Michigan waterways. In addition, EPA Region 6 officials testified before Congress that Hernandez had halted their efforts to conduct soil removals at heavily populated areas contaminated by lead.[11] The political fallout from these disclosures rendered Hernandez a liability to the administration.

Seeking to deflect the heat from the charges being aired before the six investigating subcommittees, President Reagan nominated EPA's first administrator, William D. Ruckelshaus, to return to the beleaguered agency. His selection was interpreted by some observers as a harbinger of stability and integrity in the organization.[12] Agency personnel were generally supportive and viewed the appointment as a morale booster.[13] Initial reports indicated that, as administrator, Ruckelshaus would enjoy the best of two worlds: freedom from political interference from the White House and access to the inner circle of decision makers. To solidify his acceptance, one of Ruckelshaus's first actions upon nomination was a fence-mending parley with leaders of environmental groups.[14]

Leadership changes within the agency quieted the immediate clamor, but for many weeks continued congressional investigations and persistent media attention kept the hazardous waste issue squarely in the public eye. This scrutiny yielded many disclosures as to agency relationships with the hazardous waste industry. Following are examples of four of these disclosures, and while the generalizability of these four examples may be debatable, their significance is not. First, closer examination of the once-heralded Seymour, Indiana, cleanup agreement suggested that the federal government's desire to attain a timely settlement effectively weakened its negotiating position. The cooperating generators obtained a release from all liability, including subsurface, in exchange for their limited surface removal actions.[15] Second, at the Bridgeport, New Jersey, toxic lagoon, reports indicated that EPA neglected to pursue information that would have accelerated identification of waste generators and progress toward a cleanup settlement. EPA apparently regarded this information as inconsequential, but, thus far, the agency has been able only to secure a $25,000 fine against the now-defunct site operators, despite the fact that conditions at this lagoon are so severe that EPA will incur a $200,000 expenditure of emergency funds, as a mere containment strategy, to siphon off rising wastes at the site.[16] Third, from the Midwest came assertions that EPA headquarters had forced regional personnel to drop plans to conduct on-site testing of water and fish at a Dow Chemical facility in Midland, Michigan, in 1981. Comparable tests, completed two years later, found significant concentrations of toxic organic compounds in the plant's wastewater which flows into a nearby river.[17] Fourth, at the Stringfellow Acid Pits in California, the site from which former Administrator Burford withheld Superfund monies, the Justice Department eventually sued generators for cleanup costs in April 1983. Comparing lists of the waste generators who dumped at the Stringfellow site with previous affiliations of administration and EPA personnel who were making implementation decisions raises conflict of interest questions.[18]

EPA performance in these cases also calls into question agency success in maintaining an appropriate regulatory relationship with industry. And because hazardous waste is a developing policy area, the patterns set in these early stages will mold the regulatory structure into maturity. EPA may have been unfairly castigated in the onrush of allegations, but there is little doubt that dissension impaired the agency's ability to function.

The role of subnational governments in hazardous waste management remains a vital concern. As the EPA controversy unfolds, it becomes evident that one of the developments identified in this collection—the importance of state institutions in hazardous waste regulation—is unlikely to recede. It may be that dismay over the management of hazardous wastes at the national level will increase performance pressures on state governments. Reports scoring EPA's cleanup delays often portrayed the relevant state environmental bureaucracy in a more activist light, e.g., the Michigan Department of Natural Resources' advocacy of the Dow Chemical water tests and the Ohio officials' complaints about EPA's tortuous, red-tape-laden procedures at the Chem-Dyne site.[19] These and similar instances have fostered the impression that EPA's commitment to facilitating the expenditure of Superfund monies was lukewarm at best. Defenders of the agency's operations contend that the strategy was designed to reduce confrontations between EPA and industry and to encourage upfront settlements, rather than engage in protracted legal battles. In addition, some Superfund delays were due to EPA–state disagreement over the classification of a site as privately owned (state share of costs is 10 percent) or publicly owned (state share of costs is 50 percent). Whatever the rationale, state governments find themselves inextricably involved.[20]

The critical role of states appears to rest in the permitting and licensing of hazardous waste facilities. As the EPA situation returned to a semblance of normalcy, reports of serious accusations confronting Waste Management, Incorporated, the nation's largest hazardous waste disposal firm, became public.[21] The company is facing an array of charges across the country, ranging from repeated illegal dumping of toxic chemicals, to pursuit of improper influence at EPA, to price fixing. Waste Management labels its violations "minor infractions." The resulting publicity has had wider ramifications than causing Waste Management stock prices to plummet. States with Waste Management facilities are probing the company's activities; citizen groups are forming to oppose their applications for new facility siting permits. The exercise of state government permitting and licensing responsibilities does not bode well for Waste Management, Incorporated, and raises clouds over the entire disposal industry as well.

The EPA imbroglio has had numerous effects, but a particularly significant one has been the rekindling of congressional interest in hazardous waste policy. In chapter 12 it was noted that reauthorization of RCRA would likely be a loophole-tightening venture, with little expansion of the 1976 provisions. That appears less predictable now, especially after the release of the Office of Technology Assessment's report calling for further federal action. OTA concluded that "the government is pursuing a problem it does not fully understand with too little money, vague technical standards, imperfect laws and sometimes contradictory policies."[22] Consequently, legislation has been introduced in Congress to expand governmental control over hazardous waste. Representative James J. Florio's proposal would tighten the small generator exemption and prohibit disposal of toxic wastes in injection wells located near underground water supplies. Senator Gary Hart's bill would expand Superfund monies by replacing the Superfund tax with a fee on storage or disposal of hazardous wastes.[23] Other proposals are likely to result as congressional investigations continue. Hazardous waste has become a priority item in Congress.

Hazardous waste is a congressional priority primarily because of its priority among the citizenry. A front page headline in the *Wall Street Journal* captures the sentiment: "Local Citizen Groups Take a Growing Role in Fighting Toxic

Dumps."[24] It appears that EPA's purported favoritism toward polluters has mobilized a latent environmentalism across the country. Grass roots opposition groups spring up when news of the possible construction of a hazardous waste facility, be it a landfill or not, is released.

As evidence documenting the inevitable leakage from both clay- and plastic-lined landfills is reported and their siting becomes more problematical, pressure for alternative treatment methods will intensify.[25] There is growing interest in mid-oceanic incineration of toxic wastes, in recycling and reuse, in further testing of detoxification through use of mutant bacteria dubbed "superbugs." It remains for state governments to choose the appropriate regulatory schemes that offer incentives for alternatives to land disposal, especially if EPA refuses to do so.

The hazardous waste issue has become irrevocably politicized. EPA, despite its new leadership, despite its much-delayed hiring of additional enforcement investigators, will continue to function under revenue constraints barring a significant shift in the political climate in Washington. Hazardous waste policy remains an arena ripe for experimentation. It seems that at the state level it can be accomplished most expeditiously. Whether the states will assume a significant leadership role or balk at the opportunity for positive governmental action remains an essential question for the future of responsible hazardous waste management.

Appendix A. Proposed National Priorities List for Superfund Cleanup Activities by State

City/County	Site name (Priority group)
Alabama	
Limestone & Morgan	Triana, Tennessee River (1)
Greenville	Mowbray Engineering (3)
Perdido	Perdido Groundwater Contamination (8)
American Samoa	
American Samoa	Taputimu Farms (2)
Arizona	
Globe	Mt. View Mobile Home (2)
Phoenix	19th Avenue Landfill (2)
Tucson	Tucson International Airport (2)
Goodyear	Litchfield Airport Area (4)
Kingman	Kingman Airport Industrial Area (6)
Scottsdale	Indian Bend Wash Area (6)
Arkansas	
Jacksonville	Vertac, Inc. (1)
Mena	Mid-South (4)
Edmondsen	Gurley Pit (6)
Ft. Smith	Industrial Waste Control (6)
Walnut Ridge	Fritt Industries (6)
Newport	Cecil Lindsey (7)
Marion	Crittendon Co. Landfill (8)
California	
Glen Avon Heights	Stringfellow (1)
Rancho Cordova	Aerojet (2)
Redding	Iron Mountain Mine (2)
Ukiah	Coast Wood Preserving (4)
Fresno	Purity Oil Sales, Inc. (5)
Fresno	Selma Pressure Treating (5)
Fullerton	McColl (5)
Richmond	Liquid Gold (5)

(9)(9)(9)(9)

(7)(7)(7)(7)(7)

5(7)(7)(7)

(7)(7)(7)

258 The Politics of Hazardous Waste Management

Cloverdale — MGM Brakes (7)
Hoopa — Celtor Chemical (8)
Sacramento — Jibbom Junkyard (9)

Colorado
Boulder — Marshall Landfill (2)
Leadville — California Gulch (3)
Commerce City — Woodbury Chemical (4)
Denver — Denver Radium Site (4)
Idaho Springs — Central City, Clear Creek (4)
Commerce City — Sand Creek (6)

Connecticut
Naugatuck — Laurel Park Inc. (2)
Beacon Falls — Beacon Heights (4)
Southington — Solvents Recovery System (6)
Canterbury — Yaworski (7)

Delaware
New Castle — Army Creek (1)
New Castle County — Tybouts Corner (1)
New Castle — Delaware Sand and Gravel (4)
New Castle — Tris Spill Site (6)
Delaware City — Delaware City PVC Plant (8)
Dover — Wildcat Landfill (8)
Kirkwood — Harvey Knott Drum Site (8)
New Castle — New Castle Steel Site (8)

Florida
Jacksonville — Pickettville Road Landfill (1)
Plant City — Schuylkill Metals (1)
Davie — Davie Landfill (2)
Galloway — Alpha Chemical (2)
Miami — Gold Coast Oil (2)
Hialeah — NW 58th Street (3)
Miami — Miami Drum (3)
Seffner — Taylor Road Landfill (3)
Tampa — Kassauf-Kimerling (3)
Tampa — Reeves SE Galvanizing (3)
Tampa — 62nd Street Dump (3)
Warrington — Pioneer Sand (3)
Whitehouse — Whitehouse Oil Pits (3)
Zellwood — Zellwood Groundwater Contamination (3)

Cottondale	Sapp Battery (4)
Fort Lauderdale	Hollingsworth (4)
Indiantown	Florida Steel (4)
Live Oak	Brown Wood (4)
Miami	Varsol Spill (4)
Whitehouse	Coleman Evans (4)
Pensacola	American Creosote (5)
Clermont	Tower Chemical (6)
DeLand	Sherwood Medical (6)
Mount Pleasant	Parramore Surplus (7)
North Florida	Munisport (8)

Guam

Guam	Ordot Landfill (2)

Idaho

Smelterville	Bunker Hill (2)
Caldwell	Flynn Lumber Co. (5)
Rathdrum	Arrcom (Drexler Enterprises) (9)

Illinois

Greenup	A & F Materials (2)
Waukegan	Outboard Marine Corp. (2)
Wauconda	Wauconda Sand & Gravel (3)
Marshall	Velsicol Illinois (4)
Pembroke	Cross Bros./Pembroke (5)
Waukegan	Johns-Manville (6)
Galesburg	Galesburg/Koppers (7)
Ogle County	Byron Salvage Yard (7)
La Salle	La Salle Electric Utilities (8)
Winnebago	Acme Solvent/Morristown (8)
Belvidere	Belvidere (9)

Indiana

Gary	Midco I (1)
Seymour	Seymour (2)
Kingsbury	Fisher Calo (3)
Boone County	Envirochem (4)
Bloomington	Lemon Lake Landfill (5)
Bloomington	Neal's Landfill (5)
Columbia City	Wayne Waste Oil (5)
Elkhart	Main Street Well Field (5)
Gary	Ninth Avenue Dump (5)

Gary	Lake Sandy Jo (6)
Marion	Marion (Bragg) Dump (7)
Lebanon	Wedzeb Inc. (8)
Allen County	Parrot Road (9)

Iowa

Charles City	Labounty Site (1)
Council Bluffs	Aidex Corporation (2)
Des Moines	Dico (9)

Kansas

Cherokee County	Tar Creek, Cherokee Co. (1)
Arkansas City	Arkansas City Dump (2)
Holiday	Doepke Disposal, Holiday (4)
Wichita	John's Sludge Pond (7)

Kentucky

Brooks	A. L. Taylor (2)
Louisville	Lee's Lane Landfill (6)
Newport	Newport Dump (6)
West Point	Distler Brickyard (6)
Calvert City	Airco (7)
Jefferson County	Distler Farm (7)
Calvert City	B. F. Goodrich (8)

Louisiana

Darrow	Old Inger (2)
Sorento	Cleve Reber (3)
Bayou Sorrel	Bayou Sorrel (6)
Slidell	Bayou Bonfouca (7)

Maine

Gray	McKin Company (1)
Winthrop	Winthrop Landfill (2)
Washburn	Pinette's Salvage Yard (6)
Augusta	O'Connor Site (8)
Saco	Saco Tanning (8)

Maryland

Elkton	Sand, Gravel and Stone (5)
Annapolis	Middletown Road Dump (6)
Cumberland	Limestone Road Site (8)

Massachusetts

Acton	W. R. Grace (1)
Ashland	Nyanza Chemical (1)
East Woburn	Wells G&H (1)
Holbrook	Baird & McGuire (1)
Woburn	Industri-Plex (1)
New Bedford	New Bedford (2)
Plymouth	Plymouth Harbor/Cordage (2)
Dartmouth	Re-Solve (4)
Tyngsboro	Charles-George (4)
Westborough	Hocomoco Pond (4)
Groveland	Groveland Wells (5)
Lowell	Silresim (5)
Bridgewater	Cannon Engineering (6)
Palmer	PSC Resources (6)

Michigan

Swartz Creek	Berlin & Farro (1)
Utica	Liquid Disposal Inc. (1)
Cadillac	Northernaire Plating (2)
St. Louis	Gratiot County Landfill (2)
Brighten	Spiegelburg Landfill (3)
Davisburg	Springfield Township Dump (3)
Filer City	Packaging Corp. of America (3)
Grand Rapids	Butterworth #2 Landfill (3)
Muskegon	Ott/Story/Cordova (3)
Pleasant Plains Twp.	Wash King Laundry (3)
Rose Township	Rose Township Dump (3)
Utica	G&H Landfill (3)
Albien	McGraw Edison (4)
Mancelona	Tar Lake (4)
Pennfield Township	Vernon Well Field (4)
St. Louis	Velsicol Michigan (4)
Greilickville	Grand Traverse Overall Supply Co. (5)
St. Louis	Gratiot Co. Golf Course (5)
Wyoming	Spartan Chemical Company (5)
Grand Rapids	Chem Central (6)
Ionia	Ionia City Landfill (6)
Kalamazoo	K & L Avenue Landfill (6)
Otisville	Forest Waste Products (6)
Park Township	Southwest Ottawa Landfill (6)
Temperance	Novaco Industries (6)
Whitehall	Whitehall Wells (6)

Kentwood	Kentwood Landfill (7)
Ludington	Mason County Landfill (7)
Marquette	Cliff/Dow Dump (7)
Muskegon	Duell & Gardner Landfill (7)
Muskegon	SCA Independent Landfill (7)
Petoskey	Petoskey Municipal Wells (7)
Rose Township	Cemetery Dump Site (7)
Adrian	Anderson Development (8)
Brighton	Rasmussen's Dump (8)
Charlevoix	Charlevoix Municipal Well (8)
Clare	Clare Water Supply (8)
Grandville	Organic Chemicals (8)
Kalamazoo	Auto Ion (8)
Livingston County	Shiawassee River (8)
Niles	U. S. Aviex (8)
Oden	Littlefield Township Dump (8)
Oscoda	Hedblum Industries (8)
South Ossineke	Ossineke (8)
Sparta	Sparta Landfill (8)
Buchanan	Electrovoice (9)

Minnesota

Brainerd/Baxter	Burlington Northern (1)
Fridley	FMC (1)
New Brighton/Arden	New Brighton (1)
St. Louis Park	Reilly Tar (1)
Oakdale	Oakdale (2)
St. Paul	Kopper's Coke (2)
Anoka County	Waste Disposal Engineering (3)
Lehillier/Mankato	Lehillier (3)
St. Louis Park	National Lead Taracorp (3)
Andover	South Andover Site (7)

Mississippi

Gulfport	Plastifax (2)

Missouri

Ellisville	Ellisville Site (2)
Springfield	Fulbright Landfill (5)
Verona	Syntex Facility (5)
Imperial	Arena 2: Fills 1 & 2 (7)
Moscow Mills	Arena 1 (Dioxin) (8)

Montana

Anaconda	Anaconda-Anaconda (1)
Silver Bow/ Deer Lodge	Silver Bow Creek (1)
Milltown	Milltown (4)
Libby	Libby Ground Water (6)

Nebraska

Beatrice	Phillips Chemical (9)

New Hampshire

Epping	Kes-Epping (1)
Nashua	Sylvester, Nashua (1)
Somersworth	Somersworth Landfill (1)
Kingston	Ottati & Goss (3)
Londonderry	Tinkham Site (5)
Dover	Dover Landfill (6)
Londonderry	Auburn Road Landfill (7)

New Jersey

Bridgeport	Bridgeport Rent. & Oil (1)
Fairfield	Caldwell Trucking (1)
Freehold	Lone Pine Landfill (1)
Gloucester Township	Gems Landfill (1)
Mantua	Helen Kramer Landfill (1)
Marlboro Township	Burnt Fly Bog (1)
Old Bridge Township	CPS/ Madison Industries (1)
Pittman	Lipari Landfill (1)
Pleasantville	Price Landfill (1)
Brick Township	Brick Township Landfill (2)
Carlstadt	Scientific Chemical Processing (2)
East Rutherford	Universal Oil Products (2)
Hamilton Township	D'Imperio Property (2)
Hillsborough	Krystowaty Farm (2)
Bound Brook	American Cyanamid (3)
Dover Township	Reich Farms (3)
Edison	Kin-Buc Landfill (3)
Franklin Township	Myers Property (3)
Maywood & Rochelle Pk.	Maywood Chemical Sites (3)
Parsipanny, Troy Hls	Sharkey Landfill (3)
Pedricktown	N. L. Industries (3)
Pemberton Township	Lang Property (3)

Ringwood	Ringwood Mines/Landfill (3)
South Brunswick	South Brunswick Landfill (3)
Chester	Combe Fill South Landfill (4)
Dover Township	Toms River Chemical (4)
Elizabeth	Chemical Control (4)
Plumstead	Spence Farm (4)
Plumstead Township	Goose Farm (4)
Rockaway Township	Rockaway Township Wells (4)
South Brunswick Twp.	JIS Landfill (4)
Winslow Township	King of Prussia (4)
Bayville	Denzer & Schafer X-Ray (5)
Berkley	Beachwood/Berkley Wells (5)
Dover	Dover Municipal Well 4 (5)
Edison	Renora (5)
Fair Lawn	Fair Lawn Wellfield (5)
Florence	Roebling Steel Co. (5)
Gibbstown	Hercules (5)
Howell Township	Bog Creek Farm (5)
Marlboro Township	Imperial Oil (5)
Monroe Township	Monroe Township Landfill (5)
Mt. Olive Township	Combe Fill North Landfill (5)
Piscataway	Chemsol (5)
Plumstead	Pijak Farm (5)
Rockaway Boro	Rockaway Boro Wellfield (5)
South Kearny	Syncon Resins (5)
Swainton	Williams Property (5)
Vineland	Vineland State School (5)
Jackson Township	Jackson Township Landfill (6)
Millington	Asbestos Dump (6)
Montgomery Township	Montgomery Housing Dev. (6)
Orange	US Radium (6)
Rocky Hill	Rocky Hill Municipal Well (6)
Sayreville	Sayreville Landfill (6)
Boonton	Pepe Field (7)
Evesham	Ellis Property (7)
Franklin Township	Metaltec/Aerosystems (7)
Freehold Township	Friedman Property (7)
Galloway Township	Mannheim Avenue Dump (7)
Old Bridge	Evor Phillips (7)
Pennsauken	Swope Oil and Chemical (7)
Asbury Park	M&T Delisa Landfill (8)
Jersey City	PJP Landfill (9)
Sparta	A. O. Polymer (9)

New Mexico

Albuquerque	South Valley (2)
Milan	Homestake (5)
Churchrock	United Nuclear Corp. (8)
Clovis	ATSF/Clovis (8)

New York

Oswego	Pollution Abatement Services (1)
Oyster Bay	Old Bethpage Landfill (1)
Wellsville	Sinclair Refinery (1)
Brant	Wide Beach Development (2)
Niagara Falls	Hooker-S Area (3)
Niagara Falls	Love Canal (3)
Oyster Bay	Syosset Landfill (3)
South Glens Falls	GE Moreau Site (3)
Albany	Mercury Refining (4)
Batavia	Batavia Landfill (4)
Elmira Heights	Facet Enterprises (4)
Moira	York Oil Company (4)
Olean	Olean Wellfield (4)
Port Washington	Port Washington Landfill (4)
Ramapo	Ramapo Landfill (4)
South Cairo	American Thermostat (4)
Vestal	Vestal Water Supply (5)
Brewster	Brewster Well Field (6)
Horseheads	Kentucky Ave. Wellfield (6)
Wheatfield	Niagara County Refuse (6)
Clayville	Ludlow Sand & Gravel (7)
Fulton	Fulton Terminals (7)
Lincklaen	Solvent Savers (7)
Niagara Falls	Hooker-Hyde Park (7)
Cold Springs	Marathon Battery (8)
Niagara Falls	Hooker-102nd Street (8)

North Carolina

210 miles of roads	PCB Spills (2)
Charlotte	Martin Marietta, Sodyeco (3)
Swannanoa	Chemtronics, Inc. (8)

North Dakota

Southeastern	Arsenic Trioxide Site (2)

Northern Marianas

North Marianas	PCB Warehouse (2)

Ohio

Arcanum	Arcanum Iron & Metal (1)
Hamilton	Chem Dyne (2)
Ashtabula	Fields Brook (3)
Circleville	Bowers Landfill (3)
Deerfield	Summit National (3)
Ironton	Allied Chemical (4)
Salem	Nease Chemical (4)
Ironton	E. H. Schilling Landfill (5)
Byesville	Fultz Landfill (6)
Coshocton	Coshocton City Landfill (6)
Dodgeville	New Lyme Landfill (7)
Jefferson	Poplar Oil (7)
Kingsville	Big D Campgrounds (7)
Reading	Pristine (7)
Rock Creek	Rock Creek/Jack Webb (7)
St. Clairsville	Buckeye Reclamation (7)
West Chester	Skinner Landfill (8)
Marietta	Van Dale Junkyard (9)
Zanesville	Zanesville Well Field (9)

Oklahoma

Ottawa County	Tar Creek (1)
Criner	Criner/Hardage (3)

Oregon

Albany	Teledyne Wah Chang (4)
Portland	Gould, Inc. (8)

Pennsylvania

Bruin Boro	Bruin Lagoon (1)
Grove City	Osborne (1)
McAdoo	McAdoo (1)
Douglasville	Douglasville Disposal (2)
Buffalo	Hranica (3)
Harrison Township	Lindane Dump (3)
Malvern	Malvern TCE Site (4)
Palmerton	Palmerton Zinc Pile (4)
Philadelphia	Enterprise Avenue (5)

West Ormond	Heleva Landfill (5)
Erie	Presque Isle (6)
Girard Township	Lord Shope (6)
Haverford	Havertown PCB Site (6)
Jefferson	Resin Disposal (6)
Lock Haven	Drake Chemical Inc. (6)
Lower Providence Twp.	Moyers Landfill (6)
State College	Centre County Kepone (6)
Chester	Wade (ABM) (7)
King of Prussia	Stanley Kessler (7)
Old Forge	Lackawanna Refuse (7)
Seven Valleys	Old City of York Landfill (7)
Old Forge	Lehigh Electric (8)
Philadelphia	Metal Banks (8)
Stroudsburg	Brodhead Creek (8)
West Chester Twp.	Blosenski Landfill (8)
Westline	Westline (8)
Kimberton	Kimberton (9)
Parker	Craig Farm Drum Site (9)
Upper Saucon Twp.	Voortman (9)
Warminster	Fischer & Porter (9)

Puerto Rico

Florida Afuera	Barceloneta Landfill (5)
Juana Diaz	GE Wiring Devices (5)
Rio Abajo	Frontera Creek (5)
Barceloneta	RCA Del Caribe (8)
Juncos	Juncos Landfill (8)

Rhode Island

Coventry	Picillo Coventry (1)
Burrillville	Western Sand & Gravel (3)
North Smithfield	L & RR-N. Smithfield (3)
Smithfield	Davis Liquid (4)
Cumberland	Peterson/Puritan (5)
North Smithfield	Forestdale (7)

South Carolina

Columbia	SCRDI Bluff Road (2)
Cayce	SCRDI Dixiana (5)
Fort Lawn	Carolawn, Inc. (8)

South Dakota

    Whitewood   Whitewood Creek (1)

Tennessee

    Memphis   North Hollywood Dump (2)
    Lawrenceburg   Murray Ohio Dump (4)
    Toone   Velsicol Chemical Co. (4)
    Chattanooga   Amnicola Dump (8)
    Galloway   Galloway Ponds (8)
    Lewisburg   Lewisburg Dump (8)

Texas

    Crosby   French, Ltd. (1)
    Crosby   Sikes Disposal Pits (1)
    Houston   Crystal Chemical (1)
    La Marque   Motco (1)
    Highlands   Highlands Acid Pit (6)
    Grand Prairie   Bio-Ecology (7)
    Houston   Harris (Farley Street) (7)
    Orange County   Triangle Chemical (9)

Trust Territories

    Pacific Trust Terr.   PCB Waste (2)

Utah

    Salt Lake City   Rose Park Sludge Pit (2)

Vermont

    Burlington   Pine Street Canal (2)
    Springfield   Old Springfield Landfill (7)

Virginia

    Roanoke County   Matthews (2)
    Saltville   Saltville Waste Disposal (3)
    York County   Chisman (4)
    Piney River   US Titanium (7)

Washington

    Tacoma   Commencement Bay, S. Tacoma Channel (2)
    Vancouver   Frontier Hard Chrome (2)
    Mead   Kaiser Mead (5)

Seattle	Harbor Island Lead (5)
Tacoma	Commencement Bay, near Shore Tide Flat (5)
Spokane	Colbert Landfill (6)
Kent	Western Processing (7)
Yakima	FMC Yakima (8)
Yakima	Pesticide Pit, Yakima (8)
Lakewood	Lakewood (9)

West Virginia

Point Pleasant	West Virginia Ordnance (2)
Leetown	Leetown Pesticide Pile (7)
Nitro	Fike Chemical (7)
Follansbee	Follansbee Sludge Fill (8)

Wyoming

Laramie	Baxter/Union Pacific (6)

Source: U.S. Environmental Protection Agency, "Proposed National Priorities List: As provided for in Section 105 (8) (B) of CERCLA" (Washington: Environmental Protection Agency, December 20, 1982).

Appendix B. State Hazardous Waste Funding Arrangements

State or other jurisdiction	State hazardous waste trust and spill funds	Source of fund	Major scope of fund
Alabama	Hazardous Waste Management Fund	FO	Administrative costs
	Perpetual Care Fund	FO	Monitoring beyond the active use of the site
Alaska	—	—	—
Arizona	Hazardous Waste Trust Fund	FO	Operation, maintenance, perpetual care
Arkansas	—	—	—
California	Hazardous Substances Account	TG	Match federal Superfund monies, cleanup, incident contingency fund, victim compensation fund, health studies, emergency equipment
Colorado	Hazardous Waste Disposal Fund	L, P	—
	Emergency Response Cash Fund	L	Emergency response
Connecticut	Emergency Spill Response Fund	L, R	Oil and hazardous spills
Delaware	—	—	—
Florida	Hazardous Waste Management Trust Fund	L, T, FO, R, P	Reduce hazard at abandoned sites
Georgia	Hazardous Waste Trust Fund	FO, B	Maintenance of abandoned sites
Hawaii	—	—	—
Idaho	—	—	—
Illinois	Hazardous Waste Fund	FO	Take action against long-term danger, research and development of recycling
Indiana	Hazardous Substances Emergency Trust Fund	TG	Emergency response, match under Superfund
	Environmental Management Special Fund	F, P	Multipurpose environmental response
Iowa	—	—	—
Kansas	Perpetual Care Trust Fund	FO, L	Cleanup and monitoring
Kentucky	Hazardous Waste Management Fund	FGO, R	Emergency response, post-closure, monitoring, and maintenance

State or other jurisdiction	State hazardous waste trust and spill funds	Source of fund	Major scope of fund
Louisiana	Hazardous Waste Protection Fund	B, L	Perpetual care, assure financial responsibility
	Abandoned Hazardous Waste Site Fund	Excess $, L	Match federal funds, cleanup at abandoned sites
	Environmental Emergency Response Fund	R, L, P	Environmental emergency responses, match federal funds
Maine	Hazardous Waste Fund	FG, OT	Emergency response
Maryland	Oil Disaster Containment Cleanup & Ctgy. Fund	FO	Oil and petroleum products spills
	Hazardous Substance Control Fund	FO, L	Hazardous substances in water cleanup
Massachusetts	—	—	—
Michigan	Disposal Facility Trust Fund	FO	Long-term care of closed facilities
	Hazardous Waste Service Fund	L, R	Emergency response
Minnesota	—	—	—
Mississippi	—	—	—
Missouri	Hazardous Waste Fund	FG, TO, L	Administrative costs, cleanup
Montana	—	—	—
Nebraska	—	—	—
Nevada	State Emergency Fund	L	Emergency response
New Hampshire	Hazardous Waste Cleanup Fund	F, P, L	Cleanup
New Jersey	Spill Compensation Fund	TO, L	Cleanup of spills
New Mexico	Hazardous Waste Emergency Fund	L, R, P	Cleanup, disposal, containment
New York	Hazardous Waste Remedial Fund	L	Emergency response
	Environmental Protection & Spill Comp. Fund	P	Oil spills only
North Carolina	—	—	—
North Dakota	—	—	—
Ohio	Hazardous Waste Facility Mgt. Special Account	FG	Administration, closure, abatement, grants
	Emergency Response Spill Fund	L, P	Emergency response to spills
Oklahoma	—	—	—
Oregon	Hazardous Waste Account	FG	Perpetual care
Pennsylvania	Solid Waste Abatement	P, R, B	Emergency situations, spills
Rhode Island	Hazardous Waste Substance Emergency Fund	L, BS	Abandoned site spills

State or other jurisdiction	State hazardous waste trust and spill funds	Source of fund	Major scope of fund
South Carolina	Hazardous Waste Contingency Fund	F, G	Emergencies at permitted landfills
South Dakota	—	—	—
Tennessee	Hazardous Waste Trust Fund	B	Cleanup, perpetual care
	Perpetual Care Trust Fund	FO	Containment of abandoned site
Texas	Disposal Facility Response Fund	L	Match federal Superfund monies
Utah	—	—	—
Vermont	Oil & Hazardous Spill Contingency Fund	L, R	Response to spills, hazardous substances
Virginia	—	—	—
Washington	—	—	—
West Virginia	—	—	—
Wisconsin	Hazardous Waste Fund	FO	Closing and long-term care
	Hazardous Substances Spill Fund	L, R	Cleanup and disposal
Wyoming	—	—	—
Dist. of Col.	Pending	—	—
American Samoa	—	—	—
Guam	—	—	—
No. Mariana Is.	—	—	—
Puerto Rico	—	—	—
Virgin Islands	—	—	—

Key:
F—Fees
L—Legislative appropriations
P—Penalties
R—Reimbursements
B—Bond forfeiture

T—Taxes
BS—Bond supported
O—Operator
G—Generator
OT—Out-of-state transporters

Source: Council of State Governments, *The Book of the States, 1982-1983*, (Lexington, Ky.: Council of State Governments, 1982), 611.

Appendix C. National Wildlife Federation State Toxic Substances Survey Form

NWF Survey of State Toxic Substances Programs

Instructions: The questions call for yes or no answers as a time-saving measure. Often, however, it will not be possible to answer a question with a flat yes or no. In such cases, please check the "yes/no" answer space. If you wish to explain this answer, please feel free to do so—either by cross-reference to applicable provisions of State law or regulation, or by notation in the margins or on a separate explanatory sheet. If communicating the explanations by telephone will save you time, please feel free to do so by calling [telephone number].

A. General Issues

1. Does the *State* have a specific toxic substances law (apart from general air, water and land pollution control laws)?
   Yes _____ No _____ Yes/No _____.

2. Is there a *State* agency or official with overall responsibility for coordinating or regulating toxic substances?
   Yes _____ No _____ Yes/No _____.
   If Yes, please identify the agency or official.

3. Has the *State* established by law or regulation a specific definition of "toxic" and/or "hazardous" wastes or materials—one which includes or incorporates a listing or registry of the chemicals which it encompasses?
   Yes _____ No _____ Yes/No _____.

4. Has the *State* established a 24-hour-a-day emergency telephone number for reporting hazardous waste spills and other toxic substances emergencies?
   Yes _____ No _____ Yes/No _____.
   If Yes, please provide the telephone number.

5. Does *State* law:
   a) require users or producers of toxic chemicals to furnish information on these chemicals to the *State*?
   Yes _____ No _____ Yes/No _____.
   b) specify numerical water quality criteria applicable to toxic chemicals in groundwater or limit deepwell injection and other forms of underground disposal of toxic and hazardous wastes?
   Yes _____ No _____ Yes/No _____.

c) cover the disposal of agricultural wastes (notably pesticides) by individuals?
Yes _____ No _____ Yes/No _____.

d) require or encourage the proper disposal or recycling of waste lubricating oils by individuals?
Yes _____ No _____ Yes/No _____.

e) encourage public participation in the process of regulating toxic substances?
Yes _____ No _____ Yes/No _____.

f) allow citizens to examine toxic substances information in the *State's* possession?
Yes _____ No _____ Yes/No _____.

g) authorize lawsuits by private citizens against individuals and/or *State* agencies where *State* laws regulating toxic substances have been violated or are not being enforced?
Yes _____ No _____ Yes/No _____.

h) absolve a waste generator of responsibility for toxic wastes after they are turned over to a licensed disposal or processing firm?
Yes _____ No _____ Yes/No _____.

i) exempt from permit, license, or other regulatory requirements, waste generators or processors who dispose of or store toxic wastes on their own property?
Yes _____ No _____ Yes/No _____.

B. Water Discharge of Toxics

6. *a*) Has the *State* been delegated NPDES permit authority from the Federal Government under Section 402 of the Clean Water Act?
Yes _____ No _____ Yes/No _____.

b) If Yes, do *State* controls on the water discharge of toxic pollutants go beyond those of Federal Law?
Yes _____ No _____ Yes/No _____.

7. Does *State* law or regulation:
a) impose specific discharge limits on toxic or hazardous chemicals ("T/HC") (as opposed to merely prohibiting the discharge of toxics in "harmful" amounts)?
Yes _____ No _____ Yes/No _____.

b) require toxicity testing (for example, bioassays) of proposed discharges known to contain or suspected of containing T/HC's?
Yes _____ No _____ Yes/No _____.

c) require consideration of the effects of absorption of contaminants to bottom sediments (for example, by requiring the use of "solid-phase" bioassays), in evaluating the toxicity of proposed discharges?
Yes _____ No _____ Yes/No _____.

d) explicitly limit the discharge of known or suspected carcinogens, mutagens, and teratogens?
Yes _____ No _____ Yes/No _____.

e) require chemical dischargers to participate in (or pay for) periodic or ongoing monitoring of the receiving waterway to help evaluate the effects of their discharges?
Yes _____ No _____ Yes/No _____.

f) require the *State* itself to periodically monitor *State* waters, sediments, and biota to detect the effects of waste discharges?
Yes _____ No _____ Yes/No _____.

g) require *State*-issued discharge permits to include quantitative limits on effluent levels of specified T/HC's?
Yes _____ No _____ Yes/No _____.

h) i) give the *State* the authority to require users and producers of T/HC's to conduct a material balance to account for environmentally significant losses of toxic materials?
Yes _____ No _____ Yes/No _____.

ii) If yes, is this authority frequently or routinely utilized?
Yes _____ No _____ Yes/No _____.

i) require precautions to prevent contamination of ground or surface waters in event of container rupture where T/HC's are stored or disposed of in containerized form?
Yes _____ No _____ Yes/No _____.

j) specify any limit on the maximum duration of storage of T/HC's before disposal at an approved site (or before re-examination of the containers is required to forestall corrosion and leakage)?
Yes _____ No _____ Yes/No _____.

k) regulate the discharge of T/HC's into Publicly Owned Treatment Works and/or require pretreatment of such discharges?
Yes _____ No _____ Yes/No _____; and

l) establish procedures for dealing with toxic chemical spills and other toxic chemical emergencies?
Yes _____ No _____ Yes/No _____.

C. Landfill Disposal of Toxics

8. Does *State* law or regulation require wastes proposed for landfilling to be characterized as to their content of T/HC's?
Yes _____ No _____ Yes/No _____.

9. Is a *State* permit, license, or other approval required for:
a) deposition in a landfill of T/HC's?
Yes _____ No _____ Yes/No _____.
b) initiation and operation of landfills which receive T/HC's?
Yes _____ No _____ Yes/No _____.

c) plans and specifications for the design and operation of T/HC landfills?

Yes _____ No _____ Yes/No _____.

d) methods of containerization and pretreatment of T/HC's proposed for landfilling?

Yes _____ No _____ Yes/No _____.

10. Does *State* law or regulation impose the following record-keeping and/or labeling requirements:

    a) establishment and maintenance by landfill operators of a detailed listing of T/HC's landfilled?

    Yes _____ No _____ Yes/No _____.

    b) transfer to the *State* of records identifying the T/HC's landfilled when a landfill closes down?

    Yes _____ No _____ Yes/No _____.

    c) establishment and maintenance by the *State* of a record-keeping and data retrieval system for keeping track of T/HC's disposed of at active or abandoned landfills?

    Yes _____ No _____ Yes/No _____.

    d) labeling of landfill-disposed T/HC's which are containerized?

    Yes _____ No _____ Yes/No _____.

11. Does the *State* regulate the siting of T/HC landfills by:

    a) segregating the disposal of T/HC and non-T/HC wastes at different landfills?

    Yes _____ No _____ Yes/No _____.

    b) specifying any areas in which T/HC landfills may not be established (for example, within a specified distance from a waterway)?

    Yes _____ No _____ Yes/No _____.

12. Does the *State* impose bonding requirements on T/HC landfill owners or operators to ensure proper closure/decommissioning of landfill sites?

    Yes _____ No _____ Yes/No _____.

13. *a*) Does the *State* impose groundwater and/or surface water monitoring requirements on landfill owners or operators?

    Yes _____ No _____ Yes/No _____.

    b) If yes, does this responsibility end when active use of landfill is discontinued?

    Yes _____ No _____ Yes/No _____.

    c) If no, does the *State* do such monitoring itself?

    Yes _____ No _____ Yes/No _____.

14. PLEASE ATTACH COPIES OF ALL RELEVANT STATE LAWS AND REGULATIONS.

15. Additional comments, if any (use additional sheets if desired).

Notes

1. The Process of Hazardous Waste Regulation: Severity, Complexity, and Uncertainty

1. In response to growing alarm over the hazardous waste problem, U.S. Representatives Bob Eckhardt and (later) Toby Moffett and Albert Gore, Jr., chaired several hearings to study the various dimensions of the hazardous waste issue. See, for example, U.S. House of Representatives, Committee on Interstate and Foreign Commerce, Subcommittee on Oversight and Investigation, *Hazardous Waste Disposal: Parts I and II* (Washington, D.C.: Government Printing Office, 1979); U.S. House of Representatives, Committee on Interstate and Foreign Commerce, Subcommittee on Transportation and Commerce, *Superfund* (Washington, D.C.: Government Printing Office, 1979); U.S. House of Representatives, Committee on Interstate and Foreign Commerce, Subcommittee on Oversight and Investigation and Committee on Government Operations, Subcommittee on Environment, Energy and Natural Resources, *Love Canal: Health Studies and Relocation* (Washington, D.C.: Government Printing Office, 1980); U.S. House of Representatives, Committee on Interstate and Foreign Commerce, Subcommittee on Oversight and Investigation, *Hazardous Waste Matters* (Washington, D.C.: Government Printing Office, 1980); U.S. House of Representatives, Committee on Government Operations, Subcommittee on Environment, Energy, and Natural Resources, *Toxic Chemical Contamination of Ground Water: EPA Oversight* (Washington, D.C.: Government Printing Office, 1980); and U.S. House of Representatives, Committee on Interstate and Foreign Commerce, Subcommittee on Oversight and Investigation, *Organized Crime and Hazardous Waste Disposal* (Washington, D.C.: Government Printing Office, 1980).

2. See, for example, Robert Van Den Bosch, *The Pesticide Conspiracy* (New York: Doubleday, 1978); David D. Doniger, *Law and Policy and Toxic Substances* (Baltimore: Johns Hopkins, 1979); Michael Brown, *Laying Waste: The Poisoning of America by Toxic Chemicals* (New York: Pantheon, 1980); Ralph Nader, Ronald Brownstein, and John Richard, eds., *Who's Poisoning America: Corporate Polluters and Their Victims in the Chemical Age* (San Francisco: Sierra Club, 1981); Adeline G. Levine, *Love Canal: Science, Politics, and People* (Lexington, Mass.: Lexington Books, 1982); David L. Morell and Christopher Magorian, *Siting Hazardous Waste Facilities: Local Opposition and the Myth of Preemption* (Cambridge, Mass.: Ballinger, 1982); and Samuel S. Epstein, Lester O. Brown, and Carl Pope, *Hazardous Waste in America* (San Francisco: Sierra Club, 1982).

In addition, see William Goldfarb, "The Hazards of Our Hazardous Waste Policy," *Natural Resources Journal* 19 (1979): 249–60; Malcolm Getz and Benjamin Walter, "Environmental Policy and Competitive Structure: Implications of the Hazardous Waste Management Program," *Policy Studies Journal* 9 (Winter 1980): 404–14; John A. Worthley and Richard Torkelson, "Managing the Toxic Waste Problem: Lessons from the Love Canal," *Administration and Society* 13 (1980): 145–60; Robert Eckhardt, "The Unfinished Business of Hazardous Waste Control," *Baylor Law Review* 33 (1981): 252–65; and S. Wurth-Hough, "Chemical Contamination and Governmental Policymaking: The North Carolina Experience," *State and Local Government Review* 14 (1982): 54–60.

3. U.S. General Accounting Office, *Environmental Protection: Agenda for the 1980's* (Washington, D.C.: Government Printing Office, 1982), 8; and Epstein et al., *Hazardous Waste*, 37.

4. See Kirk W. Brown and D. C. Anderson, *Effects of Organic Solvents on the Permeability of Clay Soils*, report prepared for the Office of Research and Development, U.S. Environmental Protection Agency, Grant No. R80682510 (1981); see also Peter Montague, *Four Secure Landfills in New Jersey: A Study of the State of the Art in Shallow Burial Waste Disposal Technology* (Princeton: Department of Chemical Engineering, Princeton University, 1981).

5. See testimony of Louis Harris before the Subcommittee on Health and the Environment of the House Energy and Commerce Committee, October 15, 1981, *Health Standards for Air Pollutants* (Washington, D.C.: Government Printing Office, 1981), 262–311.

6. Subcommittee on Oversight and Investigations of the House Committee on Interstate and Foreign Commerce, *Hazardous Waste Disposal Report* (Washington, D.C.: U.S. Government Printing Office, 1979), 1.

7. The Conservation Foundation, *State of the Environment, 1982: A Report from the Conservation Foundation* (Washington, D.C.: The Conservation Foundation, 1982), 145–55.

8. Ibid., 146.

9. U.S. Comptroller General, *Hazardous Waste Facilities with Interim Status May Be Endangering Public Health and the Environment* (Washington, D.C.: General Accounting Office, 1981), 11.

10. Ibid., 11.

11. U.S. Environmental Protection Agency, *Environmental News* (Washington, D.C.: Environmental Protection Agency, October 23, 1981).

12. See testimony for the record by the Office of Technology Assessment in hearings held by the U.S. House Energy and Commerce Subcommittee on Oversight and Investigation, November 16, 1981.

13. Fred C. Hart Associates, *Preliminary Assessment of Cleanup Costs for National Hazardous Waste Problems*, prepared for the U.S. Environmental Protection Agency, 1979, pp. 1 and 25.

14. See *Inside EPA* (October 30, 1981). The list of 115 sites are also found in Epstein et al., *Hazardous Waste*, 448–49.

15. "EPA Adds 45 Waste Dumps to List of Sites for Superfund Cleanup," *Washington Post* (July 24, 1982), p. A-8.

16. "EPA Unveils List of 418 Toxic Waste Sites Set for 'Superfund' Cleanup Program," *Houston Post* (December 21, 1982), p. 27-A; see also U.S. Environmental Protection Agency, Office of Water and Waste Management, *Everybody's Problem: Hazardous Waste*, SW-826 (Washington, D.C.: U.S. Government Printing Office, 1980), 14–15; Epstein et al., *Hazardous Waste*, 448–545; and Kenneth S. Kamlet, *Toxic Substances Programs in U.S. States and Territories: How Well Do They Work* (Washington, D.C.: National Wildlife Federation, 1980), iv.

17. E. J. Dionne, Jr., "Questions and Answers on Love Canal Study," *New York Times* (July 16, 1982), p. B2.

18. U.S. Comptroller General, *Hazardous Waste Sites Pose Investigation, Evaluation, Scientific, and Legal Problems* (Washington, D.C.: General Accounting Office, 1981), 1.

19. See, for example, Getz and Walter, "Environmental Policy," 404–14; William Goldfarb, "The Hazards of Our Hazardous Waste Policy," *Natural Resources Journal* 19 (1979): 249–60; and Worthley and Torkelson, "Managing Toxic Waste," 145–60.

20. Comptroller General, U.S. General Accounting Office, *Hazardous Waste Disposal Methods: Major Problems with Their Use* (Washington, D.C.: U.S. Government Printing Office, 1980).

21. Getz and Walter, "Environmental Policy," 405.

22. Ibid., 405.

23. Mary Worobee, "Analysis of the Resource Conservation and Recovery Act," *Environment Reporter: Current Developments,* vol. 11 (August 22, 1980), 637, cited in Sam A. Carnes, "Confronting Complexity and Uncertainty: Implementation of Hazardous Waste Management Policy," in Dean E. Mann, ed., *Environmental Policy Implementation* (Lexington, Mass.: Lexington Books, 1982), 37.

24. Carnes, "Confronting Complexity," 35–50.

25. Ibid., p. 46.

26. Kenneth W. Kramer and James P. Lester, "Economics, Power, and Policy: Implementing Hazardous Waste Regulation in Texas," a paper prepared for presentation at the 1982 Annual Meeting of the Southern Political Science Association, Atlanta, Georgia, October 28–30, 1982.

27. See U.S. Environmental Protection Agency, Office of Water and Waste Management, *Siting of Hazardous Waste Management Facilities and Public Opposition* (Washington, D.C., November 1979).

28. Carnes, "Confronting Complexity," 39; Morell and Magorian, *Siting Waste Facilities*; and Michael O'Hare, Lawrence Bacow, and Debra Sanderson, *Facility Siting and Public Opposition* (New York: Van Nostrand-Reinhold, 1983).

29. Carnes, "Confronting Complexity," 39.

30. Worthley and Torkelson, "Managing Toxic Waste," 158.

31. U.S. House of Representatives, Committee on Interstate and Foreign Commerce, Subcommittee on Oversight and Investigation, *Organized Crime and Hazardous Waste Disposal* (Washington, D.C.: Government Printing Office, 1980), 147–48. See also U.S. House of Representatives, Committee on Energy and Commerce, Subcommittee on Oversight and Investigation, *Hazardous Waste Enforcement* (Washington, D.C.: Government Printing Office, 1982), 1–38.

32. Regina S. Axelrod, ed., *Environment, Energy, Public Policy: Toward a Rational Future* (Lexington, Mass.: Lexington Books, 1981), 1–2.

33. See, for example, Murray L. Weidenbaum, *The Future of Business Regulation: Private Action and Public Demand* (New York: Amacom, 1979); George Stigler, *The Citizen and the State* (Chicago:

University of Chicago Press, 1975); Charles O. Jones, *Clean Air: The Policies and Politics of Pollution Control* (Pittsburgh: University of Pittsburgh Press, 1975); and James Q. Wilson, ed., *The Politics of Regulation* (New York: Basic Books, 1980), 357–94.

34. See, for example, Terry Humo, "EPA Stalls on Hazardous Waste," *Resources* 1 (Fall 1981): 20. U.S. General Accounting Office; *EPA Is Slow to Carry Out Its Responsibility to Control Harmful Chemicals* (Washington, D.C.: U.S. Government Printing Office, 1980); U.S. General Accounting Office, *Hazardous Waste Management Programs Will Not Be Effective: Greater Efforts Are Needed* (Washington, D.C.: U.S. Government Printing Office, 1979); U.S. General Accounting Office, *Hazardous Waste Facilities with Interim Status May Be Endangering Public Health and the Environment* (Washington, D.C.: U.S. Government Printing Office, 1981); "Toxic Waste Program Is Being Scuttled," *Audubon Leader* 22, no.24 (December 18, 1981), 1–2; and "The Superfund Turned Upside Down," *New York Times* (December 28, 1982), p. A22.

35. See Andy Pasztor, "Reagan Plans to Ease Toxic Waste Rules in Major Concession to Chemical Industry," *Wall Street Journal* (May 6, 1981), p. 4; and Philip Shabecoff, "E.P.A. Wants to Allow Burial of Barrels of Liquid Wastes," *New York Times* (March 1, 1982), p. A1.

36. Philip Shabecoff, "Rule on Reporting Waste Suspended," *New York Times* (March 15, 1982), p. B-2.

37. Philip Shabecoff, "U.S. Plan Offered for Cleaning Up Toxic Dump Sites," *New York Times* (March 13, 1982), p. 1; and Dale Russakoff, "Hill Has Some Withheld EPA Data," *Washington Post* (December 27, 1982), p. A1.

38. Philip Shabecoff, "U.S. Reversing Stand on Burial of Toxic Liquid," *New York Times* (March 18, 1982), p. 1.

39. "EPA Shifts on Liability Insurance," *Washington Post* (April 13, 1982), p. A-17.

40. "Hazardous Waste Rules on Burning Kept Tough," *Wall Street Journal* (June 23, 1982), p. 19; and "EPA Sets Tough Hazardous Waste Rules That May Cost Firms $500 Million a Year," *Wall Street Journal* (July 14, 1982), p. 4.

41. Conservation Foundation, *State of the Environment*, 145.

42. See testimony prepared for the House Subcommittee on Oversight and Investigation, March 21, 1979, p. 18; and *Hazardous Waste Disposal Part I*, 60–87.

43. Selim M. Senkan and Nancy W. Stauffer, "What to Do with Hazardous Waste," *Technology Review* 84 (November–December 1981), 35.

44. Gerald A. Bulanowski, *The Impact of Science and Technology on the Decision-making Process in State Legislatures: The Issue of Solid and Hazardous Waste* (Denver, Colo.: National Conference of State Legislatures, 1981).

45. U.S. House of Representatives, *Love Canal: Health Studies and Relocation*, Joint Hearing before the Subcommittee on Oversight and Investigation of the Committee on Interstate and Foreign Commerce, and the Subcommittee on Environment, Energy and Natural Resources of the Committee on Government Operations (Washington, D.C.: Government Printing Office, 1980), 5–6.

46. See Paul Goldberg, "Muzzling the Watchdog," *The Washington Monthly* 13 (December 1981): 30–35. See also Albert R. Matheny and Bruce A. Williams, "Scientific Disputes and Adversary Procedures in Policy Making," *Law and Policy Quarterly* 3 (July 1981): 341–64. A recent report prepared by twelve legal experts appointed by the American Bar Association and others, however, urged a uniform shift of the burden of proof in trials involving exposure to toxic chemicals from the plaintiff to the defendant. See *New York Times* (September 28, 1982), p. A20.

47. Ed Magnuson, "The Poisoning of America," *Time Magazine* (September 22, 1980), 58–69.

48. See footnote 4 above.

49. Montague, *Four Secure Landfills*.

50. *Federal Register* 97, no. 143 (July 26, 1982), 32284–85.

51. See U.S. Environmental Protection Agency, Office of Research and Development, *Environmental Monitoring at Love Canal*, vol. 1 (Washington, D.C.: Government Printing Office, 1982).

52. "Ring One" includes the canal and the homes immediately adjacent to it. "Ring Two" includes the area from 96th to 99th Street and "Ring Three" includes the area beyond Ring Two. Neither Rings One or Two were found to be habitable. See E. J. Dionne, Jr., "U.S. Finds Love Canal Neighborhood Is Habitable," *New York Times* (July 15, 1982), p. A1. See also Andy Pasztor, "U.S. Report That Love Canal Is Habitable Stirs Widely Conflicting Official Reaction," *Wall Street Journal* (July 15, 1982), p. 10; Josh Barbanel, "At Love Canal, Some Hope to Start Over," *New York Times* (May 17, 1982), p. 1; and E. J. Dionne, Jr., "Questions and Answers on Love Canal Study," *New York Times* (July 16, 1982), p. 2.

53. The framework, or model, used here refers to a descriptive heuristic framework rather than a formal axiomatic theory. It is used merely to organize our data collection and illustrate important linkages. For earlier work, see Donald S. Van Meter and Carl E. Van Horn, "The Policy Implementation Process: A Conceptual Framework," *Administration and Society* 6 (February 1975): 445–88; Paul Sabatier and Daniel Mazmanian, "The Implementation of Public Policy: A Framework of Analysis," *Policy Studies Journal* (special no. 2; 1980): 538–60; and George C. Edwards, III, *Implementing Public Policy* (Washington, D.C.: Congressional Quarterly Press, 1980).

54. Sabatier and Mazmanian, "Implementation of Public Policy," 545; Edwards, *Public Policy*, 10–11.

55. Van Meter and Van Horn, "Policy Implementation Process," 472.

56. J. C. Strouse and P. Jones, "Federal Aid: The Forgotten Variable in State Policy Research," *Journal of Politics* 36 (1974): 200–207.

57. Paul Sabatier, "Regulatory Policy-Making: Toward a Framework of Analysis," *Natural Resources Journal* 17 (July 1977): 415–60; see also Charles O. Jones, "Regulating the Environment," in Herbert Jacob and Kenneth N. Vines, *Politics in the American States: A Comparative Analysis* (Boston: Little, Brown and Company, 1976), 388–427. Thus, we have included such factors as the severity of chemical waste generation, the number of hazardous waste sites, and the number of waste producers, etc., as measures of the extent of the problem within different jurisdictional contexts (i.e., variation across states).

58. Sabatier and Mazmanian, "Implementation of Public Policy," 541–44.

59. Harvey Lieber, *Federalism and Clean Waters: The 1972 Water Pollution Control Act* (Lexington, Mass.: Lexington Books, 1975).

60. Van Meter and Van Horn, "Policy Implementation Process," 466.

61. Sabatier and Mazmanian, "Implementation of Public Policy," 546; and Van Meter and Van Horn, "Policy Implementation Process," 471.

62. Edwards, *Public Policy*, 11.

63. Citizens Conference on State Legislatures, *The Sometimes Governments: A Critical Study of the 50 American Legislatures* (New York: Bantam Books, 1971). See also John G. Grumm, "The Effects of Legislative Structure on Legislative Performance," in Richard I. Hofferbert and Ira Sharkansky, *State and Urban Politics* (Boston: Little, Brown, 1971), 298–322; Edward G. Carmines, "The Mediating Influence of State Legislatures on the Linkage Between Interparty Competition and Welfare Policies," *American Political Science Review* 68 (September 1974): 1118–24; Albert K. Karnig and Lee Sigelman, "State Legislative Reform: Another Look," *Western Political Quarterly* 28 (September 1975): 548–52; Lance T. LeLoup, "Reassessing the Mediating Impact of Legislative Capability," *American Political Science Review* 68 (June 1978): 616–21; and Phillip W. Roeder, "State Legislative Reform: Determinants and Policy Consequences," *American Politics Quarterly* 7 (January 1979): 51–69.

64. In addition, there is case study evidence which suggests that state legislative advisory capabilities in the science and technology area are instrumental in facilitating policy formulation involving complex issues such as hazardous waste management. See Bulanowski, "Impact of Science and Technology."

65. Van Meter and Van Horn, "Policy Implementation Process," 472; Sabatier and Mazmanian, "Implementation of Public Policy," 548–60. See also Walter A. Rosenbaum, *The Politics of Environmental Concern* (New York: Praeger, 1977); Kingsley W. Game, "Controlling Air Pollution: Why Some States Try Harder," *Policy Studies Journal* 7 (1979): 728–38; James P. Lester, "Partisanship and Environmental Policy: The Mediating Influence of State Organizational Structures," *Environment and Behavior* 12 (1980): 101–31; Carl E. Lutrin and Allen K. Settle, "The Public and Ecology: The Role of Initiatives in California's Environmental Politics," *Western Political Quarterly* 28 (1975): 352–71; Worthley and Torkelson, "Managing Toxic Waste"; John Pierce and Harvey Doerksen, eds., *Water Politics and Public Involvement* (Ann Arbor, Mich.: Ann Arbor Science Publishers, 1976); and Levine, *Love Canal*, 189–93.

66. Van Meter and Van Horn, "Policy Implementation Process," 472–74; Sabatier and Mazmanian, "Implementation of Public Policy," 553; and Edwards, *Implementing Public Policy*, 11 and 89–119.

67. Van Meter and Van Horn, "Policy Implementation Process," 472.

68. Michael A. Maggiotto and Ann O'M. Bowman, "The Impact of Policy Orientation on Environmental Regulation: A Case Study of Florida's Legislators," *Environment and Behavior* 14 (March 1982): 155–70; see also E. Constantini and Kenneth Hanf, "Environmental Concern and Lake Tahoe: A Study of the Elite Perceptions, Backgrounds, and Attitudes," *Environment and Behavior* 4 (June 1972): 209–42.

69. For an excellent discussion of the linkage between individuals' perceptions of a particular

problem and their receptivity to risk assessments of problem severity, see Paul Slovic, Baruch Fishloff, and Sarah Lichtenstein, "Rating the Risks," *Environment* 21 (April 1979): 14–20, 36–39.

70. See, for example, Heinz Eulau and Kenneth Prewitt, *Labyrinths of Democracy: Adaptation, Linkages, Representation, and Policies in Urban Politics* (Indianapolis: Bobbs-Merrill, 1973); see also Constantini and Hanf, "Environmental Concern and Lake Tahoe."

2. Toxic Substances, Hazardous Wastes, and Public Policy: Problems in Implementation

1. Cf. Donald S. Van Meter and Carl E. Van Horn, "The Policy Implementation Process: A Conceptual Framework," *Administration and Society* 6 (February 1975): 445–88; Paul Sabatier and Daniel Mazmanian, "The Implementation of Public Policy: A Framework of Analysis," *Policy Studies Journal* 8 (special no. 2; 1980): 538–60; George C. Edwards III, *Implementing Public Policy* (Washington, D.C.: Congressional Quarterly Press, 1980); and James S. Larson, *Why Government Programs Fail: Improving Policy Implementation* (New York: Praeger Publishers, 1980).

2. Analysts like Van Meter and Van Horn, Sabatier and Mazmanian, Edwards, and Larson have taken the lead advanced by Walter Williams (*Social Policy Research and Analysis: The Experience in the Federal Social Agencies* [New York: American Elsevier Publishing Co., 1971]), to explain the implementation process and why programs sometimes fail. Some of the more notable case studies of policy implementation are: Herbert Kaufman's study of the U.S. Forest Service (*The Forest Ranger* [Baltimore: Johns Hopkins Press, 1960]); Stephen K. Bailey and E. K. Mosher's study of the Elementary and Secondary Education Act of 1965 (*ESEA: The Office of Education Administers a Law* [Syracuse: Syracuse University Press, 1968]); Martha Derthick's survey of selected federal grants-in-aid (*The Influence of Federal Grants: Public Assistance in Massachusetts* [Cambridge, Mass.: Harvard University Press, 1970]); J. S. Berke and M. W. Kirst's analysis of federal education subsidies (*Federal Aid to Education: Who Benefits? Who Governs?* [Lexington, Mass.: Lexington Books, 1972]); Jeffrey Pressman and Aaron Wildavsky's seminal work on ill-fated community development in Oakland (*Implementation* [Berkeley: University of California Press, 1973]); Charles O. Jones's study of air pollution control in Pittsburgh (*Clean Air: The Policies and Politics of Pollution Control* [Pittsburgh: University of Pittsburgh Press, 1975]); George E. Rawson's look at the TVA ("Organizational Goals and Their Impact on the Policy Implementation Process," *Policy Studies Journal* 8, no. 7 [August 1980]: 1109–18); and Eugene Bardach's in-depth examination of mental health reform in California (*The Implementation Game: What Happens After a Bill Becomes a Law* [Cambridge, Mass.: The MIT Press, 1977]).

3. Cited in GAO, *EPA Is Slow to Carry Out Its Responsibility to Control Harmful Chemicals* (October 28, 1980), 1.

4. 15 U.S.C. 2601. (emphasis added)

5. 15 U.S.C. 2603.

6. 15 U.S.C. 2604.

7. 15 U.S.C. 2605.

8. 15 U.S.C. 2607.

9. EPA, Office of Toxic Substances, *Administration of the Toxic Substances Control Act, Second Annual Report, 1978* (April 1979) (emphasis added).

10. Ibid., 13.

11. EPA, *Administration of the Toxic Substances Control Act, Third Annual Report, 1979* (July 1980), 26.

12. Ibid.

13. See EPA, *Administration of the Toxic Substances Control Act, Fourth Annual Report, 1980* (April 1981), 3.

14. *Environmental Defense Fund* v. *EPA*, No. 79-1580, District of Columbia Circuit; *Octin Corporation* v. *EPA*, No. 79-1437, Fourth Circuit Court of Appeals; *Dow Chemical Co.* v. *Douglas Costle*, Civ. Action No. 79-581, D. Delaware.

15. EPA, *Third Annual Report, 1980*, 19.

16. Ibid. (emphasis added)

17. Congress, House, Committee on Interstate and Foreign Commerce, *The Toxic Substances Control Act Amendments of 1980*, Hearings before the Subcommittee on Consumer Protection and Finance on H.R. 2003, 96th Cong., 2d sess., April 15, 16, 17, 22, 1980, 16.

18. Ibid., 26.

19. Ibid., 34.

20. Ibid., 43.

21. Ibid.

22. The case involved is *Natural Resources Defense Council (NRDC)* v. *Costle*, No. 79, Civ. 2411, Southern District of New York. On February 4, 1980, the Court found against the EPA, ordering it to submit a plan for complying with Section 4(e) of TSCA. The full opinion by U.S. District Judge Laurence W. Pierce appears in *Hearings*, ibid., 56–73.

23. Ibid., 46–47.

24. The incident referred to occurred in the summer of 1979 at the Pierce Packing Company, a meat packing facility in Billings, Montana. A spare transformer had been stored in a shed at the plant and an estimated 200 gallons of PCB-containing transformer oil had leaked from the transformer into the plant's wastewater system. Solids that were recovered from the wastewater system were rendered and included in meat meal, which was then sold as an animal feed ingredient and resulted in the contamination and eventual destruction of millions of pounds of chickens, eggs, hogs, bakery products, and animals in nineteen states.

25. Ibid., 116.

26. Note testimony of Sydney J. Butler, Deputy Assistant Secretary for Food and Consumer Services, Department of Agriculture; Thomas P. Grumbly, Associate Administrator, Food Safety and Quality Service, Food and Drug Administration; and Joseph P. Hile, Associate Commissioner, Regulatory Affairs, FDA, ibid., 117–223.

27. Ibid., 226.

28. According to Bell: "Unless very stringent siting requirements are made and there is Federal preexemption once those siting requirements have been met, the possibility of siting . . . is remote. No large company or smaller company is going to expend up to $1 million for various impact statements and engineering studies with the realization that after they have done [*sic*] that they may not get even sited and after they are sited a local or state ordinance, law, or act can put them out of business." Ibid., 230.

29. Comptroller General, General Accounting Office, *EPA Slow in Controlling PCBs* (December 30, 1981), 1.

30. U.S. GAO, *EPA Responsibility to Control Harmful Chemicals.*

31. Ibid., 15.

32. Ibid., 18.

33. Ibid., 33–34.

34. Ibid., 38.

35. Ibid., 42.

36. Ibid., 44–45.

37. Malcolm Getz and Benjamin Walter, "Environmental Policy and Competitive Structures: Implications of the Hazardous Waste Management Program," *Policy Studies Journal* 9 (Winter 1980): 404–14.

38. 42 U.S.C. 6903.

39. Under RCRA, "hazardous waste" is defined as "a solid waste, or combination of solid wastes, which because of its quantity, concentration, or physical, chemical, or infectious characteristics may: (*a*) cause, or significantly contribute to an increase in mortality or an increase in serious, irreversible, or incapacitating reversible illness; or (*b*) pose a substantial present or potential hazard to human health or the environment when improperly treated, stored, transported, or disposed of, or otherwise managed." 42 U.S.C. Sec. 1004 (5) (A) (B).

40. 42 U.S.C. 6926.

41. U.S. Comptroller General, General Accounting Office, *Waste Disposal Practices: A Threat to Health and the Nation's Water Supply*, June 16, 1978.

42. Ibid., 10. General Accounting Office urged the administrator to provide Congress with necessary inventory data procedure and necessary funding requirements as soon as possible.

43. General Accounting Office, *How to Dispose of Hazardous Waste: A Serious Question That Needs to Be Resolved* (December 19, 1978), 5 (emphasis added).

44. EPA Region II—New Jersey, New York, Puerto Rico, Virgin Islands; Region III—Delaware, Virginia, Maryland, Pennsylvania, West Virginia, District of Columbia; Region V—Illinois, Indiana, Ohio, Michigan, Wisconsin, Minnesota; Region VI—Arkansas, Louisiana, Oklahoma, Texas, New Mexico.

45. GAO, *How to Dispose of Hazardous Waste*, 10–14.

46. GAO, *Hazardous Waste Management Programs Will Not Be Effective: Greater Efforts Are Needed* (January 23, 1979), 4.

47. Ibid., 6.

48. The report notes that high waste-generating states of Michigan, Texas, Pennsylvania, and New Jersey have all experienced problems of inadequate funds, insufficient disposal capacity, inadequate enforcement procedures, and insufficient definition and monitoring of hazardous waste.

49. GAO, *Hazardous Waste Management*, 11.

50. Senate, Committee on Governmental Affairs, *Oversight of Hazardous Waste Management and the Resource Conservation and Recovery Act, Hearings Before the Subcommittee on Oversight of Governmental Management*, 96th Cong., 1st sess., July 19 and August 1, 1979, 13.

51. Ibid., 18.

52. Good examples of reduced morale among staff personnel can be found in the testimony of William Sanjour, Chief, Hazardous Waste Implementation Branch, EPA (pp. 350–84); Hugh B. Kaufman, Program Manager, Hazardous Waste Assessment Program, EPA (p. 385); and Thomas C. Jorling, Assistant Administrator for Water and Waste Management, EPA (p. 475).

53. Ibid., 47.

54. Comptroller General, GAO, *Hazardous Waste Disposal Methods: Major Problems with Their Use* (November 19, 1980), 8.

55. GAO, *Hazardous Waste Sites Pose Investigation, Evaluation, Scientific, and Legal Problems* (April 24, 1981), 8–9.

56. Comptroller General, GAO, *Solid Waste Disposal Practices: Open Dumps Not Identified, States Face Funding Problems* (July 23, 1981), 9.

57. Mimeographed copy of testimony of Anne M. Gorsuch, EPA Administrator, before the Subcommittee on Commerce, Transportation and Tourism, July 29, 1981; obtained with letter of Mr. Michael B. Cook, Director, Office of Emergency and Remedial Response, EPA, August 20, 1981 (emphasis added).

58. Cited in "The 'Ice Queen' at EPA," *Newsweek* (October 19, 1981), p. 67.

59. Andy Pasztor, "Reagan Plans to Ease Toxic Waste Rules in Major Concession to Chemical Industry," *The Wall Street Journal* (May 6, 1981), p. 4.

60. "EPA Reverses Position on Waste Burial," *The Houston Post* (February 27, 1982), p. 11C; Philip Shabecoff, "EPA Wants to Allow Burial of Barrels of Liquid Wastes," *The New York Times* (March 1, 1982), p. A1.

61. Sandra Sugawara, "Landfill Safety Oozes Away Critics Charge," *The Washington Post* (March 3, 1982), p. A25.

62. Ibid.

63. Philip Shabecoff, "U.S. Reversing Stand on Burial of Toxic Liquids," *The New York Times* (March 18, 1982), p. 1.

64. Ibid., Dietrich had earlier justified EPA's decision to again allow the disposal of toxic liquids because of "a lack of capacity and uneven distributions" around the country of incinerators or other alternate disposal facilities. See *The Wall Street Journal* (March 3, 1982), p. 8.

65. *The New York Times* (March 18, 1982), p. A22.

66. Philip Shabecoff, "U.S. Plan Offered for Cleaning Up Toxic Dump Sites," *The New York Times* (March 13, 1982), p. 1.

67. Philip Shabecoff, "Rule on Reporting Waste Suspended," *The New York Times* (March 15, 1982), p. B12.

68. According to an article in *The Washington Post* detailing the relaxed reporting standard, the EPA estimated that the 308,000 hours it takes companies to complete the forms annually would be reduced by about 90 percent; *The Washington Post* (March 16, 1982), p. A21.

69. *The New York Times* (March 15, 1982), p. B12.

70. *The Washington Post* (April 13, 1982), p. A17.

71. Ibid.

72. Quoted in "EPA Requires Insurance for 10,000 Waste Sites" *The Houston Chronicle* (April 13, 1982).

73. *The Wall Street Journal* (July 14, 1982), p. 4.

74. Cf. Andy Pasztor, "Split Developing Among Reagan Advisors Over Softening Stand on the Environment," *The Wall Street Journal* (July 12, 1982).

75. See Lester et al., chapter 11 of this book.

3. Superfund: Preimplementation Planning and Bureaucratic Politics

1. ABC News-Harris Survey, Volume 11, No. 82, July 7, 1980.
2. Martin Myerson and Edward Banfield, *Politics, Planning and the Public Interest* (New York: The Free Press, 1955).
3. Ibid., 315.
4. David Braybrooke and Charles Lindblom, *A Strategy of Decision* (New York: The Free Press, 1963), 46–47.
5. Ibid, 48.
6. John Steinbrunner, *The Cybernetic Theory of Decision* (Princeton, N.J.: Princeton University Press, 1974), 57.
7. Braybrooke and Lindblom, *Strategy of Decision*, 104.
8. Yehezkel Dror, *Public Policymaking Reexamined* (Scranton, Pennsylvania: Chandler Publishing Company, 1968).
9. Steinbrunner, *Cybernetic Theory*, 57.
10. Charles O. Jones, "Speculative Augmentation in Federal Air Pollution Policy-Making," *Journal of Politics* 36 (1974): 438–64.
11. See ICF Incorporated, "Analysis of Community Involvement in Hazardous Waste Site Problems: A Report to the Office of Emergency and Remedial Response, United States Environmental Protection Agency" (Washington, D.C., July 1981).
12. EPA Internal Draft Memo, November 15, 1979.
13. Attachment to Cook Memo, April 2, 1980.
14. Harvey M. Sapolsky, *The Polaris System Development: Bureaucratic and Programmatic Success in Government* (Cambridge, Mass.: Harvard University Press, 1972), 246.

4. Federalism and Hazardous Waste Policy

1. Thomas F. Williams, *Hazardous Waste: Fifteen Years and Still Counting* (U.S. Environmental Protection Agency, Publication OPA 98/0, June 1980), 5.
2. See Kenneth Bartlett, "The Constitutional Framework for the Resource Conservation and Recovery Act," *Toxic Substances Journal* 3, no. 1 (1982): 42.
3. U.S. Council on Environmental Quality, *Ninth Annual Report* (1978), 163.
4. Environmental Protection Agency, *Waste Management Technology and Resource and Energy Recovery* (Proceedings of the Fourth National Congress, 1976), 2.
5. Kenneth Kamlet, *Toxic Substance Programs in U.S. States and Territories: How Well Do They Work?* (Washington, D.C.: National Wildlife Federation, 1979), 2.
6. Advisory Commission on Intergovernmental Relations, *Protecting the Environment: Politics, Pollution, and Federal Policy*, Publication A-83, March 1981, 9.
7. Ibid., 28.
8. John L. Whitaker, *Striking a Balance: Environment and Natural Resources Policy in the Nixon-Ford Years* (American Enterprise Institute for Public Policy Research, 1976), 111.
9. Ibid., 113.
10. U.S. Environmental Protection Agency, *Disposal of Hazardous Wastes: Report to Congress*, Publication SW-115, 1974, p. xii.
11. Williams, *Hazardous Waste*, 8.
12. Quoted in Michael Brown, *Laying Waste: The Poisoning of America by Toxic Chemicals* (New York: Pantheon, 1980), 319–20.
13. U.S. Congress, Senate, *Solid Waste Utilization Act of 1976*, Report 94–988, to accompany S. 3622, 94th Cong., 2d Sess., June 25, 1976, p. 6.
14. *Time* (September 22, 1980), 58–69.
15. Senate Committee on Governmental Affairs, Subcommittee on Oversight of Governmental Management, "Oversight of Hazardous Waste Management and the Resource Conservation and Recovery Act," *Hearings*, 96th Cong., 1st Sess., July 19, August 1, 1979, p. 374.
16. House of Representatives Committee on Interstate and Foreign Commerce, Subcommittee on Oversight and Investigation, *Hazardous Waste Disposal*, Committee Print 96-IFC 31, 96th Cong., 1st Sess., September 1979, p. 2.

17. Senate Committee on Governmental Affairs, Subcommittee on Oversight of Governmental Management, *Report on Hazardous Waste Management and the Implementation of RCRA*, 96th Cong., 2d Sess., March 1980, p. 6. See also, William L. Kovacs and John F. Klucsik, "The New Federal Role in Solid Waste Management: The Resource Conservation and Recovery Act of 1976," *Columbia Journal of Environmental Law* 3 (1977): 254.

18. U.S. Comptroller General. The CED reports and their dates are 79-14, January 23, 1979; 81-82, November 19, 1980; 81-57, April 24, 1981; 81-131, July 23, 1981; and 81-159, September 28, 1981.

19. Bartlett, "Constitutional Framework," 23. Italics added.

20. House Committee on Interstate and Foreign Commerce, *Resource Conservation and Recovery Act*, House Report 94-1491, Part I, to accompany H.R. 14496, 94th Cong., 2d Sess., September 9, 1976, pp. 24, 25, 29.

21. EPA, *Transcript*, Regional Public Meetings on the Resource Conservation and Recovery Act of 1976 (RCRA), February 18, 1977, Richmond, Virginia, 46.

22. EPA, *Transcript*, First Public Meeting on RCRA, December 16, 1976, p. 30.

23. Federal Register, Vol. 46, No. 16, January 26, 1981, p. 8299.

24. EPA had originally intended to promulgate the land disposal regulations in the fall of 1983.

25. 98 S. Ct. 2531 (1978).

26. *Hazardous Waste Disposal*, 72, 46.

27. *Report on Hazardous Waste Management*, 23.

28. See Harvey Lieber, *Federalism and Clean Water: The 1972 Water Pollution Control Act* (Lexington, Mass.: Lexington Books, 1975) and Lieber, "Institutional Arrangements for Areawide Water Resources Management in the Washington, D.C., Area Under the Federal Water Pollution Control Act Amendments of 1972," Report No. 17, Washington, D.C., Water Resources Research Center, June 1980.

29. See William Goldfarb, "The Hazards of Our Hazardous Waste Policy," *Natural Resources Journal* 19 (April 1977): 259; and Jonathan H. Steeler et al., *Hazardous Waste Management: Survey of State Legislation* (Denver: National Conference of State Legislatures, 1982).

30. *New Jersey* v. *U.S.*, D DC Cir. No. 81-0945 and *Exxon Corp. et al.* v. *Robert Hunt*, No. 71-1958m (N.J.), Docket No. SC 303A-81, SC 319A-81TC.

31. EPA, *Transcript*, Public Meeting: Strategy for Implementation of RCRA, Publication SW-33p, Arlington, Virginia, January 19, 1978, pp. 35, 47.

32. U.S. Comptroller General, *Hazardous Waste Management Programs Will Not Be Effective; Greater Efforts Are Needed*, CED 79-14 (January 23, 1979).

33. *Inside EPA* (April 23, 1982), 8.

34. U.S. Congress, House of Representatives, Committee on Interstate and Foreign Commerce, Subcommittee on Oversight and Investigation, "Hazardous Waste Disposal," *Hearings*, 96th Cong., 1st Sess., 1979, 1312–15.

35. CED 81-85, iii. Italics added.

36. Environmental Protection Agency, Office of Resources Management, *Operations/Resources Impact Analyses*, RCRA Subtitle C, September 1980, xi.

37. CEQ, *Environmental Quality, 1981*, 12th Annual Report (Washington, D.C.: Government Printing Office, 1982), 99.

38. U.S. Comptroller General, *Environmental Protection Agency's Progress Implementing the Superfund Program*, CED 82-91, June 2, 1981.

39. Steven I. Friedland, "The New Hazardous Waste Management System: Regulation of Wastes or Wastes Regulation?" *Harvard Environmental Law Review* 5, no. 128 (1981): 89–129.

40. Senate Committee on Public Works, *A Legislative History of the Solid Waste Disposal Act, As Amended*, Committee Print No. 93-22, 93rd Cong., 2d Sess., October 1974, p. 47.

41. CEQ, *10th Annual Report* (1979), 184.

42. *Report on Hazardous Waste Management*, 20.

43. Ibid., 24.

44. *Hazardous Waste Management* (September 6, 1982), 6, 11, 12.

45. *Hazardous Waste Report* (August 9, 1982), 12–13.

5. Hazardous Waste Policy In Florida: Is Regulation Possible?

1. By social regulation, we mean regulations that impose costs on a concentrated group (e.g., a particular industry) in order to obtain benefits (e.g., clean air and water) for a more diffuse group (e.g.,

the general public). David Vogel distinguishes social regulation from economic regulation by noting that the latter deals with "prices, outputs, terms of competition and entry exit." In contrast, the former deals with "the externalities and social impact of economic activity" ("The 'New' Social Regulation in Historical Comparative Perspective," in *Regulations in Perspective*, ed. Thomas K. McCraw [Cambridge, Mass.: Harvard University Press, 1981], 238).

2. Charles E. Lindblom, *Politics and Markets* (New York: Basic Books, 1977); James O'Connor, *The Fiscal Crisis of the State* (New York: St. Martin's Press, 1973); William Ophuls, *Ecology and the Politics of Scarcity* (San Francisco: W.H. Freeman and Co., 1977); K. William Kapp, *Social Costs of Business Enterprise* (New York: Asia Publishing House, 1963).

3. The two most widely cited works in this area are: Marver H. Bernstein, *Regulating Business by Independent Commission* (Princeton, N.J.: Princeton University Press, 1955); Samuel P. Huntington, "The Marasmus of the I.C.C.: The Commission, the Railroads, and the Public Interest," *Yale Law Journal* 61 (April 1952): 467–509.

4. James Q. Wilson, "The Politics of Regulation," in *The Politics of Regulation*, ed. James Q. Wilson (New York: Basic Books, 1980), 386–87.

5. While this literature is by now voluminous, some of the most thoughtful and widely cited works are Charles O. Jones, *Clean Air: The Policies and Politics of Pollution Control* (Pittsburgh: University of Pittsburgh Press, 1975); Eugene Bardach and Robert A. Kagan, *Going by the Book* (Philadelphia: Temple University Press, 1982); Eugene Bardach and Robert A. Kagan, eds., *Social Regulation: Strategies for Reform* (San Francisco: Institute for Contemporary Studies, 1982); George Stigler, *The Citizen and the State* (Chicago: University of Chicago Press, 1975); Wilson, "The Politics of Regulation"; Paul W. MacAvoy, "The Existing Condition of Regulation and Regulatory Reform," in *Regulating Business: The Search for an Optimum*, ed. Donald P. Jacobs (San Francisco: Institute for Contemporary Studies, 1978), 3–13.

6. Bardach and Kagan, *Going by the Book*, chapter 10; Eugene Bardach and Robert A. Kagan, "Liability Law and Social Regulation," in Bardach and Kagan, eds., *Social Regulation*, 237–66.

7. Bardach and Kagan, *Going by the Book*, 65.

8. Few terms have proven as resistant to adequate definition as "the public interest" and we will not resolve these problems here. For our purpose, we use the term to refer to the effects of private market operations on the general population over and above the effects that are priced by market transactions. Thus, we are assuming that the functioning of the market creates two sets of costs and benefits, and therefore interests are created. One set of costs and benefits that is priced by the market is imposed upon those that participate in private transactions and thus constitute private interests. A second set of costs and benefits (some of which may, in principle, be accurately priced, others not) falls on some portion of the wider population not participating in private transactions, and these people constitute the public interest. Regulation is usually undertaken to insure that this second set of costs and benefits is taken into account.

9. On this point see Steven Kelman, "Cost-Benefit Analysis and Environmental, Safety, and Health Regulation: Ethical and Philosophical Considerations," in *Cost-Benefit Analysis and Environmental Regulations: Politics, Ethics, and Methods*, ed. Daniel Swartzman, Richard A. Liroff, and Kevin G. Corke (Washington, D.C.: The Conservation Foundation, 1982), 137–51; Mark Sagoff, "We Have Met the Enemy and He Is Us, or Conflict and Contradiction in Environmental Law," *Environmental Law* 283 (1982): 283–315.

10. Wilson, "The Politics of Regulation"; Jones, *Clean Air*; Eugene Bardach and Robert A. Kagan, "Introduction," in Bardach and Kagan, eds., *Social Regulation*, 3–19.

11. For an example of this type of approach see Mark Green, Beverly C. Moore, Jr., and Bruce Wasserstein, *The Closed Enterprise System: Ralph Nader's Study Group on Antitrust Enforcement* (New York: Grossman Publishers, 1972).

12. Bardach and Kagan, "Introduction," 5. Reprinted with permission.

13. See, for example, Kapp, *Social Costs*; Ophuls, *Ecology*; E. F. Schumacher, *Small Is Beautiful* (New York: Harper and Row, 1973); Rufus E. Miles, Jr., *Awakening from the American Dream* (New York: Universe Books, 1976).

14. See, for example, O'Connor, *Fiscal Crisis*; Hugh Stretton, *Capitalism, Socialism and the Environment* (New York: Cambridge University Press, 1976). For a non-Marxist, but nevertheless critical perspective see Lindblom, *Politics and Markets*; Charles E. Lindblom, "The Market as Prison," *Journal of Politics* 44 (May 1982): 324–36.

15. O'Connor, *Fiscal Crisis*, chapter 2.

16. Kapp, *Social Costs*, chapters 2 and 4.

17. Lindblom, "Market as Prison."

18. See, for example, G. William Domhoff, *Who Rules America?* (Englewood Cliffs, N.J.: Prentice-Hall, Inc., 1967).

19. Edward S. Herman, *Corporate Control, Corporate Power* (New York: Cambridge University Press, 1981), 172–86.

20. Our research involved in-depth interviews with officials in relevant divisions of the Florida Department of Environmental Regulation, state senators, representatives, and their staffs, who were involved in hazardous waste legislation, staff members of the relevant House and Senate committees, lobbyists for the phosphate and electrical generating industries, lobbyists and the other representatives of the Sierra Club and the Environmental Service Center, and staff members for the Governor's Hazardous Waste Policy Advisory Council. All our interviews were conducted with a guarantee of individual anonymity. In addition, we examined the various state laws, regulations, and other documents relevant to hazardous waste management in Florida.

21. Luther J. Carter, *The Florida Experience: Land and Water Policy in a Growth State* (Baltimore: The Johns Hopkins University Press, 1974), 49–53.

22. Carter, *Florida Experience*, chapter 10; Nelson Manfred Blake, *Land into Water—Water into Land* (Tallahassee: University Presses of Florida, 1980), chapter 10.

23. R. L. Rhodes and D. H. MacLaughlin, "Federal and State Regulation of Hazardous Waste Management," *Florida Bar Journal* 54, no. 713 (1980): 713–18.

24. Governor's Hazardous Waste Policy Advisory Council, *Hazardous Waste: A Management Perspective* (Tallahassee: The Institute of Science and Public Affairs, Florida State University, 1981), 33–36.

25. Florida State House Bill 593 (1982).

26. B. D. E. Canter, "Safe Hazardous Waste Disposal: Sure, But Where?," *Florida Bar Journal* 55, no. 813 (1981): 813–17.

27. Joint Committee on Taxation, *Description of Revenue Aspects Relating to Hazardous Substance Pollution and Liability* (Washington, D.C.: Government Printing Office, September 9, 1980), 5.

28. Florida State House Bill 593 (1982); Florida State Senate Bill 559 (1982).

29. "State Battles Toxic Waste Dumps," *Miami Herald* (August 2, 1982), p. 7A.

30. Governor's Hazardous Waste Policy Advisory Council, *Hazardous Waste*, 27–28.

31. Florida Department of Environmental Regulation, "News Release" (Tallahassee: Department of Environmental Regulation, October 23, 1981); "State Battles Toxic Waste Dumps," *Miami Herald* (August 2, 1982), p. 7A.

32. Bruce A. Williams, "Bounding Behavior: Economic Regulation in the American States," in *Politics in the American States: A Comparative Analysis*, 4th ed., Virginia Gray, Herbert Jacob, and Kenneth Vines, eds. (Boston: Little, Brown and Company, 1983), 329–72.

33. DER, "News Release," no page number (emphasis added).

34. Albert R. Matheny and Bruce A. Williams, "Regulation Risk/Benefit Analysis, and the Supreme Court," paper presented at the Annual Meetings of the Law and Society Association (Toronto, June 5, 1982).

35. The following discussion is based upon Kathy Koch, "Compromise Reached on 'Superfund' Bill," *Congressional Quarterly* (November 29, 1980): 3436–37; Kathy Koch, "Compromise Superfund Proposal Cleared," *Congressional Quarterly* (December 6, 1980): 3509.

36. "Toxic Waste Rule Draws Little Applause," *Chemical Week* 126 (May 14, 1980): 15; "Superfund Compromise Wins 'Grudging' Nod," *Chemical Week* 126 (June 18, 1980): 63.

37. Governor's Hazardous Waste Policy Advisory Council, *Hazardous Waste*, 23–24.

38. The following discussion is based largely upon, the MITRE Corporation, "Working Paper" (McLean, Va.: The MITRE Corporation, June 4, 1981).

39. For a more complete discussion of the inseparability of technical and political issues in this area see, Albert R. Matheny and Bruce A. Williams, "Scientific Disputes and Adversary Procedures in Policy-Making: An Evaluation of the Science Court," *Law and Policy Quarterly* 3, no. 3 (July 1981): 341–64.

40. For an excellent discussion of the political uses of cost-benefit analysis and the financial requirements of doing such analyses adequately see George C. Eads, "White House Oversight of Executive Branch Regulation," in Bardach and Kagan, eds., *Social Regulation*, 177–97.

41. National Wildlife Federation, *Toxic Substances Programs in U.S. States and Territories: How Well Do They Work?* (Washington, D.C.: National Wildlife Federation, 1979).

42. On the importance of the relative resources of interest groups see Williams, "Bounding Behavior."

43. Wilson, "The Politics of Regulation," 370–71.

44. Jones, *Clean Air*, chapters 7–8.

45. For an article that reaches a similar conclusion in a different area of environmental regulation see Helen Ingram, Nancy Laney, and John R. McCain, "Water Scarcity and the Politics of Plenty in the Four Corners States," *Western Political Quarterly* 32 (September 1979): 298–306.

6. Intergovernmental and Public-Private Sector Relations in Hazardous Waste Management: The New York Example

1. R. C. Eckhardt, "The Unfinished Business of Hazardous Waste Control," *Baylor Law Review* 33 (April 1981): 252–65.

2. T. C. Jorling, "Hazardous Substances in the Environment," *Ecology Law Quarterly* 9 (December 1981): 520–618.

3. John A. Worthley and Richard Torkelson, "Managing the Toxic Waste Problem," *Administration and Society* 13 (August 1981): 145–60.

4. Ibid.

5. Philip Shabecoff, "Health Fears Grow as Debate Continues on Toxic Wastes," *New York Times* (January 2, 1982), pp. 1,7.

6. Hugh B. Kaufman, quoted in ibid., 1.

7. New York State Department of Environmental Conservation, *Short-Term Hazardous Waste Management Plan* (Albany: Department of Environmental Conservation, 1981), 10.

8. Ibid., 8.

9. Sloan O'Donnel, "Hazardous Wastes: A By-Product of the Good Life," *The Conservationist* 42 (January/February 1982): EQN I–EQN VI.

10. Testimony of Norman H. Nosenchuck before the Subcommittee on Commerce, Transportation and Tourism of the House Energy and Commerce Committee, April 21, 1982.

11. Lois Gibbs quoted in Miriam Pawel, "Toxic Waste Foes Get Expert Advice," *Newsday* (March 7, 1982), p. 18.

12. Sylvia Moreno, "Assembly to Probe Landfill Cleanup," *Newsday* (April 20, 1982), p. 21.

13. Judith Cummings, "No. 1 Toxic Waste Site Is a Town's No. 1 Gripe," *The New York Times* (December 9, 1981), p. A16.

14. Allison Mitchell, "Company Settles One Love Canal Suit," *Newsday* (May 1, 1982), p. 10.

15. Ed Magnuson, "The Poisoning of America," *Time* (September 22, 1980), pp. 58–69.

16. New York State Department of Environmental Conservation, *Short-Term Hazardous Waste Management Plan* (Albany: DEC, 1981), 10.

17. Richard Severo, "Toxic Waste Disposal: Search for the Best Technology Intensifies," *The New York Times* (May 19, 1981), pp. C1, 2.

18. Mark McIntyre, "A New Marketplace for Toxic Wastes," *Newsday* (April 22, 1982), p. 19.

19. Program Development Group, New York State Assembly, "Hazardous Wastes Exchange: A Proposal" (Albany: State Assembly, March 30, 1981).

20. Joseph Colby quoted in "Oyster Bay Urges Landfill Suit," *Newsday* (March 11, 1982), p. 27.

21. New York State Department of Environmental Conservation, *Hazardous Waste Program Status Update* (Albany: DEC, 1981), 8.

22. Program Development Group, New York State Assembly, "A Proposal for a New York State Mini-Superfund" (Albany: State Assembly, January 15, 1982).

23. Magnuson, "The Poisoning of America," 69.

24. Hugh L. Carey, "Governor's Press Release" (Albany: November 24, 1981).

25. Robert B. Reich, "Business Is Asking for Trouble Again," *The New York Times* (November 22, 1981), pp. 4–8.

7. Institutional Fragmentation and Hazardous Waste Policy: The Case of Texas

1. *Texas Almanac and State Industrial Guide 1980–81* (Dallas: A.H. Belo Corporation, 1979).

2. The term "solid waste," as used in the federal and state regulatory context, actually includes gaseous, liquid, semisolid, and solid waste materials.

3. Kenneth S. Kamlet, *Toxic Substances Programs in U.S. States and Territories: How Well Do

They Work? (Washington, D.C.: National Wildlife Federation, 1979), p. IV (hereinafter cited as Kamlet, *Toxic Substances Programs*).

4. Texas Department of Water Resources (TDWR), *Texas Environment* (October 1981), 9. This figure does not include wastes generated by oil and gas production activities, hazardous wastes disposed of in municipal landfills, or certain wastes exempted from regulatory requirements. Inclusion of these would probably result in a figure of at least four million metric tons.

5. Ibid., 10.

6. Izaak Walton League of America, *Waste Alert Spotlight* 2 (March 1982).

7. The Texas Solid Waste Disposal Act, Art. 4477-7, Section 2 (6) Texas Revised Civil Statutes Annotated, defines "municipal solid waste" as: "solid waste resulting from or incidental to municipal, community, trade, business and recreational activities, including garbage, rubbish, ashes, street cleanings, dead animals, abandoned automobiles, and all other solid waste other than industrial solid waste."

8. The Texas Solid Waste Disposal Act, Art. 4477-7, Section 2 (7) Texas Revised Civil Statutes Annotated, defines "industrial solid waste" as: "solid waste resulting from or incidental to any process of industry or manufacturing, or mining or agricultural operations."

9. TDWR, *Texas Environment*, 9.

10. TDWR, *Solid Waste Management Plan for Texas 1980-1986, Volume II—Industrial Solid Waste, Draft* (November, 1980), 29 (hereafter cited as TDWR, *Industrial Solid Waste Management Plan*).

11. Figures from the TDWR, *Industrial Solid Waste Management Plan*, p. 10, indicate that 88.9 percent of all "Class I" industrial wastes (the category which includes hazardous wastes) generated in Texas in 1979 were disposed of on-site. Jay Snow, Chief of the Solid Waste Section of TDWR, indicated in a panel discussion at the Region VIII Conference of the American Society for Public Administration in San Antonio, November 12, 1980, that the figure for on-site disposal of strictly hazardous wastes was probably 75 percent.

12. TDWR, *Authorized Commercial Industrial Solid Waste Sites: Solid Waste Permits and Disposal Well Permits* (revised, February 1982).

13. Texas Natural Resources Information System, *Permit Application Pending* (1982); also Texas Department of Water Resources, *Permit Applications Processed for Hazardous Wastes* (October 14, 1982).

14. North Central Texas Council of Governments, *Regional Industrial Waste Management Study* (December 1980), 5.

15. Ibid., 24 and 25.

16. Subcommittee on Hazardous Waste, Committee on Environmental Affairs, Texas House of Representatives, *Interim Report to the 67th Texas Legislature* (October 16, 1980), 5-14 (hereafter cited as Subcommittee on Hazardous Waste, *Interim Report*).

17. TDWR, *Texas Environment* (October 1981), 1.

18. Subcommittee on Hazardous Waste, *Interim Report*, 5-14.

19. Subcommittee on Oversight and Investigation, House of Representatives of the 96th Congress, *Hazardous Waste Disposal* (Washington, D.C.: Government Printing Office, 1979).

20. Testimony by TDWR Deputy Director Dick Wittington at a hearing conducted by the Natural Resources Committee, Texas House of Representatives, Austin, Texas, March 17, 1981.

21. Harold Scarlett, "Texas City Chemical Dump to Get Fence, EPA Official Says," The Houston *Post* (January 5, 1980).

22. Senate Natural Resources Committee, Texas Senate, *Interim Report to the 67th Legislature* (March 1981), 5 (hereafter cited as Senate Natural Resources Committee, *Interim Report*).

23. Carlos Byars, "Encouraging News from Chemical Dump," Houston *Chronicle* (February 1, 1982).

24. Kevin Moran, "Dallas TV Station, Reporter Sued by Texas City Firm," Houston *Chronicle* (June 20, 1981).

25. Harold Scarlett, "Waste Project Foe Accused of Faking Academic Degrees," The Houston *Post* (August 27, 1980).

26. Paul Sweeney, "Waste Disposal in the Coastal Bend: Poison Acres," The Texas *Observer* (April 11, 1980), 4-7 and 16-19.

27. Article 4477-7, Texas Revised Civil Statutes Annotated.

28. Texas Natural Resources Code, Title 3, Texas Water Code, chapter 27.

29. Texas Natural Resources Code, Title 3, Texas Water Code, chapter 21, subchapter K.

30. Texas Natural Resources Code, Title 3, Texas Water Code, chapter 26, subchapter H.
31. Texas Natural Resources Code, Title 3.
32. Kamlet, *Toxic Substances Programs*, i, indicates that fourteen states have a principal toxic regulatory agency.
33. Texas Department of Health, *Texas Hazardous Waste Management Program* (1982).
34. Ibid., and Radian Corporation (prepared for the Ozarks Regional Commission and the U.S. Geological Survey), *Permit Requirements for Development of Energy and Other Selected Natural Resources for the State of Texas* (July 1981), 77–85.
35. Sierra Club, Lone Star Chapter, *Position Paper on Hazardous Waste Management in Texas* (January 1981), 7–8.
36. Kenneth W. Kramer, "Environmental Protection and Natural Resources Management," in Wendell M. Bedichek and Neal Tannahill, eds., *Public Policy in Texas* (Glenview, Ill.: Scott, Foresman and Company, 1982), 468–69.
37. Richard Lowerre, *Supplemental Comments of the Lone Star Chapter of the Sierra Club on the State of Texas Application for Phase II Authorization* (January 29, 1982), 7–10.
38. Kamlet, *Toxic Substances Programs*, 7.
39. Ibid., i–iv. Of course, the National Wildlife Federation study derived its score on the strength of a state's regulatory program from the responses given to forty-three questions, of which the question about a single agency or official was one. However, none of the other forty-two questions were related directly to the issue of single or multiple agencies.
40. Article 6252-13a, Texas Revised Civil Statutes Annotated.
41. Gulf Coast Waste Disposal Authority, for example, in 1980 withdrew an application to site a hazardous waste management facility at a particular location due to opposition from the City of Houston and City of Pasadena and the expected costs of pursuing that site in the face of determined opposition. Another site, not opposed by those two cities, was chosen, but that application was also eventually withdrawn for a variety of reasons. *Waste Management Facilities in the Galveston Bay Area: An Overview*, Report of the Keystone Workshop on Siting Nonradioactive Hazardous Waste Management Facilities, held August 17–20, 1982, at Keystone, Colorado.
42. Texas Department of Health, *Municipal Solid Waste Management Regulations: Requirements for Solid Waste, Including Hazardous Waste* (November 1980), 32–34. In November 1981, the state Third Court of Appeals rejected a contention by a waste disposal company that these land-use regulations were too vague. "Browning-Ferris Land Use Plea Denied," Houston *Chronicle* (November 26, 1981).
43. TDWR, *Industrial Solid Waste Management, Technical Guidelines No. 2*, "Site Selection and Evaluation" (March 1, 1978), II-1.
44. Ibid. For example, the TDWR guidelines on "Site Selection and Evaluation" suggest that "flood plains, shore lands, and groundwater recharge areas should be avoided" and that sites should be "outside the paths of recurring severe storms, such as hurricanes and tornados" and have "low air pollution potential." A site which met all of those criteria would be difficult to find in the Houston-Galveston area. Only the soil characteristics of the area fit the suggested guidelines closely.
45. Pending applications (and expected applications) for industrial waste management facilities are for sites in Calhoun, Chambers, Liberty, and Matagorda counties, all of which are coastal counties subject to frequent severe storms and encompassing extensive flood plain and groundwater recharge areas.
46. Testimony of Catherine Perrine (League of Women Voters of Texas) and testimony of Joel Schroeder (Texas Farm Bureau) at a hearing conducted by the Natural Resources Committee, Texas House of Representatives, Austin, Texas, March 17, 1981. Testimony by representatives of P.A.C.E. at a hearing conducted by the Subcommittee on Toxic and Chemical Waste Disposal Siting, Texas House Committee on Environmental Affairs, The Woodlands, Texas, March 26, 1982. Sierra Club, *Position Paper*.
47. Policy Research Project, Lyndon Baines Johnson School of Public Affairs, University of Texas at Austin, *Preliminary Report on Siting of Hazardous Waste Disposal Facilities* (March 1982). The Final Draft of the *Coastal Natural Resources Report* by the Texas Energy and Natural Resources Advisory Council (June 1982), 56–62, does make a cautious argument for siting criteria also.
48. Susan Hadden, Director, *LBJ School PRP on Hazardous Waste Disposal Facility Siting: Recommendations* (May 25, 1982) (hereafter cited as LBJ School, *Recommendations*).
49. Senate Natural Resources Committee, *Interim Report*, 16–17.
50. LBJ School, *Recommendations*.

51. Ibid.; Sierra Club, *Position Paper*; Senate Natural Resources Committee, *Interim Report*, 17.

52. See, for example, the charges made in the investigative article by Paul Sweeney, "One Man's Waste Is Another Man's Business," The Texas *Observer* (February 15, 1980), pp. 3–6 and 12–13.

53. Ibid.

54. Examples of these allegations are found in such reports as "West Texas Oil Just Doesn't Mix with Water Needs," The Houston *Post* (October 29, 1978) and "Residents Claim Oil Drillers Not Properly Monitored," Odessa *American* (January 30, 1981). An organization, Southwest Soil and Water Protection Association, has been formed by many of these landowners to represent their interests and to push for more stringent regulation of oil and gas waste disposal. See Michael Hinshaw, "Water Quality: More Than a Matter of Taste," The Dallas *Morning News* (January 25, 1981).

55. Arthur Hahn, "RRC Called on to Enforce Rules," Brenham *Banner-Press* (January 1, 1982).

56. LBJ School, *Recommendations*.

57. Kirk Brown and D. C. Anderson, *Effects of Organic Solvents on the Permeability of Clay Soils*, EPA, Grant No. R80682510 (1981).

58. Peter Montague, *Four Secure Landfills in New Jersey* (Princeton, N.J.: Department of Chemical Engineering, 1981).

59. *Federal Register*, Vol. 47, no. 143 (July 26, 1982): 32284–85.

60. TDWR, *Industrial Solid Waste Management Plan*: 35-36.

61. *Texas Pollution Report* (February 24, 1982), 2.

62. Subcommittee on Toxic and Chemical Waste Disposal Siting, Texas House Committee on Environmental Affairs, *Summary of Recommendations from Subcommittee Public Hearings* (June 1982).

63. Ibid.

64. Texas Advisory Commission on Intergovernmental Relations (TACIR), *Solid Waste Management: Implementing the Federal Resource Conservation and Recovery Act of 1976 in Texas* (September 1977), 13–14 (hereafter cited as TACIR, *Solid Waste Management*).

65. Public Law 89-272.

66. TACIR, "The Federal Resource Conservation and Recovery Act of 1976: Policy Questions for Texas," *Intergovernmental Report* (March 1977), 3.

67. Ibid.

68. TDWR, *Industrial Solid Waste Management Plan*, 5.

69. Ibid., and Texas Department of Health, *Solid Waste Management Plan for Texas 1980–1986, Volume I–Municipal Solid Waste, Draft* (November 1980), 7.

70. Subcommittee on Hazardous Waste, *Interim Report*, 27.

71. For example, standards required to be set by EPA under the Clean Air Act of 1970 and the Federal Water Pollution Control Act of 1972 for discharges of toxic or other hazardous materials into the air or water.

72. Public Law 93-523.

73. The original UTC regulations were produced by EPA in August 1976. These proposed rules met with such criticisms that final rules were delayed for four years.

74. Walter A. Rosenbaum, *The Politics of Environmental Concern* (New York: Praeger Publishers, 1973), 232–51.

75. See, for example, *Environmental Quality–The Fourth Annual Report of the Council on Environmental Quality* (September 1973), 201–2.

76. TDWR, *Industrial Solid Waste Management Plan*, 5–6.

77. See, for example, Subcommittee on Hazardous Waste, *Interim Report*, 29–57; and Senate Natural Resources Committee, *Interim Report*, 21–23.

78. See, for example, the recommendations made in Texas Coastal and Marine Council, *Report to the 66th Legislature on Senate Resolution 471, Long-Term Monitoring of Industrial Solid Waste Disposal Facilities* (1979), 16–19. Senate Bill 499 to establish such a fund was passed by the Texas Senate in 1979 but did not even get a committee hearing in the House.

79. Subcommittee on Hazardous Waste, *Interim Report*, 63.

80. This was Senate Bill 758, as described in *Texas Pollution Report* (March 11, 1981), 1–2.

81. This was House Bill 8. Although it was modified after adoption of the federal Superfund to eliminate state fees on hazardous waste generators to finance the state fund (Superfund being interpreted as federal preemption of the right of states to collect those fees), the bill did not by its wording limit cleanup efforts financed by a state fund to only those associated with Superfund. *Texas Pollution Report* (March 4, 1981), 1–2.

82. See Citizens Conference on State Legislatures, *The Sometimes Governments: A Critical Study of the Fifty American Legislatures* (New York: Bantam Books, 1971).

83. Richard H. Kraemer and Charldean Newell, *Texas Politics* (St. Paul: West Publishing Company, 1979), 167–70, 175–76.

84. Citizens Conference on State Legislatures, *The Sometimes Governments*, 307.

85. James E. Anderson, Richard W. Murray, and Edward L. Farley, *Texas Politics: An Introduction* (New York: Harper and Row, 1979), 190–91.

86. Wendell M. Bedichek and Neal Tannahill, eds., *Public Policy in Texas* (Glenview, Ill.: Scott, Foresman and Company, 1982), 218–20.

87. Ann Bowman and Michael A. Maggiotto, "The Effect of Environmental Policy Orientations: Lobbying Florida's Legislators," *Social Science Quarterly* 62 (September 1981): 547–54.

88. See, for example: Bedichek and Tannahill, *Public Policy in Texas*, 139–49; Kraemer and Newell, *Texas Politics*, 55–70.

89. Kraemer and Newell, *Texas Politics*, 59–64.

90. Clifton McClesky et al., *The Government and Politics of Texas*, 7th ed. (Boston: Little, Brown and Company, 1982), 354.

91. *Texas Almanac and State Industrial Guide: 1972-1973* (Dallas: A.H. Belo Corporation, 1971), 397.

92. See, for example: testimony of W. R. Reeves, Texas Ecologist, Inc. (a waste disposal firm) and testimony of Rayford Price, Texas Institute of Chemical Waste Management, at a hearing conducted by the Natural Resources Committee, Texas House of Representatives, Austin, Texas, March 17, 1981.

93. *Texas Pollution Report* (March 25, 1981), 1.

94. Paul Sweeney, "Parker's Deadly Kidding on Hazardous Wastes," The Texas *Observer* (June 12, 1982), p. 15.

95. Kenneth W. Kramer, "Environmental Protection and Natural Resources Management," in Bedichek and Tannahill, *Public Policy in Texas*, 472.

96. For example, a recent survey of public opinion in Houston showed strong support for environmental protection. See Jim Asher, "Traditional Labels Don't Apply," The Houston *Post* (August 22, 1982).

97. Dirk Kirschten, "The New War on Pollution Is Over the Land," *National Journal* (April 14, 1979), p. 605.

98. See, for example, Subcommittee on Hazardous Waste, *Interim Report*, 5, and Carlos Byars, "Hazardous Waste Area Here Unlike Love Canal, EPA Official Says," Houston *Chronicle* (May 25, 1980).

99. Paul Sweeney, "Waste Disposal in the Coastal Bend: Poison Acres," The Texas *Observer* (April 11, 1980), pp. 3–6, 12–13.

100. See, for example, "NRC Study Reveals Todd Evaporated Nuclear Liquids," The Houston *Post* (January 18, 1980).

101. Kamlet, *Toxic Substances Program*, iv.

8. Technological Policies and Hazardous Waste Politics in California

1. Based on an EPA survey, and reported in National Wildlife Federation, *The Toxic Substances Dilemma: A Plan for Citizen Action* (Washington, D.C.: U.S. Environmental Protection Agency, 1981). Data on waste generation by region are provided in U.S. Environmental Protection Agency, *Hazardous Waste Generation and Commercial Hazardous Waste Management Capacity—An Assessment* (SW-894) (Washington, D.C.: EPA, 1980).

2. Stephen C. Jones, "Hazardous Waste Management," *California Lawyer* 3 (January 1983): 12.

3. Listed in Samuel S. Epstein, Lester O. Brown, and Carl Pope, *Hazardous Waste in America* (San Francisco: Sierra Club Books, 1982), 448–49. The new EPA list of Superfund sites issued in December 1982 includes eleven in California. See *New York Times* (December 21, 1982), p. 14.

4. California Department of Health Services, *Report to the Hazardous Waste Management Council on Current Hazardous Waste Generation in California* (Sacramento: Toxic Substances Control Division, Hazardous Waste Management Branch, August 31, 1982).

5. An additional 40 million tons per year were identified as "hazardous waste-related activities" as reported by industry to EPA on its Part A RCRA forms in 1980. A large proportion of these wastes

(60 percent) are dilute rinse waters which, after on-site treatment, are legally discharged to sewers and surface waters. And another 28 percent of the total listed as "unknown" cannot be assigned to any final disposition. These data weaknesses are inherent in the original EPA reporting form. The major study of these data done for the state by researchers at the University of California-Davis in 1981 concluded that most of this material, in liquid form, is being discharged by a combination of evaporatory losses from surface impoundments, discharges to sewers and surface waters not properly reported to EPA, and residues transported off-site for land disposal. See University of California-Davis, *Hazardous Wastes in California: On-Site Storage, Treatment, and Disposal* (Davis: Chemical Engineering Department, draft of May 20, 1982). The huge size of these numbers, however, remains of concern to planners and hazardous waste managers in the state.

6. California Department of Health Services, *Report to the Hazardous Waste Management Council*, 7–8.

7. Ibid., 6.

8. Ibid., 14.

9. Ibid., 19. Also see Michael Belliveau, *On-site Hazardous Waste Management in the San Francisco Bay Area* (San Francisco: Citizens for Better Environment, September 1981).

10. California Department of Health Services, *Report to the Hazardous Waste Management Council*, 20–21.

11. Southern California Hazardous Waste Management Project, *Hazardous Waste Generation and Facility Development Needs in Southern California* (Los Angeles: Los Angeles Sanitation Districts, draft, November 1982) (unpaginated), section II-2.

12. Ibid.

13. Ibid.

14. Ibid.

15. Ibid., section III-2.

16. Peter Steinhart, "The Spectre of Toxic Substances," *California Lawyer* 3(January 1983): 25. Also see Congressional Research Service, "Six Case Studies of Compensation for Toxic Substances Pollution;: Alabama, California, Michigan, Missouri, New Jersey, Texas" (96th Cong., 2d Sess., 1980), 365–66.

17. California Department of Health Services, *Notices of Proposed Changes in Regulations of the Department of Health Services Regarding Hazardous Waste Land Disposal Restrictions* (Sacramento: Toxic Substances Control Division, August 18, 1982) (R-32-82), Attachment: "Case Histories of Pollution Incidents from Land Disposal of Hazardous Wastes in California," p. A-9.

18. Technically, this is an existing problem rather than part of the legacy of past improper disposals in that storage of waste solvents in underground tanks is ongoing. These steel tanks were installed some years ago without adequate protection against corrosion.

19. Steinhart, "The Spectre of Toxic Substances," 25.

20. California Department of Health Services, *Notices of Proposed Changes*, A-1, A-2.

21. Chapter 1302, Statutes of 1982 (A.B. 26, Ross Johnson).

22. Epstein et al., *Hazardous Waste in America*, 349.

23. California Department of Health Services, *Notice of Proposed Changes*. Also see California Department of Health Services, "Preventing a Toxic Tomorrow: California's Program to Treat Toxic and Extremely Hazardous Wastes" (Sacramento: Toxic Substances Control Division, September 15, 1982).

24. California Governor's Office of Appropriate Technology, *Alternatives to the Land Disposal of Hazardous Wastes: An Assessment for California* (Sacramento: OAT, 1981).

25. Governor Edmund G. Brown, Jr., Executive Order B-8881, October 13, 1981, p. 2.

26. Ibid.

27. Interagency Task Force for Reduction of Land Disposal of Toxic Wastes, *Development of the Land Disposal Restriction Regulation: Discussion Papers, List of Respondents, Workshop Attendees and Meetings, and Compilation of Written Comments* (Sacramento: Department of Health Services, August 6, 1982).

28. "California Industry Prepares for Landfilling Ban," *Hazardous Materials Intelligence Report*, P.O. Box 535, Cambridge, MA 02238, 3, no. 30 (July 23, 1982), p. 4.

29. See Meredith/Boli and Associates, Inc., *Review of the OAT Reporting and Related Documents*, Prepared for the Chemical Manufacturers Association (Beverly Hills: Meredith/Boli, January 1982).

30. Testimony of Michael S. Meredith and E. Clark Boli before the Assembly Committee on Consumer Protection and Toxic Materials, Sacramento, California, February 2, 1982, 5–6.

31. Interagency Task Force, *Land Disposal Restriction Regulations.*

32. California Department of Health Services, *Notice of Proposed Changes.*

33. "California Industry Prepares . . . ," *Hazardous Materials Intelligence Report,* 5.

34. Based on a summary prepared by the California Department of Health Services for the Hazardous Waste Management Council (Sacramento, September 15, 1982).

35. California Department of Health Services, *Criteria for the Siting of Treatment Technologies for Hazardous Waste Management* (Sacramento: DHS, November 1982) (draft).

36. "Objections Delay Hazardous Waste Center Plans," *Chula Vista Star News* (November 21, 1982), 1.

37. Conversation with Rick Alexander, San Diego Association of Governments (SANDAG), October 10, 1982.

38. For a review of the buffer zone issue as it applies around nuclear power plants, see David Morell and Grace Singer, *Land-Use Controls in Communities Near Nuclear Power Plants—A Policy Analysis* (Princeton, N.J.: Center for Energy and Environmental Studies, Princeton University, 1982). Also see California Foundation on the Environment and the Economy, *Hazardous Waste Facility Siting and Permitting in California: Case Studies and Analysis* (San Francisco, February 1983), 27–39.

39. Douglas Lockwood, *California's Existing Siting and Permitting Process for Hazardous Waste Facilities: Issues and Perspectives* (Sacramento: Hazardous Waste Management Council, October 1982), 19–35.

40. This siting case also illustrates the tendency of developers to focus their attention primarily on land they already own (or on which they hold an option). The Long Beach site was probably an excellent one for expanded oily waste treatment. A treatment facility has been operating there for fifty years; local citizens were accustomed to its presence; existing sewer capacity was adequate to handle large discharges of treated effluent. The site was *not* ideal for a hazardous waste transfer station, however. It was too close to residential housing, and not well situated with respect to freeway access. Yet the developer proposed to locate a transfer station here, primarily because the land was already available for a different kind of waste management use. Resultant citizen opposition to the proposed transfer station led to defeat of the entire proposal, including the expanded oily waste-treatment facility.

41. City of Long Beach, Ordinance (5803, Section 3[c][1][aa]), 13–14, amendments adopted February 9, 1982.

42. California Department of Health Services, *Carcinogen Identification Policy* (Sacramento: DHS, July 1982).

43. Ibid., 1.

44. Ibid., 40.

45. California Department of Health Services, *Carcinogen Identification Policy—Methods for Estimating Cancer Risks from Exposures to Carcinogens* (Sacramento: DHS, October 1982).

46. California Department of Health Services, *Carcinogen Policy: A Policy for Reducing the Risk of Cancer* (Sacramento: DHS, December 1982).

47. Ibid., 1.

48. California Department of Health Services, *California Assessment Manual for Hazardous Wastes* (Sacramento: DHS, June 1981).

49. *Toxic News* (October–November 1982), 7.

50. Linda Haugsted, "Hazardous Waste Battle Lines Drawn," *San Gabriel Valley Tribune* (November 20, 1982), 1.

51. "California Setting New Hazardous Waste Rules," *Chemical and Engineering News* 60, no. 48 (November 29, 1982): 16–17.

52. California Air Resources Board, *A Proposed Policy for Reviewing New or Modified Sources of Toxic Air Contaminants* (Sacramento: ARB, September 27, 1982).

53. Richard O'Reilly, "State Moves on Toxic Chemicals—Strict Regulation of Emission into Air OK'd," *Los Angeles Times* (October 29, 1982), pp. 3, 20. Also see *Toxic News* (October–November 1982), 7.

54. Max Miller, "ARB Creates Committee to Monitor Airborne Toxins, Cancer Causes," *Sacramento Bee* (October 29, 1982), p. A11.

55. See William W. Lowrance, *Of Acceptable Risk* (Los Altos: Kaufman, 1976), 76.

56. Health and Safety Code, Sections 25100–25240 (West Supp. 1981); 22 Cal Admin Code Sections 66016–66898 (1980).

57. Water Code Sections 13000–13989; 23 Cal Admin Code Sections 1–2836.

58. Epstein et al., *Hazardous Waste in America*, 349.

59. Under Section 3006(c) of RCRA, states with "substantially equivalent" programs may be authorized by EPA to operate in lieu of the federal program.

60. California Legislature, Office of the Auditor General, *California's Hazardous Waste Management Program Does Not Fully Protect the Public from the Harmful Effects of Hazardous Waste* (Sacramento: Auditor General, October 1981), 4.

61. *Hazardous Waste Control Account*, California Health and Safety Code, Section 25174.

62. *Toxic News* (September–October 1982), 2.

63. California Health and Safety Code, Section 25330.

64. Belliveau, *On-site Hazardous Waste Management in the San Francisco Bay Area.*

65. California Auditor General, *Hazardous Waste Program.*

66. See David Morell and Christopher Magorian, *Siting Hazardous Waste Facilities: Local Opposition and the Myth of Preemption* (Cambridge, Mass.: Ballinger, 1982); and Jonathan Steeler, *A Legislator's Guide to Hazardous Waste Management* (Denver: National Conference of State Legislatures, October 15, 1980).

67. S.B. 1049, introduced on March 27, 1981; amended in Senate June 16, 1981.

68. Comments by Gary Kovall, Manager of Environmental Regulatory Affairs, ARCO, at a public workshop on hazardous waste management sponsored by the Southern California Hazardous Waste Management Project, Riverside, November 8, 1982.

69. A.B. 2196, submitted by Assemblywoman Cathie Wright (R-Los Angeles), was similar to S.B. 878. It died in Assembly committee.

70. A bill sponsored by Sen. Diane Watson in 1982—S.B. 1748—(introduced March 11, 1982) would have provided funds for county and regional hazardous waste planning activities. Hazardous waste planning committees to be established in specified counties would prepare a hazardous waste management plan. These plans would then have to be approved by the county Board of Supervisors and by a specified majority of the cities within the county. This planning effort was to be funded by increased fees on hazardous waste disposal. Even though all mention of preemption was removed, S.B. 1748 too failed to pass.

71. Chapter 89, Statutes of 1982.

72. As finally constituted in 1982, California's Hazardous Waste Management Council has the following membership: one state Senator, one member of the State Assembly, Director, Department of Health Services, Chair, Air Resources Board, Chair, Water Resources Control Board, Chair, Waste Management Board, three county supervisors, one city council member, two representatives of waste-generating industry, one representative of the waste management industry, two environmentalists, and one consulting geologist.

73. Section 25208.

74. Section 25208(c).

75. Personal communication from Kieran Bergin, Los Angeles County Sanitation Districts, December 27, 1982.

76. U.S. Environmental Protection Agency, Office of Water and Waste Management, *Siting of Hazardous Waste Management Facilities and Public Opposition*, SW-809 (Washington, D.C.: EPA, November 1979), 296–300.

77. California Department of Health Services, *Notice of Proposed Changes*, A-6, A-7.

78. Bergin, pers. comm., December 27, 1982.

79. See South Coast Air Quality Management District, "Vinyl Chloride in the South Coast Air Basin" (El Monte: SCAQMD, May 1982).

80. For an analysis of initiatives and referenda, see Carl E. Lutrin and Allen K. Settle, "The Public and Ecology: The Role of Initiatives in California's Environmental Politics," *Western Political Quarterly* 28 (1975): 352–71.

81. The first meeting took place in Los Angeles on December 18, 1980. It included representatives from the Department of Health Services, Water Resources Control Board, U.S. Environmental Protection Agency, Southern California Association of Governments, and Los Angeles Sanitation Districts.

82. Hearing, State of California, "Solving the Hazardous Waste Problem: Non-Toxic Solutions for the 1980's," Los Angeles Convention Center, November 17, 1980, 12. Bradley was referring here to the Casmalia and Kettleman Hills Class I hazardous waste landfills, located in Santa Barbara County and Kings County, respectively.

83. Southern California Hazardous Waste Management Project, *Status Report* (April 15, 1982).

84. S.B. 501, introduced March 9, 1981; amended in Senate April 30, 1981, and amended in Assembly June 18, 1981. Chapter 244, Calif. Legis. Serv. 813 (July 21, 1981), amending Cal. Health and Safety Code Section 25149, Cal. Govt. Code Section 66976(e).

85. Section 66976(e).

86. "Solving the Hazardous Waste Problem," 10, 11, 12.

87. Ibid., 13.

88. Transportation of hazardous wastes from the areas of waste generation to either Casmalia or Kettleman costs about $25/ton. Bergin, pers. comm., December 27, 1982.

89. Southern California Hazardous Waste Management Project, *Regional Goals and Policies for the Safe Management of Hazardous Wastes* (June 1982).

90. California Department of Health Services, *Criteria for the Siting of Treatment Technologies*.

91. Southern California . . . Project, *Hazardous Waste Generation*.

92. Southern California Hazardous Waste Management Project (Jean Carr, Public Participation Coordinator), *Report on November Meetings and Evaluation of Regional Workshops*, November 24, 1982.

93. California Department of Health Services, *The California Hazardous Waste Management Plan and State Program* (Sacramento: Department of Health Services, December 1981), 13.

94. Ibid.

95. Anonymous.

96. See "Panel Told All Landfills Leak, EPA Rules on Hazardous Waste Land Disposal Inadequate," *Environment Reporter* (December 3, 1982): 1276–78. William W. Lowrance deals with the general topic of uncertainty of hazardous waste landfills in *Assessment of Health Effects at Chemical Disposal Sites* (New York: Life Sciences and Public Policy Program, Proceedings of the Symposium held in New York City on June 1–2, 1981), 3–21.

97. "Election in California May Affect State's Role as Trend Setter in RCRA Area," *Hazardous Waste News* (November 15, 1982), 357.

98. Donald H. Harrison, "The Deukmejian Dilemma: Promises Meet Reality," *California Journal* 13, no. 12 (December 1982): 436.

99. "Urgent, but Not *That* Urgent," Editorial, *Los Angeles Times* (December 1, 1982), p. II-10.

9. The Politics of Public Participation in Hazardous Waste Management

1. U.S. Congress, House Committee on Interstate and Foreign Commerce, Subcommittee on Consumer Protection and Finance, *Hearings: Toxic Substances Control Act Amendments of 1980*, April 15–22, 1980, H. Rep. 96–211, p. 35.

2. Conservation Foundation, "Agencies Curtail Access to Data and Decisions," *Letter*, November 1981, p. 1.

3. Useful summaries of these provisions and the related legislation may be found in: U.S. Community Services Administration, *Citizen Participation* (San Jose, Calif.: Community Services Administration, 1978); and U.S. Advisory Commission on Intergovernmental Relations, *Citizen Participation in the American Federal System* (Washington, D.C.: GPO, 1979), chapter 4.

4. On the mixed motivations for congressional voting, see Walter A. Rosenbaum, "Public Involvement as Reform and Ritual: The Development of Federal Participation Programs," in Stuart Langron, ed., *Citizen Participation in America* (Lexington, Mass.: D.C. Heath and Co., 1978), 81–96; and Daniel P. Moynihan, *Maximum Feasible Misunderstanding* (New York: The Free Press, 1970).

5. U.S. Library of Congress, Congressional Research Service, *A Legislative History of the Water Pollution Control Act Amendments of 1972*, vols. I and II, January 1973, S. 93-1.

6. Sarah Scott, "Consumer Advocates Wait Anxiously for the Pendulum to Swing Their Way," *National Journal* 13, no. 2 (January 10, 1981): 57–58.

7. The author is indebted to Jacquelin Warren of the Natural Resources Defense Council for identifying this camouflage evident in recent budget hearings. See, for example, House Committee on Energy and Commerce, Subcommittee on Commerce, Transportation and Tourism, *Hearings: Authorization of the Toxic Substances Control Act*, March 24 and April 28, 1981, H. Rep. 97-15.

8. Richard P. Nathan, *The Plot That Failed: Nixon and the Administrative Presidency* (New York: John Wiley and Sons, 1975).

9. William W. Lowrance, *Of Acceptable Risk: Science and the Determination of Safety* (Los Altos, Calif.: William Kaufman, Inc., 1976), 105.

10. Centaur Associates, Inc., *Siting of Hazardous Waste Management Facilities and Public Opposition* (Washington, D.C.: U.S. Environmental Protection Agency, 1979), doc. No. SW-809, p. iv.

11. The differences between the newer and older forms of public involvement are emphasized in Walter A. Rosenbaum, "The Paradoxes of Participation," *Administration and Society* 8, no. 3 (November 1976): 355–84.

12. See Advisory Commission on Intergovernmental Relations, *Citizen Participation*, chapter 4, appendix A.

13. These provisions for public involvement may be found in sections 4, 5, 6, 14(b), 19, 20, and 21.

14. Section 7004(b).

15. This program is based on the act's Section 105(9) which instructs the president to prepare a "national contingency plan" for waste management that includes "specified roles for private organizations and entities in preparation for response and in responding to releases of hazardous substances, including identification of appropriate qualifications and capacity therefor."

16. The Carter actions are summarized in Advisory Commission on Intergovernmental Relations, *Citizen Participation*, 107.

17. See *Federal Register*, April 30, 1980, 28912–19.

18. The author is indebted for this information to Marsha Ramsey, EPA Office of Toxics Integration.

19. This slow program development is described in U.S. Executive Office of the President, Council on Environmental Quality, *Environmental Quality: Tenth Annual Report* (Washington, D.C.: GPO, 1979), 215–19.

20. U.S. House Committee on Appropriations, Subcommittee on the Department of Housing and Urban Development and Independent Agencies Appropriations, *Hearings*, March 13–15, 1979, 390.

21. U.S. House Committee on Interstate and Foreign Commerce, *Toxic Substances Control Act*, 35.

22. See Scott, "Consumer Advocates," 57–58. Congress, in fact, had been ambivalent about such intervenor funding almost from its inception. See, for example, U.S. Senate Committee on Governmental Affairs, *Study of Federal Regulation, Vol. III: Public Participation in Regulatory Agency Proceedings*, October 31, 1977, S. 95–71.

23. The thirteen states were selected on the basis of those ranking among the nation's ten leading states in the amount of hazardous waste generated and/or the presence of "priority" hazardous waste sites identified by EPA. The states generating the greatest hazardous waste are identified in U.S. Environmental Protection Agency, Office of Water and Waste Management, *Everybody's Problem: Hazardous Waste* (Washington, D.C.: EPA, 1980), doc. No. SW-826, p. 14. The "priority" states are those identified by the EPA under requirements of Superfund legislation as having the most dangerous abandoned waste sites. See *New York Times* (October 24, 1981). Several states appear in both categories. The thirteen states are Arkansas, California, Delaware, Florida, Illinois, Massachusetts, Michigan, New Jersey, New York, Ohio, Pennsylvania, Tennessee, and Texas.

24. Dawn P. Jackson, "Paying Lawyers to Sue the Government—An Expense That OMB Could Do Without," *National Journal* (April 17, 1982): 680–82.

25. Memorandum from William N. Hedeman, Jr., Director, Office of Emergency and Remedial Response, to regional administrators, November 18, 1981.

26. Ibid.

27. Centaur Associates, *Siting of Facilities*, 27.

28. U.S. Comptroller General, "Hazardous Waste Management Programs Will Not Be Effective: Greater Efforts Are Needed," *Report to Congress*, January 23, 1979, doc. No. CED-79-14, p. 13.

29. Centaur Associates, *Siting of Facilities*, iv.

10. State Roles in Siting Hazardous Waste Disposal Facilities: From State Preemption to Local Veto

1. Aware of the difficulties of the problem, the Speaker of the Texas House charged the Committee on Environmental Affairs to consider the need for a siting law in Texas. The Subcommittee on Siting of Hazardous Waste Disposal Facilities, Rep. Jim Turner, Chairman, in turn asked a study team at the Lyndon B. Johnson School of Public Affairs to report to it on this subject. The Lyndon B. Johnson Foundation provided funds for the travel and supplies for the research. The first author directed the study; the other two authors were students in the Policy Research Project. Other members of the

project team were Dorothy Allen, Kimberly Brown, Marc Carlisle, Vick Hines, Moira McNamara, Gwen Newman, John Schuhart, and Sue Stendebach, all of whom contributed to this chapter.

2. See *Environmental Quality: 1980* (Eleventh Annual Report of the Council on Environmental Quality) (Washington, D.C.: CEQ, 1981), 415.

3. U.S. Environmental Protection Agency, *Siting of Hazardous Waste Management Facilities and Public Opposition*, SW 809 (Washington, D.C.: EPA, November 1979).

4. See, for example, Marver Bernstein, *Regulating Business by Independent Commission* (Princeton, N.J.: Princeton University Press, 1955).

5. Author's note: in the case of HWDFs, because of the technical as well as the negatively redistributive nature of the goods.

6. Theodore Lowi, *The End of Liberalism* (New York: W.W. Norton, Inc., 1969), 96–97.

7. Although we have been focusing our discussion on siting, it is important to emphasize that siting is closely tied to all other aspects of hazardous waste policy, and that separating it out is more for analytic convenience than a representation of reality. For example, siting policy is itself dependent on the siting technology chosen. There are a variety of storage and disposal technologies available; not unexpectedly, the cost of these methods is almost inversely related to their environmental safety. Public opposition to siting of HWDFs has been closely associated with landfills. Thus one policy option for states that does not fall under the rubric of siting decisions but profoundly affects them involves provision of incentives for preferred (nonlandfill) disposal technologies. We omit discussion of these options below, recognizing that siting is inseparable from other aspects of waste policy that might serve to reduce or eliminate siting conflicts.

8. Douglas Costle, Memorandum to state governors from the Administrator, EPA, July 3, 1980, p. 4. This point is made in Donald T. Wells, "Site Control of Hazardous Waste Facilities," *Policy Studies Review* 1 (1982): 728–35.

9. Information on state laws comes from four written sources, supplemented by numerous telephone interviews. The scanty information about some laws has meant that two states, New Jersey and North Carolina, are not included in our scoring scheme, presented in table 10.4, and Tennessee is not included in any of the tables. The sources for information on state laws are: (1) Environmental Control Committee, *Legislative Responses to the Disposal of Hazardous Wastes*, January 1979; (2) Peggy Hyland, "Hazardous Waste Management Laws in the Member States of the Southern Legislative Conference," mimeo, Atlanta, Georgia, 1981; (3) National Conference of State Legislatures, *Hazardous Waste Management: A Survey of State Laws 1976–80* (update), Washington, D.C., April 1980; and (4) National Conference of State Legislatures, *Abstracts of State Hazardous Waste Siting Laws*, May 1981 (abstracts prepared jointly by several organizations).

10. EPA, *Siting of Hazardous Waste Management Facilities*, 29.

11. Ibid.

12. For a critique of Massachusetts' siting process, especially the ability of negotiation to resolve local fears, see Ken Geiser, "Clear Siting Ahead," *Exposure* 21–22 (August-September 1982): 8.

13. See, for example, Donald T. Wells, "Site Control of Hazardous Waste Facilities," *Policy Studies Review* 1, no. 4 (1982): 728-35.

11. A Comparative Perspective on State Hazardous Waste Regulation

1. The term model, as used here, refers to a descriptive heuristic framework rather than a formal theory. Much of the research is reviewed in essays by John H. Fenton and D. W. Chamberlayne, "The Literature Dealing with the Relationships Between Political Processes, Socioeconomic Conditions, and Public Policies in the American States: A Bibliographic Essay," *Polity* 1 (1969): 388–404; and Richard I. Hofferbert, "State and Community Policy Studies: A Review of Comparative Input-Output Analysis" *Political Science Annual* 3 (1973): 3–72.

2. See, for example, Charles O. Jones, "Regulating the Environment," in Herbert Jacob and Kenneth N. Vines, eds., *Politics in the American States: A Comparative Analysis* (Boston: Little, Brown and Company, 1976), 388–427; Lettie M. Wenner, *One Environmental Law: A Public Policy Dilemma* (Pacific Palisades, Calif.: Goodyear Publishing Company, 1976); and Paul Sabatier, "Regulatory Policy-Making: Toward a Framework of Analysis," *Natural Resources Journal* 4 (1977): 415–60.

3. J. Clarence Davies, *The Politics of Pollution* (New York: Pegasus, 1970); Jones, "Regulating the Environment"; Kingsley W. Game, "Controlling Air Pollution: Why Some States Try Harder," *Policy Studies Journal* 7 (1979): 728–38; and Thomas R. Dye, "Politics Versus Economics: The

Development of the Literature on Policy Determination," *Policy Studies Journal* 7 (1979): 652–62. Past attempts to measure problem severity in the environmental sphere have relied upon, for example, a single index such as "pollution potential," based on such factors as population growth, mineral production, and the type of industries found in various states (e.g., Jones, "Regulating the Environment"), or multiple indicators of demographic factors such as population density and population growth (e.g., Game, "Controlling Air Pollution"). We have employed multiple indicators of problem severity, ranging from more abstract measures (e.g., industrialization) to more functionally relevant measures (e.g., chemical waste generation and the total number of hazardous waste sites in the state). Thus, we have sought to avoid both a reliance upon single measures of problem severity or measures too far removed from the policy problem.

4. The Jones index of "pollution potential" identifies sixteen states as potentially experiencing severe pollution problems. The index ranges from a score of 10 (highest potential) to a score of 3 (less potential). In addition, thirteen states are identified as "special cases" and twenty-one states are not listed. The thirteen special cases were coded either 2 or 1 (depending on the author's estimate of pollution potential), and the remaining states were coded as a 0. See Jones, "Regulating the Environment." Other sources include U.S. Bureau of the Census, *Statistical Abstract of the United States: 1980* (Washington, D.C.: Government Printing Office, 1981); and U.S. House of Representatives, Committee on Government Operations, *Interim Report on Ground Water Contamination: EPA Oversight* (Washington, D.C.: Government Printing Office, 1980).

5. See James E. Anderson, David W. Brady, and Charles Bullock, III, *Public Policy and Politics in America* (North Scituate: Duxbury, 1978), 17–18.

6. Thomas R. Dye, *Politics, Economics, and the Public: Policy Outcomes in the American States* (Chicago, Ill.: Rand McNally, 1966).

7. J. Sacco and Edgar Leduc, "An Analysis of State Air Pollution Expenditures," *Journal of the Air Pollution Control Association* 19 (1969): 416–19; Lettie M. Wenner, "Enforcement of Water Pollution Control Laws," *Law and Society Review* 6 (1972): 481–507; Game, "Controlling Air Pollution"; and Susan E. Clarke, "Determinants of State Growth Management Policies," *Policy Studies Journal* 7 (1979): 753–62.

8. These data are taken from U.S. Bureau of the Census, *Statistical Abstract of the United States: 1980* (Washington, D.C.: Government Printing Office, 1981).

9. There are, however, some important exceptions to the findings that Democrats are more likely than Republicans to support environmental protection. See, for example, Frederick H. Buttel and William H. Flinn, "The Politics of Environmental Concern: The Impacts of Party Identification and Political Ideology on Environmental Attitudes," *Environment and Behavior* 10 (1978): 17–36; and Daniel Mazmanian and Paul Sabatier, "Liberalism, Environmentalism, and Partisanship in Public Policy-Making," *Environment and Behavior* 13 (1981): 361–84.

10. R. E. Dunlap and R. P. Gale, "Party Membership and Environmental Politics: A Legislative Roll-Call Analysis," *Social Science Quarterly* 55 (1974): 670–90.

11. L. G. Ritt and J. M. Ostheimer, "Congressional Voting and Ecological Issues," *Environmental Affairs* 3 (1974): 459–72; Dunlap and Gale, "Party Membership and Environmental Politics"; R. E. Dunlap and M. P. Allen, "Partisan Differences on Environmental Issues: A Congressional Roll-Call Analysis," *Western Political Quarterly* 29 (1976): 384–97; Jerry W. Calvert, "The Social and Ideological Bases of Support for Environmental Legislation: An Examination of Public Attitudes and Legislative Action," *Western Political Quarterly* 32 (1979): 327–37; and Henry C. Kenski and Margaret C. Kenski, "Partisanship, Ideology, and Constituency Differences on Environmental Issues in the U.S. House of Representatives: 1973–1978," *Policy Studies Journal* 9 (1981): 325–35.

12. See, for example, Calvert, "Social . . . Bases of Support"; and R. E. Dunlap and Kent D. Van Liere, "The Social Bases of Environmental Concern: A Review of Hypotheses, Explanations, and Empirical Evidence," *Public Opinion Quarterly* 44 (1980): 181–97.

13. James P. Lester, "Partisanship and Environmental Policy: The Mediating Influence of State Organizational Structures," *Environment and Behavior* 12 (1980): 101–31.

14. V. O. Key, *Southern Politics* (New York: Alfred A. Knopf, 1949); Duane Lockard, *New England State Politics* (Princeton, N.J.: Princeton University Press, 1959); Richard E. Dawson and James A. Robinson, "Interparty Competition, Economic Variables, and Welfare Policies in the American States," *Journal of Politics* 23 (1963): 265–89; John H. Fenton, *Midwest Politics* (New York: Holt, Rinehart and Winston, 1966); Richard I. Hofferbert, "The Relationship Between Public Policy and Some Structural and Environmental Variables in the American States," *American Political Science Review* 50 (1966): 73–82; Thomas R. Dye, "Politics Versus Economics"; Fenton and Chamberlayne, "The Literature with the Relationships Between Political Processes, Socioeconomic

Conditions, and Public Policies in the American States: A Bibliographic Essay"; Brian R. Fry and Richard F. Winters, "The Politics of Redistribution," *American Political Science Review* 64 (1970): 508–22; and Harvey Tucker, "It's About Time: The Use of Time in Cross-Sectional State Policy Research," *American Journal of Political Science* 26 (1982): 176–96.

15. The first measure of Democratic party strength was based on 1978 data in order to capture the impact of the 1976–77 elections in providing for increased Democratic representation in gubernatorial and legislative elections (forty-five states hold general elections in even years and five in odd years). The data from 1978 may be assumed to be the most significant *point* in time vis-à-vis hazardous waste regulation. However, an *average* of Democratic party strength for the *period* 1976–79 was also used since hazardous waste regulations were largely accomplished during this period. For elaboration on this point, see Tucker, "It's About Time."

16. Once computed, the Ranney Index of Democratic party strength may be further transformed to an index of interparty competition. To do so, all index scores above 0.500 were subtracted from 1.000 in order to transform the index from one of degree of Democratic party control to degree of interparty competition, with higher scores representing greater degrees of competition. These data are taken from Council of State Governments, *The Book of the States* (Lexington, Ky.: Council of State Governments, various years).

17. See, for example, A. E. Buck, *The Reorganization of State Governments in the United States* (New York: Columbia University Press, 1938); Committee for Economic Development, *Modernizing State Government* (New York: Committee for Economic Development, 1967); Martha W. Weinberg, *Managing the State* (Cambridge, Mass.: MIT Press, 1977); and James L. Garnett, *Reorganizing State Government: The Executive Branch* (Boulder, Colo.: Westview Press, 1980).

18. Game, "Controlling Air Pollution"; Kenneth J. Meier, "Government Reorganization for Economy and Efficiency: Some Lessons from State Government," a paper presented at the 1979 Annual Meeting of the Midwest Political Science Association; Garnett, *Reorganizing State Government*; Lester, "Partisanship and Environmental Policy"; and T. R. Carr, "State Government Reorganization: The Impact of Executive Branch Reorganization on Public Policy," a paper presented at the 1981 Annual Meeting of the Midwest Political Science Association.

19. George W. Downs, Jr., *Bureaucracy, Innovation, and Public Policy* (Lexington, Mass.: Lexington Books, 1976), 10–11.

20. This is not to suggest that this issue is necessarily settled in the literature. For competing arguments concerning the influence of centralized (versus fragmented) bureaucracies vis-à-vis environmental policy making, see Frank P. Grad, "Intergovernmental Aspects of Environmental Controls," in G. W. Rathjens et al., eds., *Environmental Control: Priorities, Policies, and the Law* (New York: Columbia University Press, 1971), 177–79; and Game, "Controlling Air Pollution."

21. Sabatier, "Regulatory Policy-Making"; Kenneth W. Kramer, *Implementing Environmental Policy*, unpublished Ph.D. Dissertation (Houston, Tex.: Rice University, 1979); Kenneth W. Kramer and James P. Lester, "Economics, Power, and Policy: Implementing Hazardous Waste Regulation in Texas," a paper presented at the 1982 Annual Meeting of the Southern Political Science Association; and Lester, "Partisanship and Environmental Policy."

22. See, for example, Grant McConnell, *Private Power and American Democracy* (New York: Vintage Books, 1966); Walter A. Rosenbaum, *The Politics of Environmental Concern* (New York: Praeger Publishers, 1973 and 1977); Anderson et al., *Public Policy and Politics*; Phillip O. Foss, *Politics and Grass* (Seattle, Wash.: University of Washington, 1960); Cynthia H. Enloe, *The Politics of Pollution in a Comparative Perspective* (New York: McKay, 1975); and John A. Worthley and Richard Torkelson, "Managing the Toxic Waste Problem: Lessons from the Love Canal," *Administration and Society* 13 (1981): 145–60.

23. Sabatier, "Regulatory Policy Making."

24. Many states have consolidated their previously fragmented environmental control programs into organized functional areas. Since 1967 there have been thirty-two major state environmental reorganizations. For the most part, the states have relied upon three basic organizational models (in terms of increasing reorganization and consolidation): (1) the health department model; (2) the state EPA model; and (3) the environmental superagency model. The data are taken from Council of State Governments, *The Book of the States: 1976–1977* (Lexington, Ky.: Council of State Governments, 1976), and updated by telephone interviews with agency officials. The states' degree of agency consolidation was computed by assigning each state a score of 3 (superagency), 2 (state EPA), 1 (health department), and 0 (no primary agency) and, for the weighted index, multiplying this score by the number of years each structure had been in place up to 1976.

25. See, for example, J. Grumm, "The Effects of Legislative Structure on Legislative Performance," in Richard I. Hofferbert and Ira Sharkansky, eds., *State and Urban Politics* (Boston, Mass.: Little, Brown and Company, 1971), 298–322; Edward G. Carmines, "The Mediating Influence of State Legislatures on the Linkage Between Interparty Competition and Welfare Policies," *American Political Science Review* 68 (1974): 1118–24; A. K. Karnig and Lee Sigelman, "State Legislative Reform: Another Look," *Western Political Quarterly* 28 (1975): 548–52; Lance T. LeLoup, "Reassessing the Mediating Impact of Legislative Capability," *American Political Science Review* 76 (1978): 616–21; and Phillip W. Roeder, "State Legislative Reform: Determinants and Policy Consequences," *American Politics Quarterly* 7 (1979): 51–69.

26. Citizens Conference on State Legislatures, *The Sometimes Governments: A Critical Study of the Fifty American Legislatures* (New York: Bantam Books, 1971).

27. Karnig and Sigelman, "State Legislative Reform"; LeLoup, "Reassessing Impact"; and Roeder, "Legislative Reform."

28. Carmines, "Mediating Influence of Legislatures."

29. Gerard A. Bulanowski, *The Impact of Science and Technology on the Decision-Making Process in State Legislatures: The Issue of Solid and Hazardous Waste* (Denver, Colo.: National Conference of State Legislatures, 1981).

30. Generally following the work of Grumm ("Effects of Legislative Structure"), the measure of legislative professionalism is based upon a factor analysis of a number of variables acknowledged to reflect varying degrees of legislative competence. The original data base on state legislatures contained over one hundred variables reflecting various facets of the legislative system. Data were reduced by inspection of correlation matrices and deletion of uncorrelated and highly (i.e., 0.95 and above) correlated variables. Finally, successive factor analyses were performed, sequentially deleting variables with low communalities. This procedure resulted in the retention of seven variables for final analysis. They were: (1) legislator's compensation; (2) average population per senate seat; (3) average population per house seat; (4) length of legislative sessions; (5) number of bills introduced; (6) general control expenditures for the legislature; and (7) the existence of a limit on the length of sessions. A single factor reflecting a general professionalism dimension was obtained. Factor scores were then assigned to each of the states. For elaboration, see Douglas Dobson and David Karns, *Public Policy and Senior Citizens: Policy Formation in the American States* (Project Report 90-A-1005, U.S. Department of Health, Education, and Welfare, Administration on Aging, 1978).

31. George W. Downs and Lawrence B. Mohr, "Toward a Theory of Innovation," *Administration and Society* 10 (1979): 379–408.

32. Helen Ingram, Nancy Laney, and John R. McCain, "Water Scarcity and the Politics of Plenty in the Four Corners States," *Western Political Quarterly* 32 (1979): 309.

33. Downs, *Bureaucracy*; Fry and Winters, "The Politics of Redistribution"; and Jacob and Vines, *Politics*.

34. Joyce M. Munns, "The Environment, Politics, and Policy Literature: A Critique and Re-formulation," *Western Political Quarterly* 28 (1975): 646–67.

35. Lester, "Partisanship and Environmental Policy."

36. Fry and Winters, "The Politics of Redistribution"; Ira Sharkansky and Richard I. Hofferbert, "Dimensions of State Politics, Economics, and Public Policy," *American Political Science Review* 63 (1969): 867–80; and Richard I. Hofferbert, *The Study of Public Policy* (New York: Bobbs-Merrill, 1974).

37. Downs, *Bureaucracy*, 71.

38. See Kenneth S. Kamlet, *Toxic Substances Programs in U.S. States and Territories: How Well Do They Work?* (Washington, D.C.: National Wildlife Federation, 1979), reprinted with permission.

39. The NWF survey is the result of inquiries and questionnaires addressed to the commissioners of state health and environmental agencies. Although the data compilers were rigorous in collecting and coding the data, they nevertheless remained at the mercy of the responding agencies. Likewise, we are at the mercy of their data collection.

The scoring procedure used by the NWF survey is described as follows: each "pro-governmental" response was awarded 1 point. Each "incorrect" answer was given no points. Each equivocal answer was given half a point. In some cases—where explanatory notes, inspection of applicable laws and regulations, and conversations with state officials made it evident that a survey question had been misinterpreted—a state's responses were re-graded to best reflect the status of its true program and to ensure uniformity in the grading of different states' answers to the same questions.

These "corrected" scores are utilized in the analysis. The maximum possible score was 43 points (the state which scored highest in the survey scored only 34 points) and the mean score for all fifty states was 20.2.

It should be understood that these survey data do not reflect state behavior after the first quarter of 1979. Indeed, many states enacted new hazardous waste legislation during 1979–81. Thus, a number of states might well receive higher rankings and scores than reflected in the present analysis if their hazardous waste programs were evaluated on the basis of more recent data.

Another more recent survey of state laws for hazardous waste management was conducted by the National Conference of State Legislatures (NCSL) in 1980. The National Wildlife Federation (NWF) survey was used in this analysis because, unlike the NCSL survey, it included multiple mailings to several state officials. In addition, due to the wording of the survey questionnaire, the NCSL survey omitted significant legislation in this area in its report. Thus, we would consider the NWF survey to be generally more reliable than the NCSL survey for the purposes of this analysis.

40. Fry and Winters, "Politics of Redistribution"; Bernard H. Booms and James R. Halldorson, "The Politics of Redistribution: A Reformulation," *American Political Science Review* 67 (1973): 924–33; and William D. Berry, "Utility Regulation in the States: The Policy of Professionalism and Salience to the Consumer," *American Journal of Political Science* 23 (1979): 263–77.

41. Jack L. Walker, "The Diffusion of Innovations Among the American States," *American Political Science Review* 63 (1969): 880–99; Virginia Gray, "Innovation in the States: A Diffusion Study," *American Political Science Review* 67 (1973): 1174–85; and Downs, *Bureaucracy*.

42. In fact, the dependent variable displays a low, but significant, fifty-state association with Walker's—now dated—innovation index.

43. Downs, *Bureaucracy*.

44. We are, of course, aware that an enumeration of the fifty states does not constitute a sample and, strictly speaking, significance testing is not appropriate. Regardless, the practice is well established in this research tradition on the grounds that it provides a guide to the importance to be attached to correlation-regression analysis results. On both these points, see Ira Sharkansky, *The Politics of Taxing and Spending* (New York: Bobbs-Merrill, 1969); and D. Morrison and R. E. Henkel, *The Significance Test Controversy* (Chicago, Ill.: Aldine, 1970).

45. The differences that do occur in the fifty-state vs. nonsouthern states' correlations are attributable to the extremely unusual characteristics of the South on three of the four independent variables to be used in the subsequent analysis. The extent of these differences is reflected by their means; Democratic party strength: south, 83.5 percent, nonsouth, 60 percent; bureaucratic consolidation: south, 5.0 percent, nonsouth, 11.0 percent; percent poor: south, 16.5 percent, nonsouth, 10.2 percent. Only in terms of legislative professionalism are the south (−0.01) and the nonsouth (0.05) means comparable.

46. The nonsouthern states were divided at the median to produce the high and low waste groupings in table 11.5. Three states that were included in the low waste category for the fifty-state analysis— Washington, Iowa, and Kansas—are now included among the high waste states. Their inclusion in the high waste category does not, however, dramatically alter the results that follow.

47. The exact nature of the effects of poverty and legislative professionalism in the high waste nonsouthern states is somewhat unusual as viewed from the perspective of the comparative state literature. Their effects upon policy are substantially independent and apparently not additive or multiplicative. While a low level of poverty and a high level of legislative professionalism both encourage policy action, they do not generally appear together in these states—their correlation is near zero—and when they do, the policy implications are not dramatic. In order to verify the correlational data, we divided these nineteen states into four categories depending upon whether they fell above or below the median on each variable. In effect, a 2 × 2 table was constructed. Average policy activity levels for each cell were then compared for each group of states. If the two phenomena—wealth and professionalism—were behaving in an additive fashion, the lowest activity level would be expected where poverty is high and professionalism is low. This expectation is met. In addition, however, we would expect the highest level where poverty is low and professionalism is high. This does not occur. Roughly equal degrees of activity occur in those states that are poor yet professional and those that are wealthy yet unprofessional. In effect, professionalism and resources are explaining different portions of the variance in hazardous waste regulation. Some states are thus reacting to the problem because they have professional legislatures that are attentive to the issue. Other states are adopting stringent hazardous waste regulations perhaps because they can afford the subsequent dislocation of industry.

48. The relationship between those two factors in the nonsouthern, low waste states does reflect additivity and the predictors are intercorrelated.

12. The Politics of Hazardous Waste Regulation: Theoretical and Practical Implications

1. "Spilled Chemicals Spread Farther," *New York Times* (September 30, 1982), p. A18.
2. "Carolinians Angry Over PCB Landfill," *New York Times* (August 11, 1982), p. D17.
3. Ralph Blumenthal, "Two More Inquiries Check Dumping by Toxics Plant," *New York Times* (October 13, 1982), p. B2.
4. Philip Shabecoff, "24 Concerns to Clean Up Waste Dump in Indiana," *New York Times* (October 27, 1982), p. A16.
5. Ibid.
6. "Philadelphia Clears Its 'Love Canal'," *New York Times* (August 25, 1982), p. A-12.
7. Richard Severo, "Jury Awards $58 Million to 47 Railroad Workers Exposed to Dioxin," *New York Times* (August 27, 1982), p. A9.
8. Edward A. Gargan, "Toxins Are Found in Bass in Hudson," *New York Times* (October 24, 1982), p. 43.
9. The extreme severity of the hazardous waste issue is discussed in chapters 1 and 13 of Samuel S. Epstein, Lester O. Brown, and Carl Pope, *Hazardous Waste in America* (San Francisco: Sierra Club, 1982).
10. As the protests against the Warren County landfill escalated, media coverage expanded. The discussion in this chapter is drawn from news accounts of the North Carolina experience appearing in the *New York Times*, the *Charlotte Observer*, and *The State* in September and October 1982.
11. Prior to a federal ban on the manufacture of PCBs, they were regularly used as coolants in electrical transformers.
12. Data on political participation in the United States indicate that greater political activity is associated with the socially and economically advantaged. See, for example, Lester Milbrath and M. L. Goel, *Political Participation*, 2nd ed. (Chicago: Rand McNally, 1977).
13. The leachate issue was raised by William Sanjour, the chief of EPA's Hazardous Waste Implementation Branch. In response, the administrator in charge of the Atlanta regional office, Charles Jeter, emphasized the safety of "secure" landfills. See "Hunt Criticizes 'Outside' Protesters," *The State* (September 24, 1982), p. 8-C.
14. As of fall 1982, reauthorization of RCRA appeared to be confined to tightening some of the provisions that existed in the original legislation. As passed by the House, RCRA coverage would be extended to small quantity generators and handlers of hazardous waste (between 100 and 1,000 kilograms per month). See "House Passes Legislation to Tighten Law Regulating Hazardous Waste Disposal," *Congressional Quarterly Weekly Report* (September 11, 1982), p. 2278.
15. See, for example, David Burnham, "Environmental Unit Accused of Seeking to Silence a Critic," *New York Times* (December 11, 1982), p. 13, and the related stories "E.P.A. Aide Accused of Ouster Attempt," *New York Times* (December 17, 1982), p. A-25; Philip Shabecoff, "Propriety of E.P.A. Aides' Talks Questioned," *New York Times* (February 16, 1983), p. A-21; Stuart Taylor, Jr., "Miss Lavelle Calls E.P.A. Badly Split," *New York Times* (February 25, 1983), p. B-8.
16. See, for example, Philip Shabecoff, "House Charges Head of E.P.A. with Contempt," *New York Times* (December 17, 1982), p. A-1; and the follow-up articles, Francis X. Clines, "President Asserts He Will Not Shield Any Fault at E.P.A.," *New York Times* (February 17, 1983), p. 1; "Accord Is Reached for House Access to Files of E.P.A.," *New York Times* (February 19, 1983), p. A1; Stuart Taylor, Jr., "3 House Panels Unhappy at Plan for E.P.A. Papers," *New York Times* (February 20, 1983), p. A1.; Philip Shabecoff, "E.P.A. Chief Urges Opening All Files to Quash Dispute," *New York Times* (March 3, 1983), p. A1.
17. Steven R. Weisman, "White House Links to E.P.A. Studied," *New York Times* (February 25, 1983), p. B-8. Amidst the glare of congressional inquiry and public arousal, EPA attempted to resolve one long-smoldering situation—the Times Beach, Missouri, dioxin contamination case. (See, for example, Nathaniel Sheppard, Jr., "Times Beach Fades Amid Delay by U.S.," *New York Times* [February 19, 1983], p. 7; and Robert Reinhold, "Missouri Dioxin Cleanup: A Decade of Little Action," *New York Times* [February 20, 1983], p. A1.) The agency released a buyout plan in which EPA could commit approximately $33 million in Superfund monies to purchase property and assist in the relocation of the Times Beach residents. See Robert Reinhold, "U.S. Offers to Buy All Homes in Town Tainted by Dioxin," *New York Times* (February 22, 1983), p. A1.

18. "Bill Is Suggested to Remake E.P.A. as Independent Unit," *New York Times* February 22, 1983), p. A12.

19. "Lawyers to Ask Changes in Toxic-Waste Suits," *New York Times* (September 1982), p. A20.

20. "Victim Compensation Legislation Proposed," *Congressional Quarterly Weekly Report* (October 23, 1982), p. 2732.

21. National Conference of State Legislatures, *Hazardous Waste Management: A Survey of State Legislation 1982* (Denver: National Conference of State Legislatures, 1982).

22. David Morell and Christopher Magorian, *Siting Hazardous Waste Facilities* (Cambridge, Mass.: Ballinger, 1982).

23. Selim M. Senkan and Nancy W. Stauffer, "What to Do with Hazardous Waste," *Technology Review* 84 (November/December 1981): 39.

24. Kenneth W. Kramer and James P. Lester, "Economics, Power, and Policy: Implementing Hazardous Waste Regulation in Texas," a paper presented at the Annual Meeting of the Southern Political Science Association, Atlanta, Georgia, 1982.

25. Harold Faber, "State Forms Unit to Investigate Dumping of Hazardous Wastes," *New York Times* (October 12, 1982), p. B1.

26. In South Carolina, for example, there has been discussion of the use of fees on hazardous waste-related industry as a funding source for the daily regulatory activities of the Bureau of Solid and Hazardous Waste Management. Industry response has been negative thus far, arguing that selected targets should not have to bear the fiscal burden for regulation that benefits the general public.

27. The Williams and Matheny research on Florida (chapter 5) expands this point.

28. Industry was able to wrench concessions from Congress in the passage of Superfund. They are mobilizing opposition to certain provisions in the RCRA reauthorization, at this writing. See "House Passes Legislation to Tighten Law Regulating Hazardous Waste Disposal," *Congressional Quarterly Weekly Report* (September 11, 1982), p. 2278.

29. Charles Davis, "Substance and Procedure in Facility Siting for the Disposal of Hazardous Wastes," a paper presented at the Annual Meeting of the Southern Political Science Association, Atlanta, Georgia, 1982.

30. Ibid.

31. Gordon A. Whitaker, "Coproduction: Citizen Participation in Service Delivery," *Public Administration Review* 40 (May/June 1980): 240–46.

32. Richard C. Rich, "Interaction of the Voluntary and Governmental Sectors: Toward an Understanding of the Coproduction of Municipal Services," *Administration and Society* 13 (May 1981): 59–76.

33. See, for example, *re:sources*, the quarterly newsletter of the Environmental Task Force.

Epilogue

1. Philip Shabecoff, "E.P.A. and Aide Reach Settlement Averting Hearings on Harassment," *New York Times*, February 15, 1983, p. A19.

2. Philip Shabecoff, "Seven Days of Decision: Why Head of E.P.A. Quit," *New York Times*, March 13, 1983, p. 1.

3. See, for example, William E. Schmidt, "Denver Lawyer's Role in E.P.A. Decisions Is Focus of Inquiries by Congress," *New York Times*, February 26, 1983, p. 8; Martin Tolchin, "Tracking the E.P.A. Investigations," *New York Times*, February 24, 1983, p. B8; Philip Shabecoff, "E.P.A. Chief Urges Opening All Files to Quash Dispute," *New York Times*, March 3, 1983, p. A1; "Reagan Expresses Faith in E.P.A. Head," *New York Times*, March 5, 1983, p. 12; "D'Amato Requests E.P.A. Chief Resign," *New York Times*, March 9, 1983, p. A1; Philip Shabecoff, "Possible Destruction of E.P.A. Data Under Inquiry," *New York Times*, March 9, 1983, p. A16; Francis X. Clines, "White House Seeks to Peer Beyond E.P.A. Smoke," *New York Times*, March 11, 1983, p. A18; Raymond Bonner, "Inquiries on E.P.A. in Congress Go On, Panels' Chiefs Say," *New York Times*, March 11, 1983, p. A1; "2 Reports Found by White House in E.P.A. Clash," *New York Times*, March 18, 1983, p. A1.

4. Philip Shabecoff, "Dismissed Official Faults E.P.A. Chief," *New York Times*, February 24, 1983, p. B14.

5. David Burnham, "2 High Officials of E.P.A. Resign, Reportedly at White House Urging," *New York Times*, February 24, 1983, p. A1.

6. Stuart Taylor, Jr., "E.P.A. Inquiries Center on Four Issues," *New York Times*, March 13, 1983, p. 36; Philip Shabecoff, "E.P.A. Aide's Memos Studied by House," *New York Times*, March 22, 1983, p. A18.

7. Taylor, "E.P.A. Inquiries Center on Four Issues."

8. Robert Reinhold, "U.S. Offers to Buy All Homes in Town Tainted by Dioxin," *New York Times*, February 22, 1983, p. A12. The effects of the dioxin experience on Times Beach are discussed in Lawrence Ingrassia and Bryan Burrough, "Toxic Waste Scare 'Killed' Small Town, Its People Complain," *Wall Street Journal*, March 1, 1983, p. 1.

9. Philip Shabecoff, "Mrs. Burford Quits at E.P.A.: Reagan Announces Accord Giving Congress All Papers," *New York Times*, March 10, 1983, p. 1; Shabecoff, "Mrs. Burford Says She Decided to Quit Before Reagan's Pact with Congress," *New York Times*, March 11, 1983, p. A18; Stuart Taylor, Jr., "Furor on E.P.A. Undercut President in His Effort to Withhold Documents," *New York Times*, March 11, 1983, p. A19.

10. Ralph Blumenthal, "Environment Unit Greets New Head," *New York Times*, March 11, 1983, p. 18; Dale Mezzacappa and Frank Greve, "Flexible Hernandez Takes Temporary Charge of EPA," *The State*, March 13, 1983, p. 16B.

11. Philip Shabecoff, "Scheuer Says E.P.A. Aide Let Dow Delete Dioxin Tie in Draft Report," *New York Times*, March 16, 1983, p. A1; "Acting Chief Contradicted," *The State*, March 17, 1983, p. 2A; Leslie Maitland, "Top E.P.A. Official Is Accused of Intervening in Behalf of Company," *New York Times*, March 24, 1983, p. A16; Richard B. Schmitt, "Some Dallas Residents Are Angry at EPA Over Agency's Handling of Alleged Hazard," *Wall Street Journal*, April 13, 1983, p. 29.

12. Steven R. Weisman, "Reagan Nominates First E.P.A. Chief to Head It Again," *New York Times*, March 22, 1983, p. A1; Hedrick Smith, "Movement to Center," *New York Times*, March 22, 1983, p. A21.

13. Philip Shabecoff, "Watt Says Status of Ruckelshaus at E.P.A. Surpasses Predecessor's," *New York Times*, March 23, 1983, p. A22.

14. Robert Lindsey, "Ruckelshaus's Ties Split Environmental Leaders," *New York Times*, March 26, 1983, p. 8; "Ruckelshaus Meets with Environmentalists," *The State*, April 15, 1983, p. 10A.

15. N. R. Kleinfeld, "$7 Million Settlement for Cleaning Up Hazardous Waste Dump Draws Fire," *New York Times*, March 3, 1983, p. B12.

16. Ralph Blumenthal, "E.P.A. Admits Not Pressing for Toxic Lagoon Cleanup," *New York Times*, April 7, 1983, p. B1.

17. Robert Reinhold, "E.P.A. Said to Bar Tests at Dow in '81," *New York Times*, March 25, 1983, p. A16; Reinhold, "E.P.A.'s Dow Tests Find High Toxicity," *New York Times*, April 1, 1983, p. 1.

18. Stuart Taylor, Jr., "U.S. Will Sue Companies for Cost of Cleanup at Waste Site on Coast," *New York Times*, March 30, 1983, p. A1; Taylor, "Meese Denies Asking for a Report on Toxic Dump," *New York Times*, March 10, 1983, p. B13.

19. Reinhold, "E.P.A. Said to Bar Tests at Dow in '81"; Stephen Kinzer, "Cleanup Delayed at Ohio Waste Site Where Chemicals Leak into Ground," *New York Times*, March 4, 1983, p. A14.

20. Taylor, "E.P.A. Inquiries Center on Four Issues"; Philip Shabecoff, "Toxic Cleanup Delay Laid to 2 Ex-Aides," *New York Times*, April 9, 1983, p. 7.

21. Raymond Bonner, "Leader in Toxic Dumps Accused of Illegal Acts," *New York Times*, March 21, 1983, p. A1; Bonner, "Big Waste Hauler Closes Site," *New York Times*, March 22, 1983, p. A1; Ralph Blumenthal, "Waste Hauler's Business Acts Faulted," *New York Times*, March 24, 1983, p. B12.

22. "Hazardous Waste Laws Called Inadequate," *The State*, March 17, 1983, p. 4A.

23. "Hazardous Waste Laws Called Inadequate"; William Kronholm, "Congress to Grapple with Toxic Waste Laws," *The State*, April 4, 1983, p. 2A.

24. Ronald Alsop, "Local Citizen Groups Take a Growing Role in Fighting Toxic Dumps," *Wall Street Journal*, April 18, 1983, p. 1.

25. Philip M. Boffey, "Experts Showing Concern on Safety of Burying Toxic Waste in Landfills," *New York Times*, March 16, 1983, p. A20.

Selected Bibliography

Books

Brown, Michael. *Laying Waste: The Poisoning of America by Toxic Chemicals*. New York: Pantheon, 1980.

Doniger, David D. *Law and Policy of Toxic Substances*. Baltimore, Md.: Johns Hopkins, 1979.

Epstein, Samuel S., Lester O. Brown, and Carl Pope. *Hazardous Waste in America*. San Francisco: Sierra Club, 1982.

Gibbs, Lois Marie. *Love Canal: My Story*. Albany, N.Y.: State University of New York Press, 1982.

Levine, Adeline G. *Love Canal: Science, Politics, and People*. Lexington, Mass.: Lexington Books, 1982.

Morell, David L., and Christopher Magorian. *Siting Hazardous Waste Facilities: Local Opposition and the Myth of Preemption*. Cambridge, Mass.: Ballinger, 1982.

Nader, Ralph, Ronald Brownstein, and John Richard, eds. *Who's Poisoning America: Corporate Polluters and Their Victims in the Chemical Age*. San Francisco: Sierra Club, 1981.

Van Den Bosch, Robert. *The Pesticide Conspiracy*. New York: Doubleday, 1978.

Whiteside, Thomas. *The Pendulum and the Toxic Cloud*. New Haven, Conn.: Yale University Press, 1979.

Published Reports

Bulanowski, Gerald. *The Impact of Science and Technology on the Decision-making Process in State Legislatures: The Issue of Solid and Hazardous Waste*. Denver, Colo.: National Conference of State Legislatures, 1981.

Carnes, Sam A. "Confronting Complexity and Uncertainty: Implementation of Hazardous Waste Management Policy," in *Environmental Policy Implementation*, ed. Dean E. Mann. Lexington, Mass.: Lexington Books, 1982. Pp. 35–50.

Conservation Foundation. *State of the Environment, 1982: A Report from the Conservation Foundation*. Washington, D.C.: The Conservation Foundation, 1982.

Council of State Governments. *Waste Management in the States*. Lexington, Ky.: The Council of State Governments, 1982.

Hart, Fred C., and Associates. *Preliminary Assessment of Cleanup Costs for National Hazardous Waste Problems*. Washington, D.C.: prepared for the U.S. Environmental Protection Agency, 1979.

ICF Incorporated. *Analysis of Community Involvement in Hazardous Waste Site Problems: A Report to the Office of Emergency and Remedial Response*. Washington, D.C.: U.S. Environmental Protection Agency, July 1981.

Kamlet, Kenneth S. *Toxic Substances Programs in U.S. States and Territories: How Well Do They Work?* Washington, D.C.: National Wildlife Federation, 1980.

Montague, Peter. *Four Secure Landfills in New Jersey: A Study of the State of the Art in Shallow Burial Waste Disposal Technology*. Princeton, N.J.: Department of Chemical Engineering, Princeton University, 1981.

National Conference of State Legislatures. *Hazardous Waste Management: A Survey of State Laws: 1976*–1980. Washington, D.C.: National Conference of State Legislatures, 1980.

Steeler, J. H. *A Legislator's Guide to Hazardous Waste Management.* Denver, Colo.: National Conference of State Legislatures, 1980.

Articles in Journals or Magazines

Canter, B. D. E. "Safe Hazardous Waste Disposal: Sure, But Where?" *Florida Bar Journal* 55 (1981): 813–17.

Eckhardt, R. C. "The Unfinished Business of Hazardous Waste Control." *Baylor Law Review* 33 (April 1981): 252–65.

Epstein, Samuel S. "The Role of the Scientist in Toxic Tort Case Preparation." *Trial* 17 (1981): 38–41, 52–53.

Friedland, Steven I. "The New Hazardous Waste Management System: Regulation of Wastes or Wasted Regulation?" *Harvard Environmental Law Review* 5 (1981): 89–129.

Getz, Malcolm, and Benjamin Walter. "Environmental Policy and Competitive Structure: Implications of the Hazardous Waste Management Program." *Policy Studies Journal* 9 (Winter 1980): 404–14.

Goldfarb, William. "The Hazards of Our Hazardous Waste Policy." *Natural Resources Journal* 19 (1979): 249–60.

Hinds, Richard. "Liability Under Federal Law for Hazardous Waste Injuries." *Harvard Environmental Law Review* 6 (1982): 1–33.

Jorling, T. C. "Hazardous Substances in the Environment." *Ecology Law Quarterly* 9 (December 1981): 520–618.

Kirschten, Dick. "The New War on Pollution Is Over the Land." *National Journal* (April 14, 1979): 603–6.

Kovacs, William L., and John F. Klucsik. "The New Federal Role in Solid Waste Management: The Resource Conservation and Recovery Act of 1976." *Columbia Journal of Environmental Law* 3 (1977): 205–61.

Landrigan, P. "Chemical Wastes: Illegal Hazards and Legal Remedies." *American Journal of Public Health* 71 (1981): 985–87.

Rhodes, R. L., and D. H. MacLaughlin. "Federal and State Regulation of Hazardous Waste Management." *Florida Bar Journal* 54 (1980): 713–18.

Wurth-Hough, S. "Chemical Contamination and Governmental Policy-making: The North Carolina Experience." *State and Local Government Review* 14 (1982): 56–60.

Worthley, John A., and Richard Torkelson. "Managing the Toxic Waste Problem: Lessons from the Love Canal." *Administration and Society* 13 (1981): 145–60.

Public Documents

Governor's Hazardous Waste Policy Advisory Council. *Hazardous Waste: A Management Perspective.* Tallahassee, Fla.: The Institute of Science and Public Affairs, Florida State University, 1981.

New York State Department of Environmental Conservation. *Hazardous Waste Program Status Update.* Albany, N.Y.: December 1981.

New York State Department of Health. *Love Canal: Public Health Time Bomb.* Special Report to the Governor and Legislature (September 1978).

Texas Department of Health, Bureau of Solid Waste Management. *Texas Hazardous Waste Management Program* (May 1981).

Texas Department of Water Resources. *Hazardous Waste Management in Texas* (July 1982).

Texas Department of Water Resources. *Solid Waste Management Plan for Texas: 1980–1986, Vol. II–Industrial Solid Waste* (January 1981).

Toxic Waste Assessment Group, Governor's Office of Appropriate Technology. *Alternatives to the Land Disposal of Hazardous Wastes: An Assessment for California.* Sacramento, 1981.

U.S. Congress. House of Representatives. Committee on Government Operations. *Interim Report on Ground Water Contamination: EPA Oversight.* Washington, D.C.: Government Printing Office, 1980.

———. *Toxic Chemical Contamination of Ground Water: EPA Oversight.* Washington, D.C.: Government Printing Office, 1980.

———. Committee on Interstate and Foreign Commerce. *Hazardous Waste Disposal: Parts I and II.* Washington, D.C.: Government Printing Office, 1979.

———. *Hazardous Waste Matters.* Washington, D.C.: Government Printing Office, 1980.

———. *Love Canal: Health Studies and Relocation.* Washington, D.C.: Government Printing Office, 1980.

———. *Organized Crime and Hazardous Waste Disposal.* Washington, D.C.: Government Printing Office, 1980.

———. *Superfund.* Washington, D.C.: Government Printing Office, 1979.

———. Senate. Committee on Public Works. *A Legislative History of the Solid Waste Disposal Act, As Amended.* Washington, D.C.: Government Printing Office, 1974.

U.S. Environmental Protection Agency. *Disposal of Hazardous Wastes: Report to Congress.* Washington, D.C.: Government Printing Office, 1974.

———. *The Effects of Organic Solvents on the Permeability of Clay Soils*, by Kirk W. Brown and D. C. Anderson. Grant No. R80682510. Washington, D.C.: Government Printing Office, 1981.

———. *Environmental Monitoring at Love Canal*, Volume I. Washington, D.C.: Government Printing Office, 1980.

———. *Hazardous Waste: Fifteen Years and Still Counting.* Washington, D.C.: Government Printing Office, 1980.

———. *Siting of Hazardous Waste Management Facilities and Public Opposition.* Washington, D.C.: Government Printing Office, 1979.

U.S. General Accounting Office. Comptroller General. *EPA Is Slow to Carry Out Its Responsibility to Control Harmful Chemicals.* Washington, D.C.: Government Printing Office, 1980.

———. *Hazardous Waste Disposal Methods: Major Problems with Their Use.* Washington, D.C.: Government Printing Office, 1980.

———. *Hazardous Waste Facilities with Interim Status May Be Endangering Public Health and the Environment.* Washington, D.C.: Government Printing Office, 1981.

———. *Hazardous Waste Management Programs Will Not Be Effective: Greater Efforts Are Needed.* Washington, D.C.: Government Printing Office, 1979.

———. *Hazardous Waste Sites Pose Investigation, Evaluation, Scientific, and Legal Problems.* Washington, D.C.: Government Printing Office, 1981.

———. *How to Dispose of Hazardous Waste: A Serious Question That Needs to Be Resolved.* Washington, D.C.: Government Printing Office, 1978.

———. *Solid Waste Disposal Practices: Open Dumps Not Identified, States Face Funding Problems.* Washington, D.C.: Government Printing Office, 1981.

———. *Waste Disposal Practices: A Threat to Health and the Nation's Water Supply.* Washington, D.C.: Government Printing Office, 1978.

Index

Acrylamide, 27
Administrative Procedures Act: public notice, comment and hearings, 182, 186
Alabama, 81, 84, 209, 221, 224
Alaska, 209, 221, 224, 225
Arizona, 198, 202–4, 205, 208, 210, 221, 224, 225
Arkansas, 68, 209, 221, 224, 234
Audubon Society, 82, 192

Banfield, Edward, 44
Bardach, Eugene, 76
Beck, Eckhart C., 44, 48–49
Bell, Melvyn, 30
Bentsen, Senator Lloyd, 133
Benzene, 114
Biglane, Kenneth, 53
Blum, Barbara, 48
Bradley, Mayor Tom, 165, 167–68
Braybrooke, David, 44, 46
Brown, Governor Edmund G., Jr., 145, 154, 157, 159, 161, 170, 172, 174
Brown, Lester O., 144
Burford, Anne. See Gorsuch, Anne

California, 209, 221, 224, 225; Citizens for a Better Environment, 159; environmental groups, 161; facility permitting process, 148–55; hazardous waste generation, 140–43; industry groups, 155, 161, 175; institutional fragmentation, 157, 172; land disposal restrictions, 144–48; manifest system, 158; state cancer policy, 154–57; Stringfellow Quarry, 144
California Air Resources Board, 146, 155–57, 158
California Assessment Manual, 155
California Council on Environmental and Economic Balance, 147, 161
California Department of Health Services, 140, 144, 146, 148, 149, 151, 154–57, 158, 159, 160, 162, 166, 167, 168, 172, 173; Division of Toxic Substances Control 159, 162; Hazardous Materials Management Section, 158
California Environmental Quality Act, 150
California Hazardous Control Act, 157
California Hazardous Waste Disposal Facility Projects: cement kiln incineration, Lebec, 148–49, 154; comprehensive treatment and land disposal facility, Sand Canyon, 151–53, 154, 160–61; demonstration scale waste treatment unit, West Covina, 150–51, 164, 166, 167, 168, 170, 172; industrial waste treatment, Wilmington, 149–51, 154, 170; transfer and treatment facility, Otay, 151, 152, 154, 164;

waste treatment and transfer station, Long Beach, 153–54, 160, 161; wet oxidation, Casmalia, 149, 154, 168
California Hazardous Waste Management Council, 154, 160, 161–63, 172
California Office of Appropriate Technology, 145–48, 159, 160, 172, 173, 174
California Water Resources Control Board, 146, 157, 158, 166
CAPONE (Citizens Against Polluting Our Neighborhood Environment), 133
Carey, Governor Hugh, 15
Carnes, Sam, 10
Carter administration: hazardous waste policy, 12–13, 38–39, 57, 64, 65, 75, 133, 239; support of citizen participation programs, 176, 181, 182, 184, 185–87
Chemical industry, 26, 27, 29, 32, 38–39, 41, 105; Florida, 81, 90, 91; Texas, 133–35
Chemical Manufacturers' Association, 91
CHEMTREC, 105
Chlorinated benzenes, 27
Chloromethane, 27
Citizen Participation Fund, 192–93
Citizen participation programs: role of the states, 177, 179, 189–91, 193–95; workshops, 185, 186, 188, 239
Citizens Clearinghouse for Hazardous Wastes, 104, 193–94, 250
Clean Air Act of 1970, 3, 4, 63, 183, 184
Clean Water Act, 63, 124, 184. See also Federal Water Pollution Control Act
Clements, Governor William, 131
Colorado, 209, 221, 224, 225
Common Cause, 192
Comprehensive Environmental Response, Compensation, and Liability Act. See Superfund
Connecticut, 198, 202–4, 205, 207, 208, 210, 221, 224, 225, 234
Conservation Foundation, 176, 193
Cook, Michael, 49, 50, 52, 53, 55
Cooperative agreements, 57, 58, 67, 83, 85–86
Costle, Douglas, 184, 188
Council on Environmental Quality, 70–71, 128, 196

D'Amato, Senator Alfonse, 15
Decision-making models: incremental 44, 45, 46; rational, 44, 45
Delaware, 209, 221, 224, 225
Deukmejian, Governor George, 159, 170, 171, 174
Dibromochloropropane (DBCP), 143–44

About the Contributors

Ann O'M. Bowman is Assistant Professor of Government and International Studies at the University of South Carolina. Professor Bowman received her Ph.D. in political science from the University of Florida in 1979. She has contributed articles to *Western Political Quarterly, Urban Affairs Quarterly, Social Science Quarterly*, and *Environment and Behavior*, and is interested in the process and politics of environmental policy making at the state level. She is currently serving on South Carolina's Interim Study Committee on Hazardous Waste.

Thomas Brandt is currently a graduate student at the Lyndon B. Johnson School of Public Affairs of the University of Texas at Austin and is a candidate for the M.A. degree.

Steven Cohen is Assistant Professor of Political Science at Columbia University. He was formerly an environmental protection specialist in EPA's Office of Hazardous Emergency Response. In this capacity, he was primarily responsible for developing Superfund community relations policy.

James L. Franke is Assistant Professor of Political Science at Texas A&M University. He is interested in energy and environmental policy making at the state and local levels and has coauthored an article in the *Western Political Quarterly*.

Susan G. Hadden is Assistant Professor of Political Science at the Lyndon B. Johnson School of Public Affairs of the University of Texas at Austin. She is interested in risks to health and the environment and is the coeditor of a recent symposium on "Public Policy Toward Risk," published by the *Policy Studies Journal*.

Kenneth W. Kramer, formerly a Visiting Assistant Professor of Political Science at Texas A&M University, is now the legislative policy chairman for the Lone Star Chapter of the Sierra Club. He is a contributing author (or coauthor) of *Realignment in American Politics: Toward a Theory* (1980), *Water Quality Administration: A Focus on Section 208* (1980), and *Public Policy in Texas* (1982).

James P. Lester is a Visiting Associate Professor at Colorado State University and was formerly an Assistant Professor of Political Science at Texas A&M University. Professor Lester received his Ph.D. in political science from the George Washington University in 1980. He is a contributor to *Environmental Policy Implementation* (1982) and *Technology Transfer and U.S. Foreign Policy* (1976), as well as the author (or coauthor) of articles on environmental politics and policy in *Western Political Quarterly, Environment and Behavior*, and *Ocean Development and International Law*. His current research concerns the impacts of institutional reforms upon environmental policy making and implementation.

Harvey Lieber is Associate Professor of Government and Public Administration at American University. He is the author of *Federalism and Clean Waters: The 1972 Water Pollution Control Act* (1975), as well as numerous articles on environmental policy making and implementation.

Albert R. Matheny is Assistant Professor of Government and Public Administration at the University of Florida. He is a contributor to *The Political Science of Criminal Justice* (1982), as well as the author (or coauthor) of articles in *Law and Society Review* and *Law and Policy Quarterly*. He is particularly interested in the use of scientific information in the policy process.

David L. Morell is research political scientist at Princeton University's Center for Energy and Environmental Studies and a lecturer in the Department of Politics. He was on leave during 1982–83 to serve as director of policy for the hazardous waste program within the California Department of Health Services. He is the coauthor of *Siting Hazardous Waste Facilities: Local Opposition and the Myth of Preemption* (1982).

Richard Riley is Assistant Professor of Political Science at Baylor University. He is interested in a broad range of topics relating to regulatory policy and state-local relations.

Walter A. Rosenbaum is Professor of Political Science at the University of Florida. He is the author of several books including *The Politics of Environmental Concern* (1973 and 1977), *Coal and Crisis* (1978), and *Energy, Politics, and Public Policy* (1981), as well as numerous articles on environmental politics and policy.

Marc Tipermas is a project manager with ICF Incorporated, a private consulting group in Washington, D.C. He received his Ph.D. from Harvard University in political science in 1976, and was formerly director of EPA's Superfund Policy and Program Management Office.

Richard Torkelson is Assistant Commissioner for Administration at the New York State Department of Environmental Conservation. He served eight years with the New York State Division of the Budget, and is on the adjunct faculty of Russell Sage College.

Joan Veillette is currently a graduate student at the Lyndon B. Johnson School of Public Affairs, University of Texas at Austin, and is a candidate for the M.A. degree.

Bruce A. Williams is Assistant Professor of Political Science at Florida Atlantic University. He is a contributor to *The Organizational Politics of Criminal Justice* (1980), *The Determinants of Public Policy* (1980), and *Politics in the American States* (1982), as well as several articles in professional journals dealing with public policy issues.

John A. Worthley is Professor of Public Administration at Seton Hall University. He has authored several books, including the forthcoming *Computers and Information Systems in Health Care Management*, and has published in several journals, including *Public Administration Review, Western Political Quarterly*, and the *Journal of Higher Education*.

of related interest

Duke Press Policy Studies

International Environmental Policy
Emergence and Dimensions
Lynton Keith Caldwell

Global Deforestation and the Nineteenth-Century World Economy
Edited by Richard P. Tucker and J. F. Richards

What Role for Government?
Lessons from Policy Research
Edited by Richard J. Zeckhauser and Derek Leebaert

The Making of Federal Coal Policy
Robert H. Nelson

other books of interest

The Beaches Are Moving
The Drowning of America's Shoreline
Wallace Kaufman and Orrin H. Pilkey, Jr.

Living with the Shore series